AF361935

# Wearables for Movement Analysis in Healthcare

# Wearables for Movement Analysis in Healthcare

Editors

**Paolo Capodaglio**
**Veronica Cimolin**

MDPI • Basel • Beijing • Wuhan • Barcelona • Belgrade • Manchester • Tokyo • Cluj • Tianjin

*Editors*

Paolo Capodaglio
Ospedale San Giuseppe,
Istituto Auxologico Italiano,
IRCCS
Piancavallo (VB), Italy

Veronica Cimolin
Department of Electronics,
Information and
Bioengineering, Politecnico di
Milano, Piazza Leonardo da
Vinci 32
Milano, Italy

*Editorial Office*
MDPI
St. Alban-Anlage 66
4052 Basel, Switzerland

This is a reprint of articles from the Special Issue published online in the open access journal *Sensors* (ISSN 1424-8220) (available at: https://www.mdpi.com/journal/sensors/special_issues/ anal_health).

For citation purposes, cite each article independently as indicated on the article page online and as indicated below:

LastName, A.A.; LastName, B.B.; LastName, C.C. Article Title. *Journal Name* **Year**, *Volume Number*, Page Range.

**ISBN 978-3-0365-4019-1 (Hbk)**
**ISBN 978-3-0365-4020-7 (PDF)**

# Contents

# About the Editors

**Paolo Capodaglio** received his MD degree and his specialty in Physical and Rehabilitation Medicine from the University of Pavia, Italy. At present, he is a Professor in Physical and Rehabilitation Medicine at the University of Torino and the Director of the Rehabilitation Department (80 in-patients) and the Laboratory for Research in Biomechanics, Rehabilitation, and Ergonomics at the Istituto Auxologico Italiano IRCCS in Piancavallo, Italy. He has devoted most of his research to functional evaluation in ageing and pathological conditions (spinal cord injuries, musculoskeletal disorders, obesity, and metabolic conditions), published more than 250 papers in peer-reviewed journals and 8 monographic books, and serves as a reviewer for several indexed papers.

**Veronica Cimolin** received her Master's degree in Bioengineering from the Politecnico di Milano in 2002 and obtained her PhD degree in Bioengineering in 2007 at the Politecnico di Milano based on her work of the quantitative analysis of movement for the assessment of functional limitation in children with cerebral palsy. At present, she is an Associate Professor at the Department of Electronics, Information, and Bioengineering of Politecnico di Milano. Since 2007, she has carried out research activities at "Luigi Divieti" Posture and Movement Analysis Laboratory. Her research is in the field of quantitative movement analysis for clinical and rehabilitative applications, and she is involved in the activities of several movement analysis laboratories of national and international institutes. She is also the teacher of the courses "Impianti ospedalieri e sicurezza" and "Analisi e organizzazione di sistemi sanitari", and she is the author of more than 150 papers in international journals indexed in Scopus/Isi Web of Knowledge.

# Preface to "Wearables for Movement Analysis in Healthcare"

Recent advances in technology offer one solution to innovate healthcare and meet demand at a low cost. Wearable sensors are the most promising technology when it comes to: (a) supporting health and social care providers in delivering safe, more efficient, and cost-effective care, (b) improving people's ability to self-manage their health and well-being, (c) alerting healthcare professionals about changes in their condition, and (d) supporting adherence to the prescribed intervention. A variety of compact wearable sensors available today have allowed researchers and clinicians to pursue applications in which individuals are monitored at home and in community settings. Wearable sensors can help to reduce time devoted to assessment and provide objective, quantifiable data on patients' capabilities, unobtrusively and continuously. These technologies provide the opportunity not only to study motor function while patients perform daily-life activities but also to provide timely, meaningful feedback to patients and their physiotherapists.

**Paolo Capodaglio and Veronica Cimolin**
*Editors*

*Article*

# Computation of Gait Parameters in Post Stroke and Parkinson's Disease: A Comparative Study Using RGB-D Sensors and Optoelectronic Systems

Veronica Cimolin [1], Luca Vismara [2,3], Claudia Ferraris [4], Gianluca Amprimo [4,5], Giuseppe Pettiti [4], Roberto Lopez [1,6], Manuela Galli [1], Riccardo Cremascoli [2,3], Serena Sinagra [2], Alessandro Mauro [2,3] and Lorenzo Priano [2,3,*]

[1] Department of Electronics, Information and Bioengineering, Politecnico di Milano, Piazza Leonardo da Vinci 32, 20133 Milano, Italy; veronica.cimolin@polimi.it (V.C.); roberto.lopez@biomedica.udec.cl (R.L.); manuela.galli@polimi.it (M.G.)
[2] Istituto Auxologico Italiano, IRCCS, Department of Neurology and Neurorehabilitation, S. Giuseppe Hospital, 28824 Piancavallo, Italy; lucavisma@hotmail.com (L.V.); riccardo.cremascoli@unito.it (R.C.); serenasinagra@gmail.com (S.S.); alessandro.mauro@unito.it (A.M.)
[3] Department of Neurosciences, University of Turin, Via Cherasco 15, 10100 Torino, Italy
[4] Institute of Electronics, Computer and Telecommunication Engineering, National Research Council, Corso Duca degli Abruzzi 24, 10129 Torino, Italy; claudia.ferraris@ieiit.cnr.it (C.F.); gianluca.amprimo@ieiit.cnr.it (G.A.); giuseppe.pettiti@ieiit.cnr.it (G.P.)
[5] Department of Control and Computer Engineering, Politecnico di Torino, Corso Duca degli Abruzzi 24, 10129 Torino, Italy
[6] Department of Electrical Engineering, Universidad de Concepción, Víctor Lamas 1290, Concepción 4030000, Chile
* Correspondence: lorenzo.priano@unito.it; Tel.: +39-0323-514-392

**Citation:** Cimolin, V.; Vismara, L.; Ferraris, C.; Amprimo, G.; Pettiti, G.; Lopez, R.; Galli, M.; Cremascoli, R.; Sinagra, S.; Mauro, A.; et al. Computation of Gait Parameters in Post Stroke and Parkinson's Disease: A Comparative Study Using RGB-D Sensors and Optoelectronic Systems. *Sensors* **2022**, *22*, 824. https://doi.org/10.3390/s22030824

Academic Editor: Jeffrey M. Hausdorff

Received: 22 December 2021
Accepted: 20 January 2022
Published: 21 January 2022

**Publisher's Note:** MDPI stays neutral with regard to jurisdictional claims in published maps and institutional affiliations.

**Abstract:** The accurate and reliable assessment of gait parameters is assuming an important role, especially in the perspective of designing new therapeutic and rehabilitation strategies for the remote follow-up of people affected by disabling neurological diseases, including Parkinson's disease and post-stroke injuries, in particular considering how gait represents a fundamental motor activity for the autonomy, domestic or otherwise, and the health of neurological patients. To this end, the study presents an easy-to-use and non-invasive solution, based on a single RGB-D sensor, to estimate specific features of gait patterns on a reduced walking path compatible with the available spaces in domestic settings. Traditional spatio-temporal parameters and features linked to dynamic instability during walking are estimated on a cohort of ten parkinsonian and eleven post-stroke subjects using a custom-written software that works on the result of a body-tracking algorithm. Then, they are compared with the "gold standard" 3D instrumented gait analysis system. The statistical analysis confirms no statistical difference between the two systems. Data also indicate that the RGB-D system is able to estimate features of gait patterns in pathological individuals and differences between them in line with other studies. Although they are preliminary, the results suggest that this solution could be clinically helpful in evolutionary disease monitoring, especially in domestic and unsupervised environments where traditional gait analysis is not usable.

**Keywords:** RGB-D sensors; optoelectronic system; movement analysis; gait; Parkinson's disease; hemiparesis; spatio-temporal parameters

## 1. Introduction

The world population is aging rapidly as a consequence of the longer life expectancy. According to an OECD analysis, by the middle of the 21st century, more than 20% of the world population will be over 65, and this demographic change will affect both industrialized and developing countries [1]. In addition, the World Health Organization estimates an increase in the neurological diseases linked to the aging trend, particularly those ones characterized by chronic or progressive disabilities such as Parkinson's disease and stroke, with a consequent exponential growth of healthcare costs [2]. It is, therefore, possible to understand the importance of therapeutic assistance interventions which, in addition to improving the psycho-physical conditions of the individual, can reduce the costs of health care.

Stroke is one of the principal causes of morbidity and mortality in adults and one of the leading causes of disability in industrialized countries [3,4]. Hemiplegia/hemiparesis following stroke is one of the consequences of the acute loss of focal brain functions and it is clinically characterized by a deficit of voluntary motor activity in one half of the body, contralateral to the ictal lesion involving the primary motor cortex [5]. Six months after the event, 50–70% of patients still present sensory and motor deficits, such as paresis and spasticity of the limbs, which affect the ability to perform functional tasks, lead to reduced quality of life, and reduce participation in activities of daily living. The chronic motor disability then imposes significant challenges for prolonged treatment and patient care [6]. Focusing on post-stroke with hemiparesis condition, patients' gait can be variably altered by the impairments of motor functions (weakness and spasticity of lower limbs) and by alterations in posture and balance controls [7].

Parkinson's disease (PD) is the second most common neurodegenerative disorder, with a prevalence that increases with age [8]. The main characteristic is the progressive worsening of motor control and coordination capabilities induced by the death of dopaminergic neurons [9]. The expression of motor dysfunctions varies among subjects and over time. Nevertheless, subjects mainly exhibit some typical symptoms, including tremor, bradykinesia (slowness of movements), rigidity, postural instability and abnormalities, gait disorders, alterations in speech, mimicry, and writing [10]. As for stroke, the progression of motor disabilities imposes significant efforts for prolonged patient care and rehabilitation programs aiming to control symptom severity and enhance muscle strength, balance, gait, and mobility [11]. Focusing on gait, the alterations in walking patterns such as reduced speed, reduced step length, and increased step variability, combined with postural instability, lead to limited mobility and increased falling risk [12].

As gait impairments are experienced to be particularly disabling by patients, the walking recovery is a major objective in stroke and PD rehabilitation programs. Therefore, in the last decades, walking capability has been the object of study for the development of gait analysis methods in stroke survivors [13] and subjects affected by PD [14], both for diagnostic and rehabilitation purposes.

In the literature, the gait pattern in post-stroke subjects was quantitatively described using three-dimensional instrumented gait analysis (3D-GA), evidencing walking speed reduction, asymmetric postural behavior during walking and standing [15], and altered kinematics and reduced ankle push-off ability during terminal stance [16]. 3D-GA was also used in PD subjects to quantify the efficacy of rehabilitation [17] and to estimate spatio-temporal parameters [18,19].

3D-GA is widely used in clinical practice and research to investigate gait disorders, as it provides complete and objective information regarding specific gait features, including joint motion (kinematics), time–distance variables (spatio-temporal data), and joint moments and powers (kinetics). Conventionally, body segment kinematic and kinetic parameters are measured in gait laboratories, using marker-based optoelectronic systems and force plates. 3D-GA is considered accurate [20], but the availability of specific laboratories, high costs, and dependency on trained users [21,22] limit its use in clinical practice. Additionally, optoelectronic 3D-GA generally requires few clothes to be worn, and this condition could cause anxiety and embarrassment to patients [21,23].

Over the last decade, low-cost optical body-tracking sensors (i.e., RGB-D cameras) have been introduced in the gaming market as innovative devices for a new paradigm of human–computer interaction based on body movement. In particular, Microsoft Kinect© was the first device developed and released for this purpose. Although RGB-D cameras have not achieved significant success in the gaming market, they have subsequently found widespread use in other contexts, such as developing new rehabilitation systems for gait analysis, limb motion tracking, and gesture and posture classification [24,25]. Kinect-based solutions have become progressively more used in computer-assisted medical care and treatment thanks to lower costs and non-invasive motion capture techniques. Consequently, clinicians gained the possibility to explore the fields of body recognition and analysis of motor function as promising tools to monitor patients also in out-of-hospital settings. This need is considered extremely important for long-term care, for National Health System (NHS) costs, and the hospital service management and resource policies.

In recent years, several studies considered the use of RGB-D cameras for the assessment of the motor condition [26,27], for action and activity recognition [28], and motor rehabilitation [29], especially combined with virtual environments [30,31]. Regarding gait analysis [32], the first model of the Kinect sensor was widely used to analyze gait patterns in young adults [33,34], healthy adults [33,35], and children [36]. The characteristics of the next model include improvements of the on-board sensors, a wider field of view, stability of the body-tracking algorithm, and higher resolution of color and depth streaming, leading to higher performance and greater accuracy of motion capture than the older model [26,37–39]: the enhanced features allowed more detailed investigations of gait characteristics in healthy and pathological subjects. As for traditional gait analysis, some studies implemented a multi-camera-based system in order to cover greater walking distances (generally up to 10 m). However, this entails an increasing complexity of the implemented solution for the management, flow synchronization, and calibration procedures [40,41]. Other studies combined optical sensors and treadmills as an alternative approach to capture a more significant number of steps for the gait assessment [42,43]. On the other hand, the high cost and size of these approaches must be considered, which means that their implementation and use are still limited to motion analysis laboratories, such as for 3D-GA. Thus, conducting a walking test using these approaches could become impractical in smaller and unsupervised environments such as domestic settings.

A more practical solution is to use a single camera approach, which makes it possible to carry out walking tests in more confined spaces. Solutions based on a single camera could be easily used in any environment and adapted to specific needs, thus addressing a relevant clinical requirement, that is, the availability of a valid and straightforward method for quantifying gait patterns in pathological subjects that is suitable for home settings.

Single-camera approaches have been used to evaluate gait patterns in people with different health conditions, including cerebral palsy [44], ataxia [45], Parkinson's disease [46], and polyneuropathy [47], or to analyze young and older individuals [48,49]. More generally, RGB-D sensors (i.e., Kinect-like optical sensors) have been used in PD subjects to evaluate upper limb tasks [50], lower limb and posture [51], and gait and postural stability [46]. Concerning post-stroke, RGB-D sensors have been used to predict the risk of falls [52], evaluate the upper limb function [53], analyze balance recovery [54], rehabilitate upper limbs [55], estimate gait features [56], and evaluate the reliability of gait assessment and correlation with balance tests [57]. However, validating the accuracy of these non-invasive devices in capturing gait outcomes appears to be mandatory.

Studies on Parkinson's Disease [58,59], individuals with hemiplegia and stroke [56,60,61], subjects with other pathologies and disorders [62–65], and healthy people [35,48] indicate that some spatio-temporal gait parameters could be appropriately estimated using a single-camera approach. Concerning the gait validation, some studies are available using treadmills [43,66,67], multi-camera approaches [40,41], or wearable sensors [12,68–70]. However, there is scarce information for comparing the spatio-temporal gait parameters estimated by a single RGB-D sensor and an optoelectronic system on PD subjects [46,71], post-stroke individuals [61], and those with both pathological conditions [60], especially through a solution that is suitable for unsupervised or semi-supervised environments. In fact, from a clinical point of view and to ensure the sustainability of the health system, it is essential to have a simple method to quantify gait strategies in these pathological subjects. However, at the same time, it is necessary to define the level of accuracy and any limitations to allow clinicians to target its use properly; for example, for diagnostic or rehabilitation purposes.

Along this line of research, this study aims to extend our previous work [72] by validating the spatio-temporal gait parameters in level walking for both post-stroke individuals and subjects with Parkinson's disease. A single-camera approach on a reduced walking path, which is thus suitable for the limited spaces in domestic environments, is proposed to this end. The objective measurements estimated with this solution are compared with the corresponding ones obtained from the instrumented 3D-GA and other validation studies available in the literature, albeit on inhomogeneous subsets of spatio-temporal parameters, correlation criteria, and populations. This solution could become, for example, a valid method for evaluating and monitoring walking strategies over time, also integrating into more complex solutions [51] that assess the patient's general condition frequently and in unsupervised environments, enabling the follow-up of patients even outside of healthcare facilities. The possibility of home monitoring would allow the clinicians to constantly assess the evolution of the disease and to target interventions according to real and current patient needs, with the patient being able to feel "cured" and the national health care system being able to save money through the careful management of hospitalization periods, when required, and personalized rehabilitation programs in both supervised and unsupervised settings.

Furthermore, the non-invasiveness and versatility to different clinical needs could also favor an increasingly important role in future ecological neuro-rehabilitative contexts by stimulating patients at home and in daily activity [73,74] and ensuring the continuity of those services currently provided only from health facilities.

In conclusion, the main goals of this study are summarized as follows: (i): evaluate the agreement and robustness of the estimated gait parameters obtained from a single-camera approach based on Microsoft Kinect v.2 versus an instrumented standard 3D-GA system; (ii): objectively characterize the gait patterns on a cohort of parkinsonian and post-stroke subjects; (iii): compare the estimated gait parameters with several studies in the literature.

## 2. Materials and Methods

In this section, the study design is described. Since the study aimed to characterize the gait patterns in pathological subjects using a technological solution suitable for the home setting, a validation phase was necessary to check accuracy and robustness versus an instrumented 3D-GA system. To this end, two groups of subjects were enrolled to participate in the experimental study. Then, a dedicated setup and an acquisition protocol were defined to allow for the simultaneous motion capture and the estimation of gait parameters through the two systems. Finally, statistical analysis was performed to compare parameters and evaluate the correlation, reliability, and significant differences between the two systems and the pathological groups.

### 2.1. Patients

The recruitment procedure of PD and post-stroke participants to the experimental campaign, planned for this study and performed in a supervised setting, considered the potential end-users of the proposed system in home settings. In particular, we included only post-stroke (PS) and PD subjects who had already been evaluated and selected for standard rehabilitation in the hospital setting, and furthermore, who would have benefited from continuous and prolonged neurorehabilitation at home. Enrolment was then performed at the Division of Neurology and Neurorehabilitation, San Giuseppe Hospital, Istituto Auxologico Italiano, Piancavallo (Verbania), Italy. The inclusion criteria for post-stroke subjects were: minor disability of the lower limbs (possibility of walking), ability to walk 10 m without the assistance of another person or aids, ability to understand the instructions for performing the gait analysis test. The inclusion criteria for PD subjects were: tremor severity <=1, Hohen and Yahr (H&Y) score in 1–3 range. The exclusion criteria for both groups were: cognitive impairment with Mini-Mental State Examination (MMSE) < 27/30, previous neurosurgical procedures, history of other neurological or musculoskeletal disorders unrelated to stroke and PD. The exclusion criteria did not include age, sex, side dominance, or therapy.

For the experimental test, we recruited eleven post-stroke subjects who presented partial anterior circulation infarcts (PACI) according to Bamford's classification [75], and ten PD subjects who satisfied the inclusion criteria.

The clinical supervisor explained the experimental procedure in detail and instructed all participants accurately about the systems and the acquisition protocol, so all participants performed the walking test under the same conditions. The experimental campaign was carried out following the ethical standards of Istituto Auxologico Italiano, whose local ethical committee approved the study, and the latest amendments of the Helsinki declaration (1964). The enrollment procedure required all participants to sign informed consent forms to participate in the study.

### 2.2. Characteristics of the RGB-D and 3D-GA Systems

A single RGB-D sensor was used to implement a non-invasive motion capture system that includes hardware and software elements. The hardware relies on the Microsoft Kinect© v2 sensor (Microsoft Corporation, Redmond, WA, USA) and an elaboration unit consisting of a laptop running Windows 10 to which the RGB-D sensor connects through a dedicated USB port. The optical sensor produces color and depth streams at about 30 frame/s, using the time-of-flight technology to estimate the depth information: these features are adequate for the real-time motion capture and 3D reconstruction of the human body movement. Human body movements map on a skeletal model consisting of 25 joints that approximately correspond to specific anatomical points of the body: the tracking algorithm recognizes body patterns via the depth streaming and identifies the spatial regions associated with the joints of the skeletal model through a random forest classifier trained on thousands of images [76]. For each joint, the relative 3D position to the origin of the sensor reference system is available and returned for the 3D reconstruction of the body movements. In our previous studies [51,72], the 3D trajectories of joints and segments were compared to a gold reference system, verifying the accuracy and robustness of angular and linear measurements.

The RGB-D motion capture system includes custom-written software consisting of MATLAB® scripts (Mathworks Inc, Natick, MA, USA) that runs on the elaboration unit. The software component implements access, saving, and analysis procedures of the raw information provided by the RGB-D sensor through the Software Development Kit (SDK), including color images, depth images, and structured skeletal model data. The analysis procedure works on the collected 3D trajectories of joints to automatically segment every step and estimate the gait parameters that will be compared with the 3D-GA system: the Data Processing section (Section 2.4) describes the analysis procedure in detail. In addition, the RGB-D system has a Graphical User Interface (GUI), managed by MATLAB scripts, to simplify motion capture through specific utilities to check device operation, start and stop acquisition, and check data acquisition correctness [72].

To satisfy the validation aims, the study compared the RGB-D motion capture system to a standard 3D-GA system consisting of an optoelectronic system with six cameras (VICON, Oxford Metrics Ltd., Oxford, UK; sample rate: 50 Hz) and two force platforms (Kistler, Einterthur, CH). Optoelectronic systems represent the gold standard of technologies used in motion analysis for the evaluation of kinematics. After taking some anthropometric measurements, the operator placed the passive markers on the subject's skin at specific key points [77]. The reference systems for each segment of the lower limbs were calculated starting from the 3D coordinates of the markers positioned on the pelvis, thigh, leg, and foot: the angles of flexion–extension, abdominal-adduction and intra–extra rotation of the joints of the lower limbs were computed. The kinematic (angles) and kinetic (moments and powers) data from the 3D-GA system were not used for this study.

Before starting the experimental test, the 3D-GA system was calibrated to ensure the system's accuracy and to allow the estimation of the 3D marker coordinates. The average measurement error was computed based on the difference between the estimated and actual distances of two passive markers fixed on the extremities of a rigid bar (actual distance was 600 mm): the calibration procedure ended with an average error within 0.3 mm (standard deviation: 0.2 mm). In this condition, the calibrated working volume was 5 m in length ($x$-axis of the laboratory reference system), 2 m in height ($y$-axis of the laboratory reference system), and 2 m along the $z$-axis of the laboratory reference system.

### 2.3. Setup and Data Acquisition

In the movement analysis laboratory of the Institute, traditional gait analysis was performed using a 10-m long walkway. However, considering that the body tracking of the RGB-D system begins approximately 4.5 m from the sensor, it was necessary to define a setup that allowed for a peer comparison between the two systems. The RGB-D sensor was mounted on a tripod to ensure stability and placed at the end of the walkway. In this way, it was possible to capture the participants' gait as they moved towards the optical sensor that was preferable [41].

Then, a gait analysis path (GAP) was defined inside the walkway to ensure the total body tracking with the necessary accuracy: the start line was 4.2 m away from the sensor's position, while the end line was about 1.5 m away. This setup seemed to limit the analysis to a restricted area of the walkway, which was, in any case, sufficient to capture at least one complete gait cycle (or full stride) for each leg based on the motor conditions of the participants [57]. Moreover, GAP was adequate to guarantee the maximum depth accuracy on which the body-tracking accuracy and robustness also depend: according to [78], the vertical and lateral scattering of light pulses affect the depth accuracy in the working volume. The greater precision was along the central visual cone of the RGB-D system, where the depth accuracy was less than 2 mm, increasing to 4 mm up to 3.5 m and over 4 mm beyond 4.0 m from the sensor. It was also necessary to consider that the GAP was set approximately across the walkway mid-zone and around the force platforms, thus enabling the simultaneous gait analysis using the 3D-GA system. Figure 1 shows the experimental setup. Particular attention was paid to avoiding the presence, on the scene, of reflective surfaces and light sources entering the RGB-D sensor: these elements could interfere with the light pulses emitted by the device, generating artifacts in the depth map and, consequently, errors in the detection of the body map and the reconstruction of the 3D skeletal model. These measures can become constraints, especially for domestic environments. In this scenario, further constraints concerned clothing: in particular, baggy and too dark clothing and reflective objects needed to be avoided as they could interfere with the accurate motion capture. All these requirements will be part of the domestic experimental protocol, and all participants will have to be trained and supported in the system's configuration.

The acquisition protocol required participants to have completed two or more practice trials across the walkway to understand and be comfortable with the experimental procedure. After an initial familiarization, participants had to stand up and maintain an upright posture for a few seconds and then walk straight at their normal walking pace, from the beginning of the walkway to the RGB-D sensor. In this way, each subject entered and then left the GAP at his/her maximum walking ability. Only forward walking was considered for the gait analysis [79]. According to the experimental protocol, at least five trials were performed by each participant to guarantee the reproducibility of the results.

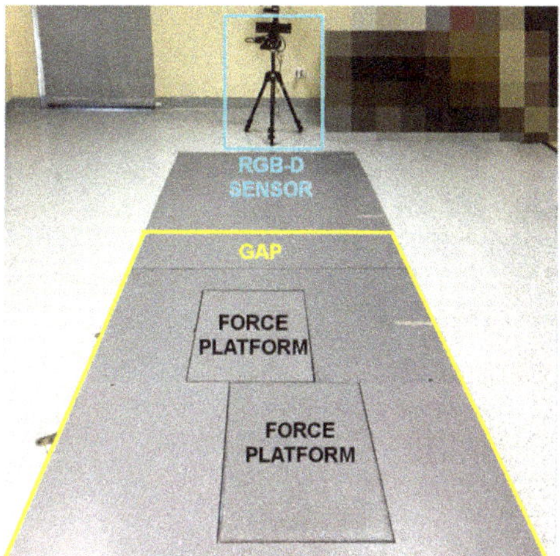


**Figure 1.** Experimental setup: position of the RGB-D sensor (light blue square); position of the force platform for 3D-GA system; approximate position of the final area of the Gait Analysis Path on the walkway for RGB-D system (yellow lines).

*2.4. Data Processing and Estimation of Gait Parameters*

For the analysis of 3D-GA data, the passive markers were used to identify the main walking events from which to estimate spatio-temporal parameters. In particular, the first heel-strike event was detected for each foot from the force platforms. Then, the previous and following similar events were identified by comparing the first heel-strike kinematic configuration (consisting of foot position, hip/knee/ankle flexion angles, and other specific features) over the 3D-GA data collected during the entire walking test. These events allowed the detection of at least two or three complete gait cycles inside the GAP, according to the walking ability and health condition of the subject, which were the same as those recognized by the RGB-D system.

For the data collected by the RGB-D system, the analysis procedure used dedicated MATLAB scripts that worked on the 3D joints recorded during the walking test. The analysis procedure consisted of three phases: data preprocessing, step segmentation, and estimation of gait features. The preprocessing phase applied resampling and filtering techniques to clean and align the data to the 3D-GA ones before using them for the step segmentation phase. In particular, a 50 Hz resampling with cubic interpolation was performed to ensure uniformity of the time baseline and avoid timestamp jitter due to a variable device framerate at around 30 FPS. A 10 Hz low-pass filter was then applied to the resampled data in order to exclude high frequency signal noise from the analysis.

The second phase relied on a custom-written step segmentation algorithm explicitly designed to identify each step when the subject was inside the GAP zone. The step segmentation algorithm worked on the 3D trajectories of the ankle joints to avoid the loss of accuracy of foot joints [80]. When the subject entered the sensor's field of view, the body-tracking algorithm started to provide the skeletal model. However, for the gait analysis with RGB-D system, it was essential to detect the time when the subject entered and left the GAP zone to identify the time window to be analyzed. To this end, the algorithm firstly estimated the 3D body center of mass ($COM_{BODY}$) as in [51] and then used its 3D Euclidean distance from the sensor to determine the time when the subject entered and left the GAP, as in [72]. After identifying the time window, the step segmentation algorithm used the 3D trajectories of left ($ANK_L$) and right ($ANK_R$) ankles to analyze each leg individually inside the GAP and estimate the related gait parameters. In particular, the z-component of the 3D joint was used to apply a binary thresholding in order to identify the "stationary" and "in movement" ankle periods. The stationary ankle period was when the difference between two consecutive z-component values was less than the prefixed threshold; the "in movement" ankle period was when the difference was over the prefixed threshold. As in [72], the threshold was 2 cm: the aim was to verify that the threshold defined for post-stroke subjects was also valid for PD subjects and that the step segmentation algorithm worked exactly the same way without tuning parameters. The results of the step segmentation algorithm were two binary arrays that allowed the extraction of some traditional gait parameters per leg and overall, as in other reference studies using RGB-D approaches [37,46,56,61,66]. Figure 2 shows an example of the sequence of steps and some temporal parameters estimated from the binary arrays.

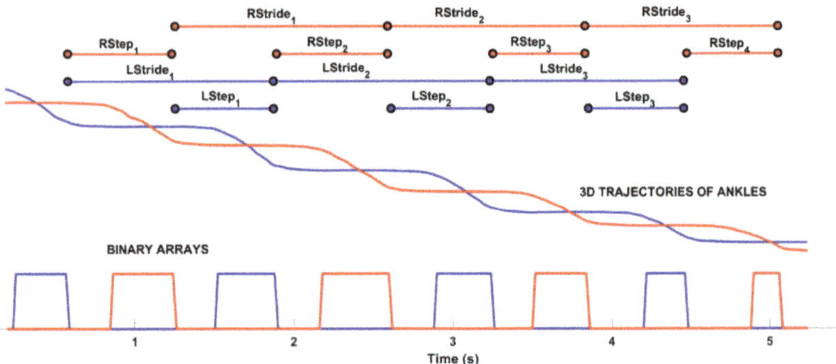


**Figure 2.** Results of the step segmentation algorithm: binary arrays for left (square line in blue) and right (square line in red) ankles estimated from 3D trajectories. At the top, some of the spatio-temporal parameters are shown; in particular, the step and gait cycle duration.

According to standard biomechanical protocols, the level walking analysis with 3D-GA commonly estimates many parameters, but only a subset is traditionally considered clinically relevant to highlight gait disorders [81]. For this reason, this study focused only on some spatio-temporal parameters that typically characterize gait patterns in post-stroke and PD subjects. It is important to note that the RGB-D and 3D-GA systems work on detecting different and specific events to estimate the same gait parameters, as in [72]: the two systems present differences in the physical location of passive markers and skeletal model joints, other than in the data processing algorithms. Nevertheless, starting from the events detected, it was possible to estimate the same spatial and temporal parameters and their average value inside the GAP, both for each side (i.e., step length, speed, time, and so on) and the overall walking test (i.e., cadence).

Another relevant feature of impaired gait pattern is related to the body sway during walking [82,83]: in fact, evident body sways are the consequence of the attempt to correctly balance walking in the presence of impairment and altered postural attitude while walking, and this is true both for post-stroke and PD subjects. For this investigation, the 3D-GA system estimated the position of the body center of mass from the 3D trajectories of the passive markers [84] and its peak-to-peak sways along the medio-lateral and vertical directions. Again, to limit the differences in locations of the body's center of mass, the RGB-D system computed the 3D midpoint of $HIP_L$-$HIP_R$ segment ($COM_{HIP}$): its peak-to-peak sways along the medio-lateral and vertical directions were computed and compared with the same parameters estimated by the 3D-GA system. Parameters estimated from the center of mass were relative to the overall walking test.

### 2.5. Statistical Analysis

Three consistent trials were selected and considered for the analysis. For the estimation of the spatio-temporal parameters, the two body sides were analyzed separately, while the COM parameters were estimated as a single value for each trial.

After applying the Kolmogorov–Smirnov test, a non-parametric analysis was considered as the parameters were not normally distributed; thus, the median and quartile values of all the parameters were computed. The statistical analysis included two phases.

Firstly, all the statistical tests were performed on the entire group of patients (PS and PD subjects). The Wilcoxon test was used to compare the measurements obtained by the RGB-D system and those obtained by 3D-GA. The research for correlation between the two methods was performed using Spearman's rank-order correlation, while, to assess the absolute agreement, the Intra Class Correlation (ICC) was used [85].

Then, the Bland–Altman plot was created to display the level of agreement between the RGB-D and the optoelectronic systems. This is a graphical method for comparing two measurements of the same variable in which the $x$-axis represents the mean of two measurements, and the $y$-axis represents the difference between. The plot can then highlight anomalies. For example, if one method always gives too high a result, then all points are above or below the zero line. It can also reveal that one method overestimates high values and underestimates low values. Otherwise, if the points on the Bland–Altman plot are scattered all over the place, above and below zero, then it suggests that there is no consistent bias of one approach versus the other.

Then, the analysis was performed by considering the two groups of patients (PS and PD subjects) separately. A repeated measure ANOVA was performed on the parameters with the "within-subject" factor of systems (3D-GA vs. RGB-D system) and the "between-subject" factor of groups (PD vs. PS group). Post hoc tests were performed, where appropriate, for the significance threshold. A significance level of 0.05 was implemented throughout. The statistical analysis was performed using Minitab® (version 18.1, State College, PA, USA).

## 3. Results

In this section, we present the results of our study, which were determined from the data collected on two groups of subjects during the experimental test. In particular, we enrolled eleven post-stroke subjects (average age: $53.3 \pm 13.9$ years; 3 females and 8 males; weight: $84.0 \pm 21.7$ kg; height: $1.8 \pm 0.1$ m; BMI: $25.6 \pm 4.2$ kg/m$^2$), with the following characteristics: four subjects with left and seven with right hemiparesis; $4.36 \pm 1.54$ years from stroke event. Sensory deficits were present in nine patients, homonymous hemianopia in five patients, dysphasia in four patients, and visuospatial disorders in five patients. In addition, we enrolled ten PD subjects (average age: $66.4 \pm 13.7$ years; five females and five males; weight: $85.6 \pm 15.4$ kg; height: $1.7 \pm 0.1$ m; BMI: $30.1 \pm 5.8$ kg/m$^2$) with the following features: 2.1 average H&Y score; $11.2 \pm 7.5$ years from disease diagnosis.

### 3.1. Statistical Analysis and Correlation Results

Firstly, the comparison results between the measurements obtained by the RGB-D system and those obtained by 3D-GA are reported in Table 1, where the median values, with the first and third quartiles, for each parameter, estimated on all participants' walking performance (eleven PS and ten PD individuals) captured simultaneously by the two systems, are shown. The analysis indicated no statistical differences between the two systems' groups ($p \geq 0.05$), showing the agreement between the two systems. The only exception is related to the step width for which a statistical difference was found ($p < 0.001$).

All the parameters indicated good reliability between the measurements (V sway showed moderate reliability), with the exclusion of step width, which displayed poor reliability (ICC = 0.44). The Spearman's correlation values between measures from the 3D-GA and RGB-D systems were all statistically significant ($p < 0.05$) and generally good for the measurements; only step width and V sway showed moderate correlation values, confirming the results obtained for ICC.

These results indicate, in general, a good agreement between the two methods.

**Table 1.** Median (first and third quartile) values for spatio-temporal and COM parameters estimated for the two systems, ICC and values of Spearman's correlation coefficient between the 3D-GA and RGB-D system (*: $p < 0.05$).

| Spatio-Temporal Parameters (Unit) | 3D-GA System | RGB-D System | ICC | Spearman's Correlation |
|---|---|---|---|---|
| Step length (m) | 0.43 (0.37, 0.56) | 0.45 (0.32, 0.56) | 0.81 | 0.67 * |
| Stance duration (%) | 66.50 (62.65, 71.60) | 67.00 (63.35, 71.19) | 0.78 | 0.59 * |
| Double support duration (s) | 0.46 (0.31, 0.91) | 0.53 (0.34, 0.77) | 0.80 | 0.71 * |
| Mean velocity (m/s) | 0.74 (0.59, 0.99) | 0.76 (0.54, 0.95) | 0.95 | 0.90 * |
| Cadence (step/min) | 97.20 (85.50, 105.18) | 98.36 (87.77, 102.99) | 0.97 | 0.82 * |
| Step width (m) | 0.23 (0.19, 0.24) | 0.17 (0.14, 0.20) * | 0.44 | 0.45 * |
| **Center of Mass Parameters (Unit)** | **3D-GA System** | **RGB-D System** | **ICC** | **Spearman's Correlation** |
| ML sway (m) | 0.09 (0.07, 0.10) | 0.09 (0.06, 0.12) | 0.94 | 0.61 * |
| V sway (m) | 0.05 (0.04, 0.06) | 0.04 (0.04, 0.05) | 0.60 | 0.48 * |

In Figure 3, the Bland–Altman plot is displayed. It is a scatterplot of the mean of the RGB-D system and instrumented 3D-GA method plotted against the difference between the two methods [86]. It is possible to observe that the Bland–Altman graphs globally display a good agreement between the two measurement systems as most of the points fell within the interval.

### 3.2. Gait Patterns in PD and PS Subjects

In order to demonstrate the ability of the system in the qualitative characterization of the walking performance, the following Figure 4 shows two examples of gait patterns. In particular, Figure 4a represents the walking pattern of one PD participant, while Figure 4b represents the walking pattern of one post-stroke (PS) participant. The aim of Figure 4 is not to highlight the differences between the walking scheme of the two pathologies but to verify the system's ability to detect gait features in every circumstance, that is, in different subjects affected by distinct pathologies with their own peculiar gait characteristics. The PD subject (Figure 4a) performed five complete steps inside the GAP zone, whose length was short and variable on both sides: the first and the last steps were at the limits of the GAP zone, so the analysis procedure did not consider them. The graphical representation of the $COM_{HIP}$ trajectory points out the small lateral sways during the walking test. This qualitative information suggests impairment during gait. On the contrary, the post-stroke subject (Figure 4b) performed four steps inside the GAP zone. The left steps seemed longer than the right ones, denoting a symmetry deterioration in the walking scheme. In addition, lateral sways seemed more accentuated than the PD subject, indicating greater instability and compensatory strategies during gait.

However, quantitative and objective measurements should support the qualitative evidence derived from the simple graphical representation. As reported in Table 2, the estimated spatio-temporal and COM parameters confirmed the preliminary qualitative indications. On average, PD performance showed an asymmetric gait, in length and speed, characterized by short steps (step length), slight slowness (mean velocity and cadence), and high stance phase (percentage of the gait cycles). In addition, the data confirmed the low lateral sways of the body's center of mass. On the other hand, PS performance also showed asymmetry during walking, both in length and speed, but longer steps and higher slowness than PD performance, slightly greater than double support duration, and less stance duration. The data also confirmed more significant medio-lateral sway than PD performance. In contrast, the step width and vertical sway were comparable, confirming the lesser significance of these parameters. Finally, as expected, the best performance was relative to the left side in both cases, which was in line with the subjects' conditions: the

PS subject showed a right-side residual hemiparesis, while the PD subject exhibited more important symptoms on the right-side.

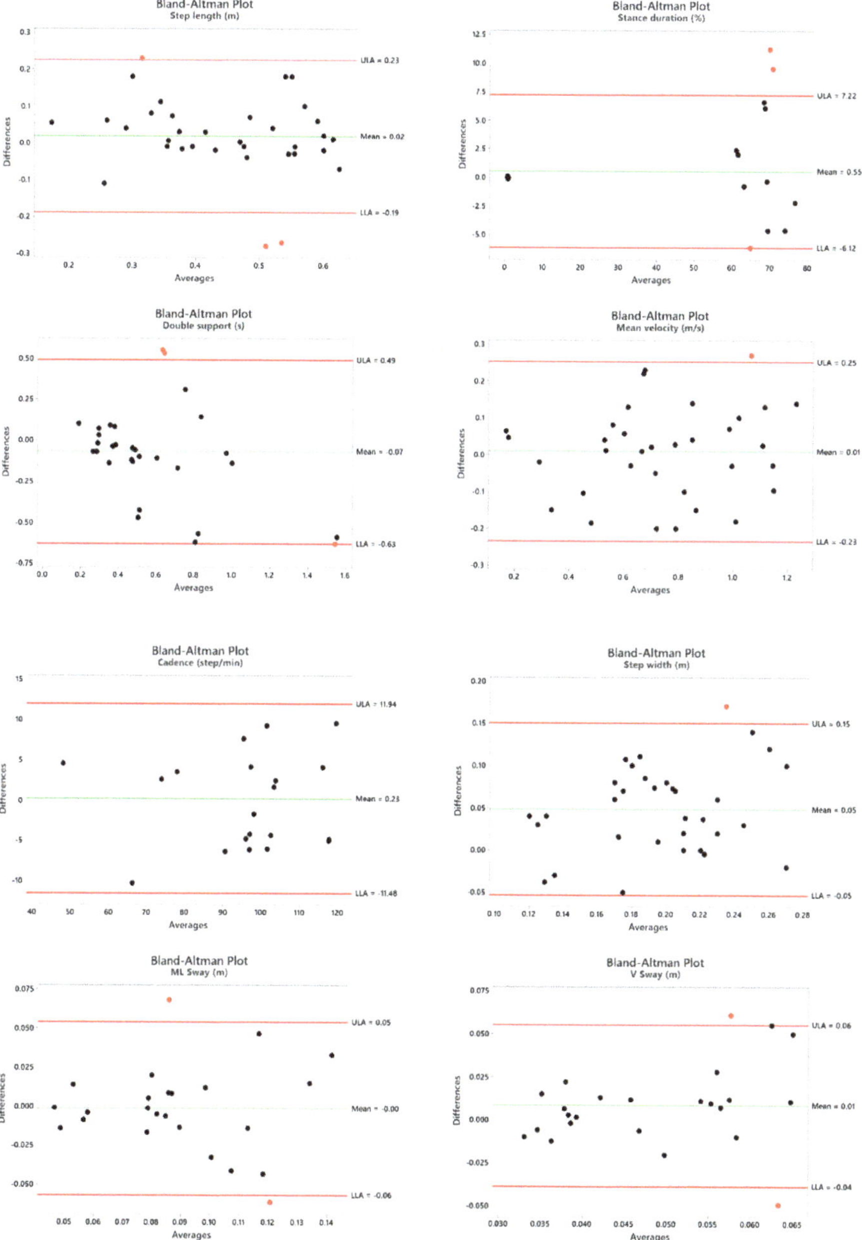

**Figure 3.** Bland–Altman plots of the mean of 3D-GA and the RGB-D systems against the difference between the two methods for the spatio-temporal parameters and the COM Sway.

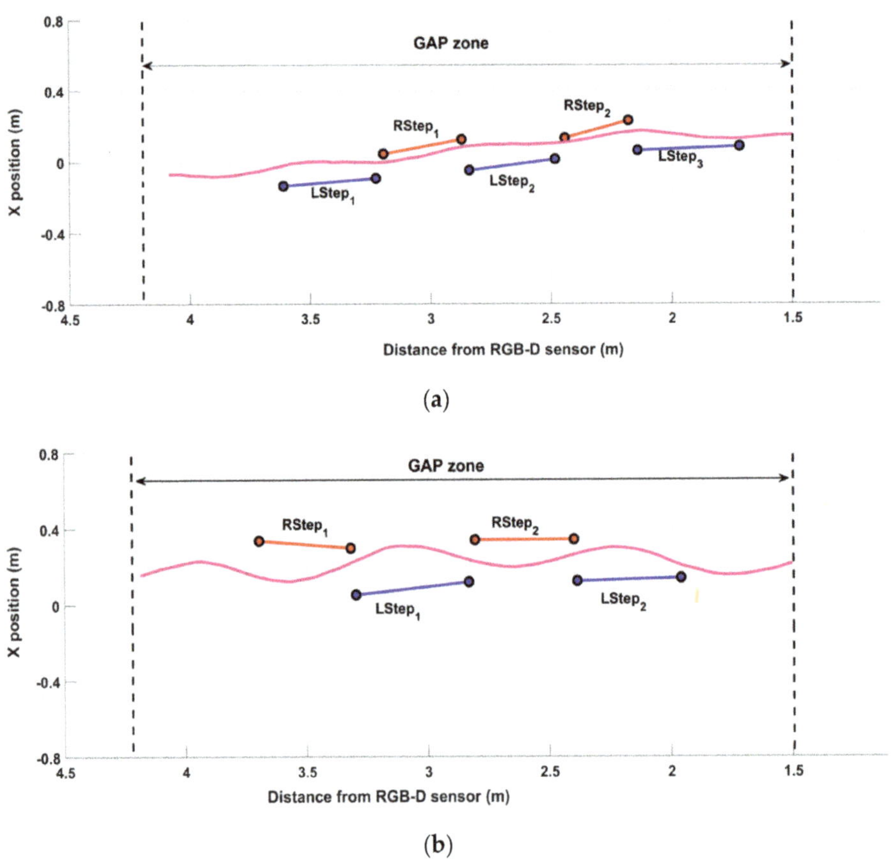


**Figure 4.** Gait patterns of one PD (**a**) and one post-stroke (**b**) participant. The black dotted lines delimit the GAP zone. The *x*-axis represents the 3D distance of the subject from the RGB-D sensor, while the *y*-axis represents the position relative to the sensor's *x*-axis. The magenta lines represent the trajectory of COM$_{\text{HIP}}$.

**Table 2.** Spatio-temporal and COM parameters for the left and right sides of PD and PS performance shown in Figure 4.

| Parameters | PD #4–Figure 4a | | PS #7–Figure 4b | |
|---|---|---|---|---|
| | **Left Side** | **Right Side** | **Left Side** | **Right Side** |
| Step length (m) | 0.39 | 0.29 | 0.45 | 0.39 |
| Stance duration (%) | 76.20 | 77.69 | 72.20 | 72.65 |
| Double support duration (s) | 0.60 | 0.60 | 0.70 | 0.70 |
| Mean velocity (m/s) | 0.77 | 0.56 | 0.59 | 0.55 |
| Step width (m) | 0.13 | 0.19 | 0.19 | 0.23 |
| Cadence (steps/min) | 114.65 | | 79.47 | |
| ML Sway (m) | 0.05 | | 0.10 | |
| V Sway (m) | 0.03 | | 0.04 | |

The following radar graphs summarize the differences in the most significant spatio-temporal parameters between PD and PS performance. Since the gait parameters represent different physical quantities, they were normalized before representing them graphically. In particular, a minimum–maximum normalization was used by considering the estimated parameters for all the PD and PS walking tests that were collected and analyzed. Then, each parameter value was scaled to be more effectively represented in a range 0–1, using a minimum–maximum normalization. To this end, we considered the average normative values for healthy adult subjects [81] as the best walking performance, which practically would be associated with the maximum values (i.e., 1). On the contrary, the lowest parameter values estimated on the PD and PS groups were considered to represent the worst walking performance associated with the minimum values (i.e., 0). It is important to note that a direct relationship with the gait impairment characterized some parameters (stance duration, double support duration, and step width): in this case, the values of the parameters increased with increasing gait dysfunctions. Therefore, according to our choices regarding the graphical representation of the radar graphs, the worsening of these parameters (i.e., higher parameter values) corresponded to normalized values closer to zero.

Conversely, other parameters (step length, mean velocity) were characterized by an inverse relationship with the gait impairment: in this case, the values of the parameters decreased with increasing gait dysfunctions. The worsening of these parameters (i.e., lower parameter values) corresponded to normalized values closer to zero. In this way, all the normalized parameters were within the 0–1 range, producing radar charts that expanded outward when the gait performance was good, and collapsed inward when the gait performance was impaired. The radar graphs corresponding to Figure 4 and Table 2 are shown in Figure 5.

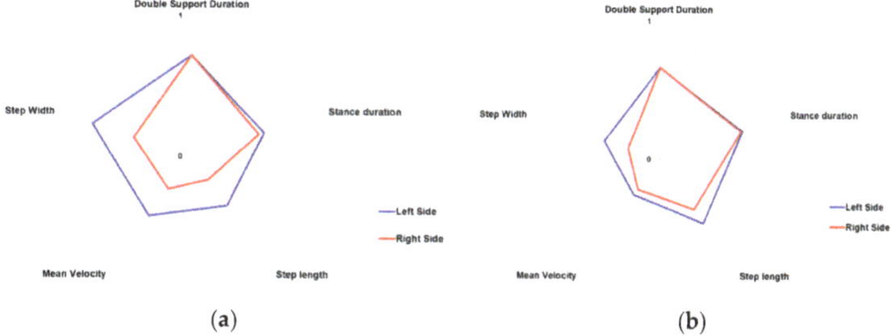


**Figure 5.** Radar charts of the relevant gait parameters for the left and right sides of the PD (**a**) and PS (**b**) performance.

### 3.3. Analysis of the PD and PS Groups

The median and quartiles for each parameter are presented in Table 3 for each group of subjects and both systems. Measurements, commonly attributable to the left and right sides, were averaged for this analysis.

**Table 3.** Median (first and third quartile) values for spatio-temporal and COM parameters estimated for the two systems of the PD and PS group.

| Parameters | RGB-D System | | 3D-GA System | |
| --- | --- | --- | --- | --- |
| | PD Group | PS Group | PD Group | PS Group |
| Step length (m) | 0.36 (0.29, 0.58) | 0.46 (0.39, 0.56) | 0.39 (0.37, 0.42) | 0.49 (0.38, 0.64) |
| Stance duration (%) | 65.75 (63.67, 71.14) | 67.50 (62.75, 71.75) | 70.30 (62.55, 74.53) | 65.00 (62.52, 69.50) |
| Double support duration (s) | 0.66 (0.37, 0.77) | 0.51 (0.31, 0.69) | 0.73 (0.28, 0.91) | 0.42 (0.33, 0.57) |
| Mean velocity (m/s) | 0.61 (0.54, 1.14) | 0.80 (0.54, 0–94) | 0.71 (0.54, 1.11) | 0.74 (0.62, 0.94) |
| Step width (m) | 0.15 (0.12, 0.16) | 0.19 (0.16, 0.22) | 0.23 (0.14, 0.24) | 0.24 (0.21, 0.26) |
| Cadence (steps/min) | 103.55 (98.90, 113.92) | 95.24 (72.58, 99.17) | 102.20 (97.20, 118.00) | 93.90 (75.20, 99.40) |
| ML Sway (m) | 0.08 (0.06, 0.11) | 0.09 (0.08, 0.12) | 0.08 (0.06, 0.09) | 0.10 (0.08, 0.14) |
| V Sway (m) | 0.04 (0.04, 0.07) | 0.05 (0.04, 0.09) | 0.05 (0.04, 0.08) | 0.04 (0.03, 0.04) |

The results of the "between-subject" analysis of Group (that is, PD and PS groups), the "within-subject" analysis of System (that is, 3D-GA and RGB-D systems), and the Group $\times$ System interaction for each parameter are reported in Table 4.

**Table 4.** Results of the "between-subject" analysis of Group (PD and PS groups), the "within-subject" analysis of System (3D-GA and RGB-D systems), and the Group $\times$ System interaction for spatio-temporal and COM parameters. Statistically significant data are highlighted in bold.

| Parameter | Factor | F | *p*-Value | Partial $\eta^2$ |
|---|---|---|---|---|
| Step length (m) | Group | 2.20 | 0.143 | 0.0333 |
|  | System | 0.26 | 0.613 | 0.0040 |
|  | Group $\times$ System | 0.03 | 0.869 | 0.0004 |
| Stance duration (%) | Group | 0.66 | 0.090 | 0.0995 |
|  | System | 0.91 | 0.344 | 0.0140 |
|  | Group $\times$ System | 0.96 | 0.332 | 0.0147 |
| Double support duration (s) | Group | 0.61 | 0.438 | 0.0094 |
|  | System | 0.41 | 0.525 | 0.0063 |
|  | Group $\times$ System | 0.18 | 0.677 | 0.0027 |
| Mean velocity (m/s) | Group | 0.45 | 0.502 | 0.0071 |
|  | System | 0.01 | 0.903 | 0.0002 |
|  | Group $\times$ System | 0.01 | 0.981 | 0.00001 |
| Step width (m) | **Group** | **10.75** | **0.002** | **0.1438** |
|  | **System** | **17.96** | **<0.001** | **0.2192** |
|  | Group $\times$ System | 0.001 | 0.982 | 0.0001 |
| Cadence (steps/min) | **Group** | **9.50** | **0.003** | **0.1293** |
|  | System | 0.02 | 0.885 | 0.0003 |
|  | Group $\times$ System | 0.10 | 0.752 | 0.0016 |
| ML Sway (m) | **Group** | **6.17** | **0.017** | **0.1281** |
|  | System | 0.05 | 0.819 | 0.0013 |
|  | Group $\times$ System | 0.40 | 0.533 | 0.0093 |
| V Sway (m) | Group | 0 | 0.958 | 0.0001 |
|  | System | 0.05 | 0.824 | 0.0012 |
|  | Group $\times$ System | 3.15 | 0.083 | 0.0698 |

The analysis confirms that, in general, there were no significant differences between the two systems in the estimation of the spatio-temporal parameters and the sways of the center of mass. The only exception seemed to be related to step width, where slightly significant differences were highlighted both between groups and systems (but not in the Group $\times$ System interaction): this suggests that step width is probably a very challenging and sensible parameter related to the different positions of passive markers and joints of the skeletal model.

Regarding the gait cadence, the analysis showed a significant difference between groups (i.e., PD and PS) but not between systems (i.e., 3D-GA and RGB-D): the two systems seemed to agree in the estimation of the gait cadence; therefore, it could be an effective discriminatory parameter between the two groups of subjects. The same also occurred for ML sway, which showed a significant difference between groups (i.e., PD and PS) but not between systems (i.e., 3D-GA and RGB-D): again, this suggests that ML sway could be a valuable parameter to differentiate the two groups of subjects.

### 3.4. Comparison Versus Other Studies

Some studies in the literature have analyzed and validated technologies and methodologies against the "gold standard" 3D-GA. Nevertheless, the direct comparison of the results is generally problematic because not all studies consider the same spatio-temporal parameters, correlation methods, populations, or technological approaches.

An approach based on a single RGB-D sensor was used in [37] on healthy young adults: this study reported a slightly greater agreement with 3D-GA than our findings, in particular for step length (ICC= 0.93 [37] vs. ICC = 0.81 [our]) and average velocity (ICC = 0.96 [37] vs. ICC = 0.95 [our]) under normal pace conditions. In [40], a multi-RGB-D sensors approach (four cameras along the 10-m walkway) was used on healthy subjects: in this study, the step width showed the lowest correlation (ICC = 0.65 [40]) compared to the other spatio-temporal parameters. The same occurred in our study, in which we obtained an even lower agreement value (ICC = 0.44 [our]).

Concerning the step width, our results for the Spearman correlation (r = 0.45 [our]) and agreement (ICC = 0.44 [our]) were comparable with the findings in [41], where values for the Pearson correlation (r = 0.52 [41]) and agreement (ICC = 0.40 [41]) were reported.

However, the latter results for step width seem to disagree with [43] that reports a high Pearson's correlation coefficient (r = 0.85 [43]): in this study, the first model of Microsoft Kinect was used, combined with a treadmill, on healthy participants. Again, with respect to [43], our estimated Spearman correlation (r = 0.59 [our]) was globally lower than the Pearson correlation of stride time and stance time (r = 0.85 [43] and r = 0.77 [43], respectively). On the contrary, the result for double support seemed much better (r = 0.24 [43] vs. r = 0.71 [our]).

The same controversial results for step width are present in [66] both as agreement (ICC = 0.84 [66]) and Pearson's correlation coefficient (r = 0.73 [66]): also, in this case, the first model of the Microsoft Kinect was used in conjunction with a treadmill on healthy participants. Regarding the other parameters, the results in [66] are worse for step length (ICC = 0.76 [66] vs. ICC = 0.81 [our]) and mediolateral sway (ICC = 0.84 [66] vs. ICC = 0.94 [our]), referring to the lowest speed setting on the treadmill.

In [36], two Kinect v.1 and an automatic algorithm for step segmentation were used to compare gait parameters on healthy children: the reported findings on average correlation and agreement are slightly greater than our results for step length (r = 0.79 [36] vs. r = 0.67 [our]; ICC = 0.85 [36] vs. ICC = 0.81 [our]), but lower for average velocity (r = 0.79 [36] vs. r = 0.90 [our]; ICC = 0.77 [36] vs. ICC = 0.95 [our]) and cadence (r = 0.79 [36] vs. r = 0.82 [our]; ICC = 0.79 [36] vs. ICC = 0.97 [our]).

Regarding the COM sway parameters during walking, objective results on correlation or agreement do not appear to be available; nevertheless, the high average sway along mediolateral direction suggested a concordance with the conclusions on pathological gaits in [83].

The direct comparison between parameter values is even more difficult because only a few validation studies specifically work on post-stroke and PD populations. Limiting the comparison to the spatio-temporal parameters of Table 3 and to the RGB-D sensor approaches, our average values were in line with the normative data (age 50–59 years) of post-stroke subjects in [61], as regards the step length (mean = 0.46 [61] vs. mean = 0.47 [our]), double support (mean = 0.47 [61] vs. mean = 0.52 [our]), and step width (mean = 0.21 [61] vs. mean = 0.19 [our]). On the contrary, the average values for walking speed (mean = 0.81 [61] vs. mean = 0.72 [our]) and cadence (mean = 94.34 [61] vs. mean = 89.52 [our]) were lower in our study, but this was probably related to the different composition of the PS group. For the same reason, the average value for step length was in line with [56] (mean = 0.51 [56] vs. mean = 0.47 [our]), while the walking speed (mean = 0.87 [56] vs. mean = 0.72 [our]) and double support (mean = 0.40 [56] vs. mean = 0.52 [our]) were slightly different.

Regarding studies on subjects with PD, our average values were comparable with those reported in [46] (H&Y 2.5, single task condition), in particular for cadence (mean = 98.8 [46] vs. mean = 101.81 [our]) and step width (mean = 0.10 [46] vs. mean = 0.15 [our]). However, there were some differences regarding double support (mean = 0.4 [46] vs. mean = 0.65 [our]) and walking speed (mean = 0.9 [46] vs. mean = 0.77 [our]), probably due to the different motor conditions of subjects included in our group. The average values were also directly comparable with [66] for cadence (mean = 101.16 [66]), walking speed (mean = 0.83 [66]), and stance duration (mean = 73.0% [66] vs. mean = 67.29% [our]).

The mean values of parameters were also comparable with studies that used different technological approaches, particularly inertial and wearable sensors, to analyze gait and compare spatio-temporal parameters with gold reference systems. For example, [68] and [12] employed inertial sensors and presented results on PD and post-stroke subjects. In [68], the mean values obtained for walking speed are greater than our findings, both for PD (mean = 0.85 [68] vs. mean = 0.77 [our]) and for post-stroke (mean = 0.61 [68] vs. mean = 0.72 [our]). On the contrary, the mean values obtained for stance duration are very similar, both for PD (mean = 66.70% [68] vs. mean = 67.29% [our]) and for post-stroke subjects (mean = 66.70% [68] vs. mean = 66.91% [our]). We also compared the mean values with [12]: in this case, the differences for stance duration were slightly higher than with the previous study. In particular, the most evident difference referred to the PD group (mean = 55.9% [12] vs. mean = 67.29% [our]) rather than post-stroke subjects (mean = 64.5% [12] vs. mean = 66.91% [our]): this probably depended on some differences in age, disease onset, and severity of the PD subjects involved.

However, despite some minor discrepancies due to different technological and methodological approaches in the referenced studies, the results globally demonstrated the reliability of the proposed solution and segmentation algorithm to estimate relevant spatio-temporal gait parameters that agree, concerning range and correlation, with studies that rely on different approaches and populations.

## 4. Discussion

In order to test whether an RGB-D sensor is capable of identifying gait characteristics and deviations during level walking in neurological diseases (in particular, post-stroke and PD), the walking abilities of two small cohorts of subjects were analyzed simultaneously with a single-camera solution and a 3D-GA system. In particular, the standard spatio-temporal parameters and the estimation of the center of mass excursion during gait were compared. The single-camera solution relies on an RGB-D sensor (i.e., Microsoft Kinect v2) and its body-tracking algorithm is able to track, in real-time, the 3D movements of the human body during gait on a reduced walking path (GAP) that is suitable for domestic environments. For the experimental procedure, the GAP was set along the traditional walking path used by the instrumented 3D-GA in order to allow a fair comparison and validation of the two systems. A custom-written step segmentation algorithm extracted some standard spatio-temporal parameters from the 3D trajectories of the skeletal model, provided by the optical sensor, by detecting every step performed inside the GAP. The computed COM information was used to determine when to perform the gait analysis inside the GAP (i.e., when the subject entered and left the GAP) and then to measure any dynamic balance anomalies during gait.

The results show that, using the implemented step segmentation algorithm, the RGB-D system provided an estimation of the same standard parameters measured with the optoelectronic system during the gait analysis test. Any slight differences between some measurements are probably attributable to the different reference points (i.e., the 3D position of markers on body and joints constituting the skeletal model) or to the different algorithms used by the two systems to estimate gait parameters. Consequently, this could have affected the agreement between the two systems and make it complex to compare results directly with other studies.

Despite this, our results exhibited a good agreement between the two systems in all the analyzed parameters, with the exception of the step width. No statistical differences were found between the two systems, and good reliability between the measurements was displayed according to the values of ICC. In addition, the research on the correlation between the 3D-GA and RGB-D systems exhibited values that were statistically significant, demonstrating, in general, a fair proportionality between the two methods. It is important to underline that, in general, the correlation values and ICC presented high and satisfactory values for all the parameters, with the exception of step width, as previously anticipated, and the V sway parameters. Unlike the other parameters, the results related to these two parameters showed, in fact, only moderate reliability (ICC = 0.44 for step width, ICC= 0.60 for V sway) and correlation values (r = 0.45 for step width, r = 0.48 for V sway) between the two measurements. The results were also confirmed by the Bland–Altman plots, which globally displayed a good association between the two measurement systems as most of the points fell within the defined interval.

The lower agreement obtained for the step width could depend on the close position of the two feet in some of the tests, which resulted in a lower precision in the estimation of the $x$-axis distance by the RGB-D system.

Similar considerations could be made for the V sway parameter. Although there were no particular concerns about this index, it is evident that the obtained results, in terms of ICC and correlation values, were lower than those of the other parameters. These differences could be related to the different 3D COM positions estimated by the two systems, which could have slightly interfered with the reliability of the results; or, more likely, to the lesser relevance of vertical swaying during walking, which involves less marked up–down body movements.

However, the results generally confirmed those obtained in previous studies carried out on post-stroke and parkinsonian subjects, even if it is difficult to directly compare our data with the literature.

The main difficulties for a fair comparison rely on the use of different correlation methods (ICC, Spearman's correlation, and Pearson's correlation), technologies (optical sensors, wearable sensors, and optoelectronic systems), experimental setup (single-camera, multi-camera, and treadmills), and populations (healthy, pathological subjects, adults, and young people): several studies in the literature adopt different methodological and technological approaches and analyze gait in different groups of subjects, so the comparison of results with other researches is not easy.

Concerning the agreement between vision-based and 3D-GA systems, as previously highlighted in Section 3.4, in [37], a slightly higher ICC agreement for mean velocity and step length was reported, while a greater Pearson's correlation was estimated for cadence in [40]. On the contrary, our results showed a greater agreement for step length and mediolateral sway than [66]; and for step length, mean velocity, and cadence than [66]. The results for step width seem to be more controversial in the literature: for example, while [40,41,72] obtained the lowest correlation and agreement for step width as our analysis, other studies [43,66] presented satisfactory agreement data both in terms of ICC and Pearson's correlation. The Spearman's correlation for the Stance duration, expressed as a percentage of stride time, seemed to be in agreement with the Pearson's correlation of stride time and stance time in [43], while the agreement seemed to be greater in our study for double support. The COM excursion results displayed high deviations, especially in the mediolateral direction, as expected in pathological gait patterns [83]. The weak results related to the vertical direction could be related to the different 3D COM positions ($y$-axis) considered by the two systems, as discussed and confirmed in our previous research [72].

Considering the PD and PS groups separately, the ranges of values estimated by the two systems were found to be in line, as evidenced by the median and the quartiles in Table 3, except for minor discrepancies due to the different landmarks and analysis algorithms. These data demonstrates that the RGB-D system is able to extract relevant features of gait with an accuracy comparable to the 3D-GA for both pathological conditions and this aspect is the most important finding of our study, according to its aim. The graphical representation of gait patterns and COM trajectories provides an immediate and intuitive qualitative indication about the global gait performance of PD and PS subjects (Figure 4): this is, however, supported by objective measures (Table 2) and by the summary representation of normalized parameters in the form of radar charts (Figure 5), which makes it possible to easily quantitatively compare performance over time, to reveal differences between body sides (as in Table 2, where more impairment is evidenced for the right side), and to highlight different gait strategies for PD and PS subjects (as in Table 2, where there are relevant dissimilarities in mean velocity, cadence and ML sway between the two evaluated subjects).

Specifically, the overall comparison between PS and PD groups confirmed that the two systems did not show significant differences in estimating the spatio-temporal parameters and the sways of the center of mass, as highlighted by the System and Group × System analysis (Table 4).

The data analysis showed slightly significant differences in step width between populations and systems. Furthermore, the same data were found to not be relevant for the Group × System interaction: this confirmed that step width is probably a challenging parameter to compare due to the different placement of passive markers and skeletal model joints.

Regarding the gait cadence, there were significant differences between groups (i.e., PD and PS) but not between systems (i.e., 3D-GA and RGB-D): this suggests that the two systems agreed in the estimation of the gait cadence, which could, therefore, be an effectively discriminating parameter for different gait schemes due to different neurological impairments. The analysis also showed a significant difference between groups (i.e., PD and PS) but not between systems (i.e., 3D-GA and RGB-D) concerning ML sway: this suggested that the two systems agreed in the estimation of ML sway that could be a valuable parameter to differentiate the two populations. It is, in fact, highly probable that the gait impairment due to hemiplegia in post-stroke subjects is associated with a more dangling walking to compensate for gait imbalance.

According to the overall analysis, the RGB-D system seemed able to extract important features of the gait patterns, in line with the 3D-GA system and regardless of the complexity of gait of the two neurological diseases considered: this suggested the suitability of the RGB-D system to perform objective measures of walking strategies where 3D-GA is not applicable, such as domestic environments.

Particularly relevant is the fact that the obtained measures were found to be comparable with those obtained by means of the optoelectronic system, which represents the gold standard for movement analysis.

These results are also significant from a clinical and rehabilitative point of view. The monitoring of gait parameters using an easy and non-invasive testing system may be helpful in clinical settings to expand the number of patients examined. The RGB-D system could be considered a valid means for a preliminary, quick, easy, and low-cost evaluation of gait variables. However, it is important to underline that the proposed system does not intend to replace gait laboratories based on optoelectronic systems. The optoelectronic systems should continue to be used in clinical settings for a more in-depth assessment of gait strategies and clinical decision-making. On the other hand, the more relevant advantages of the RGB-D solution could be in unsupervised settings, overcoming the limitations of the traditional gait analysis in laboratory settings and—not of secondary importance—without undressing patients for marker placement, which is frequently a psychological barrier for frail patients [21,23]. Furthermore, outpatient rehabilitation facilities could benefit from the information obtained from the proposed system, which could provide quantitative gait parameters even outside of gait-analysis laboratories, without excessively interfering with the subject's usual activities.

For rehabilitation purposes, the application of these kinds of solutions could be exploited both in the hospital (at entrance and discharge of patients), in the outpatient ambulatory, and in-home settings. In addition to objective evaluation of gait patterns, solutions based on optical devices, such as the RGB-D sensors, could be helpful also for remote rehabilitation, which has proved to be a promising intervention to reduce the effect of impairment and improve quality of life both for post-stroke and parkinsonian subjects.

In both pathologies, continuous rehabilitation and physical activity are required to reduce motor and cognitive decadence. The possibility of simple clinical monitoring would allow a more specific and targeted therapeutic action in wider cohorts of patients.

Several systematic reviews [87–90] agree on the effectiveness of remote rehabilitation interventions for recovery from the motor, cognitive, and psychological dysfunctions linked to neurological events. This aspect could be crucial, especially in countries with a scarcity of socio-economic resources. In addition, the recent COVID pandemic restrictions have also made it clear that telemedicine could become a valid and desirable means to continue health care at home, especially in frail subjects, even in the advanced economies. The current pandemic scenario has highlighted the need to offer telemedicine, remote assistance, and monitoring services when access to health facilities is limited or impossible. In this way, it would be possible to follow patients remotely, especially in cases of chronic and disabling diseases, avoiding the worsening of psycho-physical conditions and the onset of a sense of abandonment, as recent studies on the effects of the pandemic have shown. Several systems are now available for the remote monitoring of subjects with specific pathologies in the real world, including diabetes [91], osteoarthritis [92], respiratory dysfunctions [93], amyotrophic lateral sclerosis [94], and cognitive rehabilitation [95].

In this context, the system we described could permit the real-time integration of remote objective motor evaluation with exergames and ecological exercises, allowing more intensive rehabilitation training activities to be carried out at home with lower costs. This possibility represents a strength of the system: the specialists in the rehabilitation sector, in fact, claim greater effectiveness for the exercises defined as "ecological", that is, referring to the behaviors, habits, and activities of daily life. This result could be achieved through easy-to-use and low-cost solutions, such as the one we proposed, that constantly and intensively analyze, monitor, evaluate, stimulate, and rehabilitate the patients, developing their motor and cognitive potential to the maximum.

Nevertheless, the study has some limitations. At present, the small sample size could limit the generalization of the results to a broader range of disease severity and other neurological pathologies. In subsequent studies, we will extend the sample size by including more PD and PS subjects with wider disability scores to confirm our preliminary results, and we are planning to also collect data in-home settings. In addition, the following point must be made regarding the use of Microsoft Kinect© v2 as the RGB-D sensor of the proposed solution: although the device was discontinued a few years ago, it is still widely used for clinical research, as evidenced by several recent studies in the literature. In any case, the availability on the market of other RGB-D sensors will allow us to improve the current analysis and results using non-invasive technologies and vision-based approaches with higher performance and accuracy, as recent studies have shown [96,97].

It is important to underline that, in this study, only the spatio-temporal parameters and the COM sways were analyzed, even if other kinematic variables were calculated. However, it is well-known that spatio-temporal parameters are among the measurements most used by clinicians for evaluating gait patterns in post-stroke and parkinsonian patients. Effectively, the spatio-temporal parameters accurately represent the patient's ability to satisfy the general gait requirements, such as weight acceptance, support on a single limb, and progression of the swing [98]. They are also able to highlight the presence of asymmetrical gait strategies, due, for example, to anomalous stance phase, double support duration, slow walking speed, or shorter steps: these are among the main features considered in the clinical practice for the characterization of pathological gait patterns on which to evaluate, for example, the effects and benefits of rehabilitation treatments. By considering only reliable measurements, we can still obtain a significant clinical picture of the gait. In future studies, other relevant information on walking strategies could be analyzed more in-depth, such as upper limb movements and postural attitude during gait, thereby improving the overall characterization of pathological gait and focusing on the specific neurological condition of interest.

In conclusion, it is important to underline that even if this study focuses only on gait analysis, one of the most challenging motor tasks for vision-based solutions, it is part of a broader project on parkinsonian and post-stroke subjects. The project aims, in fact, to automate and comprehensively assess the motor conditions of pathological subjects through

RGB-D sensors and computer vision techniques, thus adopting non-invasive and easy-to-use approaches that are suitable for unsupervised contexts, such as the home setting.

## 5. Conclusions

This study proposed a vision-based solution for assessing gait patterns and dysfunctions due to neurological impairment associated with post-stroke and Parkinson's disease comorbidities by means of a single RGB-D sensor capable of capturing the 3D trajectories of body movements on a reduced walking path that is also suitable for domestic settings.

The qualitative analysis of the spatio-temporal parameters and COM demonstrated the capability of the RGB-D system to extract gait characteristics and differentiate specific gait patterns in agreement with the more complex and demanding gold standard 3D-GA. In fact, the data analysis confirmed that no difference between the two systems was statistically significant, except for step width, the results of which were in line with other reference studies: this indicates that the two systems measure the same quantities. It is noteworthy that some parameters, particularly cadence and ML sway, showed statistically significant differences between the two groups of subjects, suggesting that the system is also able to catch discriminatory parameters in populations with different gait patterns. However, a more in-depth analysis is necessary to confirm this finding by enlarging the sample size.

The ability to estimate gait parameters in agreement with standard 3D-GA systems is essential to monitor gait parameters in unsupervised settings, such as domestic environments, where traditional 3D-GA systems are not applicable. At this stage of development, immediate clinical implications relate to the possibility to schedule rehabilitation programs at home, both exclusively or in combination with hospitalization periods. In fact, the non-invasiveness, portability, and easy-to-use features of the proposed solution allow the combining of evaluation and rehabilitation tasks for home settings, thus defining new strategies for the management of neurological diseases and frail subjects. Future developments could be focused to the potential diagnostic capability of the system: in this case, the high sensibility of the system for detecting minimal alterations of movements could be exploited as a support for clinical judgments and differential diagnoses.

**Author Contributions:** Conceptualization, V.C., L.V., C.F. and L.P.; methodology, L.V., C.F., V.C. and L.P.; software, C.F., V.C., and G.A.; validation, V.C., C.F., and L.V.; formal analysis, V.C., C.F., G.A., L.V. and. R.C.; investigation, C.F., V.C., G.A., L.V. and R.C.; resources, L.V., C.F., V.C., and S.S.; data curation, V.C., C.F., L.V., and S.S.; writing—original draft preparation, V.C., C.F., L.V., R.L. and L.P.; writing—review and editing, C.F., V.C., L.V., G.A., and G.P.; visualization, G.A. and V.C.; supervision, L.P., M.G. and A.M.; project administration, G.P. and A.M.; funding acquisition, G.P. and A.M. All authors have read and agreed to the published version of the manuscript.

**Funding:** This work was supported by "ReHome—Soluzioni ICT per la tele-riabilitazione di disabilità cognitive e motorie originate da patologie neurologiche", Grant POR FESR 2014/2020 - Piattaforma Tecnologica Salute e Benessere from Regione Piemonte (Italy) and by the Department of Excellence Grant of the Italian Ministry of Education, University and Research to the 'Rita Levi Montalcini' Department of Neuroscience, University of Torino, Italy.

**Institutional Review Board Statement:** The study was conducted according to the guidelines of the Declaration of Helsinki, and approved by the Ethics Committee of COMITATO ETICO DELL'ISTITUTO AUXOLOGICO ITALIANO (Project identification code: ReHOME - 21A702; Codice CE 2020_02_18_01 and date of approval: February 18th, 2020).

**Informed Consent Statement:** Informed consent was obtained from all subjects involved in the study.

**Data Availability Statement:** Data available on request due to restrictions e.g., privacy or ethical.

**Acknowledgments:** L.V. would like to thank the PhD Programme in Experimental Medicine and Therapy of University of Turin.

**Conflicts of Interest:** The authors declare no conflict of interest.

## References


1. Patrick, L. (Ed.) *Ageing: Debate the Issues*; OECD Insights; OECD Publishing: Paris, France, 2015; ISBN 978-92-64-20660-1.
2. World Health Organization. *Neurological Disorders: Public Health Challenges*; World Health Organization: Geneva, Switzerland, 2006; ISBN 978-92-4-156336-9. Available online: https://apps.who.int/iris/handle/10665/43605 (accessed on 23 November 2021).
3. Avan, A.; Digaleh, H.; Di Napoli, M.; Stranges, S.; Behrouz, R.; Shojaeianbabaei, G.; Amiri, A.; Tabrizi, R.; Mokhber, N.; Spence, J.D.; et al. Socioeconomic status and stroke incidence, prevalence, mortality, and worldwide burden: An ecological analysis from the Global Burden of Disease Study 2017. *BMC Med.* **2019**, *17*, 191. [CrossRef] [PubMed]
4. Langhorne, P.; Bernhardt, J.; Kwakkel, G. Stroke rehabilitation. *Lancet* **2011**, *377*, 1693–1702. [CrossRef]
5. Takashima, R.; Murata, W.; Saeki, K. Movement changes due to hemiplegia in stroke survivors: A hermeneutic phenomenological study. *Disabil. Rehabil.* **2016**, *38*, 1578–1591. [CrossRef] [PubMed]
6. Aprile, I.; Piazzini, D.B.; Bertolini, C.; Caliandro, P.; Pazzaglia, C.; Tonali, P.; Padua, L. Predictive variables on disability and quality of life in stroke outpatients undergoing rehabilitation. *Neurol. Sci.* **2006**, *27*, 40–46. [CrossRef]
7. Chen, G.; Patten, C.; Kothari, D.H.; Zajac, F.E. Gait differences between individuals with post-stroke hemiparesis and non-disabled controls at matched speeds. *Gait Posture* **2005**, *22*, 51–56. [CrossRef]
8. Pringsheim, T.; Jette, N.; Frolkis, A.; Steeves, T.D. The prevalence of Parkinson's disease: A systematic review and meta-analysis. *Mov. Disord.* **2014**, *29*, 1583–1590. [CrossRef]
9. Balestrino, R.; Schapira, A.H.V. Parkinson disease. *Eur. J. Neurol.* **2020**, *27*, 27–42. [CrossRef]
10. Armstrong, M.J.; Okun, M.S. Diagnosis and Treatment of Parkinson Disease: A Review. *JAMA* **2020**, *323*, 548–560. [CrossRef]
11. Mak, M.; Wong-Yu, I.S.; Shen, X.; Chung, C.L.-H. Long-term effects of exercise and physical therapy in people with Parkinson disease. *Nat. Rev. Neurol.* **2017**, *13*, 689–703. [CrossRef]
12. Chang, H.; Hsu, Y.; Yang, S.; Lin, J.; Wu, Z. A Wearable Inertial Measurement System With Complementary Filter for Gait Analysis of Patients with Stroke or Parkinson's Disease. *IEEE Access* **2016**, *4*, 8442–8453. [CrossRef]
13. Boudarham, J.; Roche, N.; Pradon, D.; Bonnyaud, C.; Bensmail, D.; Zory, R. Variations in kinematics during clinical gait analysis in stroke patients. *PLoS ONE* **2013**, *8*, e66421. [CrossRef] [PubMed]
14. Chen, P.H.; Wang, R.L.; Liou, D.J.; Shaw, J.S. Gait Disorders in Parkinson's Disease: Assessment and Management. *Int. J. Gerontol.* **2013**, *7*, 189–193. [CrossRef]
15. Beyaert, C.; Vasa, R.; Frykberg, G.E. Gait post-stroke: Pathophysiology and rehabilitation strategies. *Neurophysiol. Clin.* **2015**, *45*, 335–355. [CrossRef] [PubMed]
16. Wonsetler, E.C.; Bowden, M.G. A systematic review of mechanisms of gait speed change post-stroke. Part 2: Exercise capacity, muscle activation, kinetics, and kinematics. *Top. Stroke Rehabil.* **2017**, *24*, 394–403. [CrossRef] [PubMed]
17. Peppe, A.; Chiavalon, C.; Pasqualetti, P.; Crovato, D.; Caltagirone, C. Does gait analysis quantify motor rehabilitation efficacy in Parkinson's disease patients? *Gait Posture* **2007**, *26*, 452–462. [CrossRef] [PubMed]
18. Pau, M.; Corona, F.; Pili, R.; Casula, C.; Guicciardi, M.; Cossu, G.; Murgia, M. Quantitative Assessment of Gait Parameters in People with Parkinson's Disease in Laboratory and Clinical Setting: Are the Measures Interchangeable? *Neurol. Int.* **2018**, *10*, 69–73. [CrossRef]
19. Pistacchi, M.; Gioulis, M.; Sanson, F.; de Giovannini, E.; Filippi, G.; Rossetto, F.; Marsala, S.Z. Gait analysis and clinical correlations in early Parkinson's disease. *Funct. Neurol.* **2017**, *32*, 28–34. [CrossRef]
20. McGinley, J.L.; Baker, R.; Wolfe, R.; Morris, M.E. The reliability of three-dimensional kinematic gait measurements: A systematic review. *Gait Posture* **2009**, *29*, 360–369. [CrossRef]
21. van den Noort, J.C.; Ferrari, A.; Cutti, A.G.; Becher, J.G.; Harlaar, J. Gait analysis in children with cerebral palsy via inertial and magnetic sensors. *Med. Biol. Eng. Comput.* **2013**, *51*, 377–386. [CrossRef]
22. Buganè, F.; Benedetti, M.G.; Casadio, G.; Attala, S.; Biagi, F.; Manca, M.; Leardini, A. Estimation of spatial-temporal gait parameters in level walking based on a single accelerometer: Validation on normal subjects by standard gait analysis. *Comput. Methods Programs Biomed.* **2012**, *108*, 129–137. [CrossRef]
23. Cimolin, V.; Capodaglio, P.; Cau, N.; Galli, M.; Santovito, C.; Patrizi, A.; Tringali, G.; Sartorio, A. Computation of spatio-temporal parameters in level walking using a single inertial system in lean and obese adolescents. *Biomed. Tech.* **2017**, *62*, 505–511. [CrossRef] [PubMed]
24. Pogrzeba, L.; Wacker, M.; Jung, B. Potentials of a Low-Cost Motion Analysis System for Exergames in Rehabilitation and Sports Medicine. In *E-Learning and Games for Training, Education, Health and Sports*; Göbel, S., Müller, W., Urban, B., Wiemeyer, J., Eds.; Lecture Notes in Computer Science Book Series; Springer: Berlin/Heidelberg, Germany, 2012; Volume 7156, pp. 125–133. [CrossRef]
25. Gonzalez-Ortega, D.; Diaz-Pernas, F.J.; Martinez-Zarzuela, M.; Anton-Rodriguez, M. A Kinect-based system for cognitive rehabilitation exercises monitoring. *Comput. Methods Programs Biomed.* **2014**, *113*, 620–631. [CrossRef] [PubMed]
26. Clark, R.A.; Mentiplay, B.F.; Hough, E.; Pua, Y.H. Three-dimensional cameras and skeleton pose tracking for physical function assessment: A review of uses, validity, current developments and Kinect alternatives. *Gait Posture* **2019**, *68*, 193–200. [CrossRef] [PubMed]
27. Sathyanarayana, S.; Satzoda, R.; Sathyanarayana, S.; Thambipillai, S. Vision-based patient monitoring: A comprehensive review of algorithms and technologies. *J. Ambient. Intell. Humaniz. Comput.* **2018**, *9*, 225–251. [CrossRef]

28. Zhang, H.-B.; Zhang, Y.-X.; Zhong, B.; Lei, Q.; Yang, L.; Du, J.-X.; Chen, D.-S. A Comprehensive Survey of Vision-Based Human Action Recognition Methods. *Sensors* **2019**, *19*, 1005. [CrossRef]

29. Da Gama, A.; Fallavollita, P.; Teichrieb, V.; Navab, N. Motor Rehabilitation Using Kinect: A Systematic Review. *Games Health J.* **2015**, *4*, 123–135. [CrossRef]

30. Saenz-de-Urturi, Z.; Garcia-Zapirain Soto, B. Kinect-Based Virtual Game for the Elderly that Detects Incorrect Body Postures in Real Time. *Sensors* **2016**, *16*, 704. [CrossRef]

31. Capece, N.; Erra, U.; Romaniello, G. A Low-Cost Full Body Tracking System in Virtual Reality Based on Microsoft Kinect. Augmented Reality, Virtual Reality, and Computer Graphics. AVR 2018. In *Augmented Reality, Virtual Reality, and Computer Graphics*; De Paolis, L., Bourdot, P., Eds.; Lecture Notes in Computer Science Book Series; Springer: Cham, Switzerland, 2018; Volume 10851. [CrossRef]

32. Springer, S.; Seligmann, G.Y. Validity of the Kinect for Gait Assessment: A Focused Review. *Sensors* **2016**, *16*, 194. [CrossRef]

33. Gabel, M.; Gilad-Bachrach, R.; Renshaw, E.; Schuster, A. Full body gait analysis with Kinect. In Proceedings of the IEEE International Conference on Engineering in Medicine and Biology Society (EMBC), San Diego, CA, USA, 28 August–1 September 2012; pp. 1964–1967. [CrossRef]

34. Narayan, J.; Pardasani, A.; Dwivedy, S.K. Comparative Gait Analysis of Healthy Young Male and Female Adults using Kinect-Labview Setup. In Proceedings of the International Conference on Computational Performance Evaluation (ComPE), Shillong, India, 2–4 July 2020; pp. 688–693. [CrossRef]

35. Clark, R.A.; Bower, K.J.; Mentiplay, B.F.; Paterson, K.; Pua, Y.H. Concurrent validity of the Microsoft Kinect for assessment of spatiotemporal gait variables. *J. Biomech.* **2013**, *46*, 2722–2725. [CrossRef]

36. Motiian, S.; Pergami, P.; Guffey, K.; Mancinelli, C.A.; Doretto, G. Automated extraction and validation of children's gait parameters with the Kinect. *Biomed. Eng. OnLine* **2015**, *14*, 112. [CrossRef]

37. Dolatabadi, E.; Taati, B.; Mihailidis, A. Concurrent validity of the Microsoft Kinect for Windows v2 for measuring spatiotemporal gait parameters. *Med. Eng. Phys.* **2016**, *38*, 952–958. [CrossRef]

38. Kuan, Y.W.; Ee, N.O.; Wei, L.S. Comparative Study of Intel R200, Kinect v2, and Primesense RGB-D Sensors Performance Outdoors. *IEEE Sens. J.* **2019**, *19*, 8741–8750. [CrossRef]

39. Gonzalez-Jorge, H.; Rodríguez-González, P.; Martínez-Sánchez, J.; González-Aguilera, D.; Arias, P.; Gesto, M.; Díaz-Vilariño, L. Metrological comparison between Kinect I and Kinect II sensors. *Meas. J. Int. Meas. Confed.* **2015**, *70*, 21–26. [CrossRef]

40. Geerse, D.J.; Coolen, B.H.; Roerdink, M. Kinematic Validation of a Multi-Kinect v2 Instrumented 10-Meter Walkway for Quantitative Gait Assessments. *PLoS ONE* **2015**, *10*, e0139913. [CrossRef] [PubMed]

41. Müller, B.; Ilg, W.; Giese, M.A.; Ludolph, N. Validation of enhanced kinect sensor based motion capturing for gait assessment. *PLoS ONE* **2017**, *12*, e0175813. [CrossRef]

42. Auvinet, E.; Multon, F.; Meunier, J. New Lower-Limb Gait Asymmetry Indices Based on a Depth Camera. *Sensors* **2015**, *15*, 4605–4623. [CrossRef]

43. Xu, X.; McGorry, R.W.; Chou, L.S.; Lin, J.H.; Chang, C.C. Accuracy of the Microsoft Kinect™ for measuring gait parameters during treadmill walking. *Gait Posture* **2015**, *42*, 145–151. [CrossRef]

44. Ma, Y.; Mithraratne, K.; Wilson, N.C.; Wang, X.; Ma, Y.; Zhang, Y. The Validity and Reliability of a Kinect v2-Based Gait Analysis System for Children with Cerebral Palsy. *Sensors* **2019**, *19*, 1660. [CrossRef]

45. Summa, S.; Tartarisco, G.; Favetta, M.; Buzachis, A.; Romano, A.; Bernava, G.; Vasco, G.; Pioggia, G.; Petrarca, M.; Castelli, E.; et al. Spatio-temporal parameters of ataxia gait dataset obtained with the Kinect. *Data Brief* **2020**, *32*, 106307. [CrossRef]

46. Álvarez, I.; Latorre, J.; Aguilar, M.; Pastor, P.; Llorens, R. Validity and sensitivity of instrumented postural and gait assessment using low-cost devices in Parkinson's disease. *J. Neuroeng. Rehabil.* **2020**, *17*, 149. [CrossRef]

47. Vilas-Boas, M.D.C.; Rocha, A.P.; Choupina, H.M.P.; Cardoso, M.N.; Fernandes, J.M.; Coelho, T.; Cunha, J.P.S. Validation of a Single RGB-D Camera for Gait Assessment of Polyneuropathy Patients. *Sensors* **2019**, *19*, 4929. [CrossRef] [PubMed]

48. Dubois, A.; Bresciani, J.P. Validation of an ambient system for the measurement of gait parameters. *J. Biomech.* **2018**, *69*, 175–180. [CrossRef] [PubMed]

49. Morgavi, G.; Nerino, R.; Marconi, L.; Cutugno, P.; Ferraris, C.; Cinini, A.; Morando, M. An Integrated Approach to the Well-Being of the Elderly People at Home. In *Ambient Assisted Living. Biosystems & Biorobotics*; Andò, B., Siciliano, P., Marletta, V., Monteriù, A., Eds.; Springer: Cham, Switzerland, 2015; Volume 11. [CrossRef]

50. Albani, G.; Ferraris, C.; Nerino, R.; Chimienti, A.; Pettiti, G.; Parisi, F.; Ferrari, G.; Cau, N.; Cimolin, V.; Azzaro, C.; et al. An Integrated Multi-Sensor Approach for the Remote Monitoring of Parkinson's Disease. *Sensors* **2019**, *19*, 4764. [CrossRef] [PubMed]

51. Ferraris, C.; Nerino, R.; Chimienti, A.; Pettiti, G.; Cau, N.; Cimolin, V.; Azzaro, C.; Priano, L.; Mauro, A. Feasibility of Home-Based Automated Assessment of Postural Instability and Lower Limb Impairments in Parkinson's Disease. *Sensors* **2019**, *19*, 1129. [CrossRef] [PubMed]

52. Bower, K.; Thilarajah, S.; Pua, Y.-H.; Williams, G.; Tan, D.; Mentiplay, B.; Denehy, L.; Clark, R. Dynamic balance and instrumented gait variables are independent predictors of falls following stroke. *J. Neuroeng. Rehabil.* **2019**, *16*, 3. [CrossRef] [PubMed]

53. Kim, W.S.; Cho, S.; Baek, D.; Bang, H.; Paik, N.J. Upper Extremity Functional Evaluation by Fugl-Meyer Assessment Scoring Using Depth-Sensing Camera in Hemiplegic Stroke Patients. *PLoS ONE* **2016**, *11*, e0158640. [CrossRef]

54. Lloréns, R.; Alcañiz, M.; Colomer, C.; Navarro, M.D. Balance recovery through virtual stepping exercises using Kinect skeleton tracking: A follow-up study with chronic stroke patients. In *Annual Review of Cybertherapy and Telemedicine*; Wiederhold, B., Riva, G., Eds.; IOS Press: Amsterdam, The Netherlands, 2012; Volume 181, pp. 108–112.
55. Aşkın, A.; Atar, E.; Koçyiğit, H.; Tosun, A. Effects of Kinect-based virtual reality game training on upper extremity motor recovery in chronic stroke. *Somatosens. Mot. Res.* **2018**, *35*, 25–32. [CrossRef]
56. Latorre, J.; Llorens, R.; Colomer, C.; Alcaniz, M. Reliability and comparison of Kinect-based methods for estimating spatiotemporal gait parameters of healthy and post-stroke individuals. *J. Biomech.* **2018**, *72*, 268–273. [CrossRef]
57. Clark, R.A.; Vernon, S.; Mentiplay, B.F.; Miller, K.J.; McGinley, J.L.; Pua, Y.H.; Paterson, K.; Bower, K.J. Instrumenting gait assessment using the Kinect in people living with stroke: Reliability and association with balance tests. *J. Neuroeng. Rehabil.* **2015**, *12*, 15. [CrossRef]
58. Rocha, A.P.; Choupina, H.; Fernandes, J.M.; Rosas, M.J.; Vaz, R.; Silva Cunha, J.P. Kinect v2 based system for Parkinson's disease assessment. In Proceedings of the IEEE International Conference on Engineering in Medicine and Biology Society (EMBC), Milan, Italy, 25–29 August 2015; pp. 1279–1282. [CrossRef]
59. Galna, B.; Barry, G.; Jackson, D.; Mhiripiri, D.; Olivier, P.; Rochester, L. Accuracy of the Microsoft Kinect sensor for measuring movement in people with Parkinson's disease. *Gait Posture* **2014**, *39*, 1062–1068. [CrossRef]
60. Cao, Y.; Li, B.Z.; Li, Q.N.; Xie, J.D.; Cao, B.Z.; Yu, S.Y. Kinect-based gait analyses of patients with Parkinson's disease, patients with stroke with hemiplegia, and healthy adults. *CNS Neurosci. Ther.* **2017**, *23*, 447–449. [CrossRef]
61. Latorre, J.; Colomer, C.; Alcaniz, M.; Llorens, R. Gait analysis with the Kinect v2: Normative study with healthy individuals and comprehensive study of its sensitivity, validity, and reliability in individuals with stroke. *J. NeuroEng. Rehabil.* **2019**, *16*, 97. [CrossRef]
62. Salonini, E.; Gambazza, S.; Meneghelli, I.; Tridello, G.; Sanguanini, M.; Cazzarolli, C.; Zanini, A.; Assael, B.M. Active video game playing in children and adolescents with cystic fibrosis: Exercise or just fun? *Respir. Care* **2015**, *60*, 1172–1179. [CrossRef]
63. Zoccolillo, L.; Morelli, D.; Cincotti, F.; Muzzioli, L.; Gobbetti, T.; Paolucci, S.; Iosa, M. Video-game based therapy performed by children with cerebral palsy: A cross-over randomized controlled trial and a cross-sectional quantitative measure of physical activity. *Eur. J. Phys. Rehabil. Med.* **2015**, *51*, 669–676.
64. Vukićević, S.; Đorđević, M.; Glumbić, N.; Bogdanović, Z.; Đurić Jovičić, M. A Demonstration Project for the Utility of Kinect-Based Educational Games to Benefit Motor Skills of Children with ASD. *Percept. Mot. Skills* **2019**, *126*, 1117–1144. [CrossRef]
65. Gonsalves, L.; Campbell, A.; Jensen, L.; Straker, L. Children With Developmental Coordination Disorder Play Active Virtual Reality Games Differently Than Children With Typical Development. *Phys. Ther.* **2015**, *95*, 360–368. [CrossRef]
66. Eltoukhy, M.; Oh, J.; Kuenze, C.; Signorile, J. Improved kinect-based spatiotemporal and kinematic treadmill gait assessment. *Gait Posture* **2017**, *51*, 77–83. [CrossRef]
67. Pfister, A.; West, A.M.; Bronner, S.; Noah, J.A. Comparative abilities of Microsoft Kinect and Vicon 3D motion capture for gait analysis. *J. Med. Eng. Technol.* **2014**, *38*, 274–280. [CrossRef]
68. Trojaniello, D.; Ravaschio, A.; Hausdorff, J.M.; Cereatti, A. Comparative assessment of different methods for the estimation of gait temporal parameters using a single inertial sensor: Application to elderly, post-stroke, Parkinson's disease and Huntington's disease subjects. *Gait Posture* **2015**, *42*, 310–316. [CrossRef]
69. Schlachetzki, J.C.M.; Barth, J.; Marxreiter, F.; Gossler, J.; Kohl, Z.; Reinfelder, S.; Gassner, H.; Aminian, K.; Eskofier, B.M.; Winkler, J.; et al. Wearable sensors objectively measure gait parameters in Parkinson's disease. *PLoS ONE* **2017**, *12*, e0183989. [CrossRef]
70. Perumal, S.V.; Sankar, R. Gait and tremor assessment for patients with Parkinson's disease using wearable sensors. *ICT Express* **2016**, *2*, 168–174. [CrossRef]
71. Eltoukhy, M.; Kuenze, C.; Oh, J.; Jacopetti, M.; Wooten, S.; Signorile, J. Microsoft Kinect can distinguish differences in over-ground gait between older persons with and without Parkinson's disease. *Med. Eng. Phys.* **2017**, *44*, 1–7. [CrossRef] [PubMed]
72. Ferraris, C.; Cimolin, V.; Vismara, L.; Votta, V.; Amprimo, G.; Cremascoli, R.; Galli, M.; Nerino, R.; Mauro, A.; Priano, L. Monitoring of Gait Parameters in Post-Stroke Individuals: A Feasibility Study Using RGB-D Sensors. *Sensors* **2021**, *21*, 5945. [CrossRef] [PubMed]
73. Jagos, H.; David, V.; Reichel, M.; Kotzian, S.; Schlossarek, S.; Haller, M.; Rafolt, D. Tele-monitoring of the rehabilitation progress in stroke patients. *Stud. Health Technol. Inform.* **2015**, *211*, 311–313. [PubMed]
74. Garcia-Agundez, A.; Folkerts, A.K.; Konrad, R.; Caserman, P.; Tregel, T.; Goosses, M.; Göbel, S.; Kalbe, E. Recent advances in rehabilitation for Parkinson's Disease with Exergames: A Systematic Review. *J. Neuroeng. Rehabil.* **2019**, *16*, 17. [CrossRef] [PubMed]
75. Bamford, J.; Sandercock, P.; Dennis, M.; Warlow, C.; Burn, J. Classification and natural history of clinically identifiable subtypes of cerebral infarction. *Lancet* **1991**, *337*, 1521–1526. [CrossRef]
76. Shotton, J.; Fitzgibbon, A.; Cook, M.; Sharp, T.; Finocchio, M.; Moore, R.; Kipman, A.; Blake, A. Real-time Human Pose Recognition in Parts from Single Depth Images. Machine learning for Computer Vision. In *Studies in Computational Intelligence*; Cipolla, R., Battiato, S., Farinella, G., Eds.; Springer: Berlin/Heidelberg, Germany, 2013; Volume 411, pp. 119–135. [CrossRef]
77. Davis, R.B.; Ounpuu, S.; Tyburski, D.; Gage, J.R. A gait analysis data collection and reduction technique. *Hum. Mov. Sci.* **1991**, *10*, 575–587. [CrossRef]

78. Wang, Q.; Kurillo, G.; Ofli, F.; Bajcsy, R. Evaluation of Pose Tracking Accuracy in the First and Second Generations of Microsoft Kinect. In Proceedings of the International Conference on Healthcare Informatics, Dallas, TX, USA, 21–23 October 2015; pp. 380–389. [CrossRef]

79. Gianaria, E.; Grangetto, M. Robust gait identification using Kinect dynamic skeleton data. *Multimed. Tools Appl.* **2019**, *78*, 13925–13948. [CrossRef]

80. Da Gama, A.E.F.; Chaves, T.M.; Fallavollita, P.; Figueiredo, L.S.; Teichrieb, V. Rehabilitation motion recognition based on the international biomechanical standards. *Expert Syst. Appl.* **2019**, *116*, 396–409. [CrossRef]

81. Perry, J.; Burnfield, J.M. Gait Analysis: Normal and Pathological Function. *J. Sports Sci. Med.* **2010**, *9*, 353. [CrossRef]

82. do Carmo, A.A.; Kleiner, A.F.; Barros, R.M. Alteration in the center of mass trajectory of patients after stroke. *Top. Stroke Rehabil.* **2015**, *22*, 349–356. [CrossRef] [PubMed]

83. Tesio, L.; Rota, V. The Motion of Body Center of Mass During Walking: A Review Oriented to Clinical Applications. *Front. Neurol.* **2019**, *10*, 999. [CrossRef] [PubMed]

84. Tisserand, R.; Robert, T.; Dumas, R.; Chèze, L. A simplified marker set to define the center of mass for stability analysis in dynamic situations. *Gait Posture* **2016**, *48*, 64–67. [CrossRef] [PubMed]

85. Koo, T.K.; Li, M.Y. A guideline of selecting and reporting intraclass correlation coefficients for reliability research. *J. Chiropr. Med.* **2016**, *15*, 155–163. [CrossRef]

86. Giavarina, D. Understanding Bland Altman analysis. *Biochem. Med.* **2015**, *25*, 141–151. [CrossRef]

87. Shankaranarayana, A.M.; Gururaj, S.; Natarajan, M.; Balasubramanian, C.K.; Solomon, J.M. Gait training interventions for patients with stroke in India: A systematic review. *Gait Posture* **2021**, *83*, 132–140. [CrossRef]

88. Sarfo, F.S.; Ulasavets, U.; Opare-Sem, O.K.; Ovbiagele, B. Tele-Rehabilitation after Stroke: An Updated Systematic Review of the Literature. *J. Stroke Cerebrovasc. Dis.* **2018**, *27*, 2306–2318. [CrossRef]

89. Schwamm, L.; Holloway, R.G.; Amarenco, P.; Audebert, H.J.; Bakas, T.; Chumbler, N.R.; Handschu, R.; Jauch, E.; Knight, W.A.; Levine, S.R.; et al. A review of the evidence for the use of telemedicine within stroke systems of care: A scientific statement from the American Heart Association/American Stroke Association. *Stroke* **2009**, *40*, 2616–2634. [CrossRef]

90. Johannson, T.; Wild, C. Telerehabilitation in stroke care–a systematic review. *J. Telemed. Telecare* **2011**, *17*, 1–6. [CrossRef]

91. Rodriguez-León, C.; Villalonga, C.; Munoz-Torres, M.; Ruiz, J.R.; Banos, O. Mobile and Wearable Technology for the Monitoring of Diabetes-Related Parameters: Systematic Review. *JMIR Mhealth Uhealth* **2021**, *9*, e25138. [CrossRef]

92. Cudejko, T.; Button, K.; Willott, J.; Al-Amri, M. A pplications of Wearable Technology in a Real-Life Setting in People with Knee Osteoarthritis: A Systematic Scoping Review. *J. Clin. Med.* **2021**, *10*, 5645. [CrossRef] [PubMed]

93. Angelucci, A.; Kuller, D.; Aliverti, A. A Home Telemedicine System for Continuous Respiratory Monitoring. *IEEE J. Biomed. Health Inform.* **2021**, *25*, 1247–1256. [CrossRef] [PubMed]

94. De Marchi, F.; Cantello, R.; Ambrosini, S.; Mazzini, L.; CANPALS Study Group. Telemedicine and technological devices for amyotrophic lateral sclerosis in the era of COVID-19. *Neurol. Sci.* **2020**, *41*, 1365–1367. [CrossRef] [PubMed]

95. Mantovani, E.; Zucchella, C.; Bottiroli, S.; Federico, A.; Giugno, R.; Sandrini, G.; Chiamulera, C.; Tamburin, S. Telemedicine and Virtual Reality for Cognitive Rehabilitation: A Roadmap for the COVID-19 Pandemic. *Front. Neurol.* **2020**, *11*, 926. [CrossRef] [PubMed]

96. Albert, J.A.; Owolabi, V.; Gebel, A.; Brahms, C.M.; Granacher, U.; Arnrich, B. Evaluation of the Pose Tracking Performance of the Azure Kinect and Kinect v2 for Gait Analysis in Comparison with a Gold Standard: A Pilot Study. *Sensors* **2020**, *20*, 5104. [CrossRef] [PubMed]

97. Yeung, L.F.; Yang, Z.; Cheng, K.C.C.; Du, D.; Tong, R.K.Y. Effects of camera viewing angles on tracking kinematic gait patterns using Azure Kinect, Kinect v2 and Orbbec Astra Pro v2. *Gait Posture* **2021**, *87*, 19–26. [CrossRef]

98. Gouelle, A.; Mégrot, F. Interpreting spatiotemporal parameters, symmetry, and variability in clinical gait analysis. In *Handbook of Human Motion*; Muller, B., Wolf, S.I., Brueggemann, G.P., Deng, Z., McIntosh, A., Miller, F., Selbie, W.S., Eds.; Springer: Cham, Switzerland, 2016; pp. 1–20. [CrossRef]

*Article*

# Evaluation and Application of a Customizable Wireless Platform: A Body Sensor Network for Unobtrusive Gait Analysis in Everyday Life

**Markus Lueken** [1,*]**, Leo Mueller** [1]**, Michel G. Decker** [1]**, Cornelius Bollheimer** [2]**, Steffen Leonhardt** [1] **and Chuong Ngo** [1]

[1]    Medical Information Technology, RWTH Aachen University, Pauwelsstr. 20, 52074 Aachen, Germany; leo.mueller@rwth-aachen.de (L.M.); michel.georges.decker@rwth-aachen.de (M.G.D.); leonhardt@hia.rwth-aachen.de (S.L.); ngo@hia.rwth-aachen.de (C.N.)

[2]    Department of Geriatrics, RWTH Aachen University Hospital, Pauwelsstr. 30, 52074 Aachen, Germany; cbollheimer@ukaachen.de

*    Correspondence: lueken@hia.rwth-aachen.de; Tel.: +49-241-8023515

Received: 12 November 2020; Accepted: 14 December 2020; Published: 20 December 2020

**Abstract:** Body sensor networks (BSNs) represent an important research tool for exploring novel diagnostic or therapeutic approaches. They allow for integrating different measurement techniques into body-worn sensors organized in a network structure. In 2011, the first Integrated Posture and Activity Network by MedIT Aachen (IPANEMA) was introduced. In this work, we present a recently developed platform for a wireless body sensor network with customizable applications based on a proprietary 868 MHz communication interface. In particular, we present a sensor setup for gait analysis during everyday life monitoring. The arrangement consists of three identical inertial measurement sensors attached at the wrist, thigh, and chest. We additionally introduce a force-sensitive resistor integrated insole for measurement of ground reaction forces (GRFs), to enhance the assessment possibilities and generate ground truth data for inertial measurement sensors. Since the 868 MHz is not strongly represented in existing BSN implementations, we validate the proposed system concerning an application in gait analysis and use this as a representative demonstration of realizability. Hence, there are three key aspects of this project. The system is evaluated with respect to (I) accurate timing, (II) received signal quality, and (III) measurement capabilities of the insole pressure nodes. In addition to the demonstration of feasibility, we achieved promising results regarding the extractions of gait parameters (stride detection accuracy: 99.6 ± 0.8%, Root-Mean-Square Deviation (RMSE) of mean stride time: 5 ms, RMSE of percentage stance time: 2.3%). Conclusion: With the satisfactory technical performance in laboratory and application environment and the convincing accuracy of the gait parameter extraction, the presented system offers a solid basis for a gait monitoring system in everyday life.

**Keywords:** body sensor network; gait analysis; inertial sensors; ground reaction force

## 1. Introduction

Healthcare costs have risen massively in recent years and the trend does not seem to be reversed. According to the United States (US) National Health Expenditure, spending has tripled in the US since 1995 and amounted to 18% of GDP in 2018 [1]. With an increasing life expectancy together with a stagnating birth rate in the corresponding countries, an aging society is being observed. The actual effects of this development are difficult to estimate and are the subject of current studies. It can be assumed that the burden on health care systems will continue to increase. Therefore, new cost-effective concepts for medical care are needed to maintain the standard of living and the quality of care.

Body sensor network (BSN), alternatively wireless body area network (WBAN), technology could make a decisive contribution to relieving the burden on health systems and thus keeping them viable for society in the future [2,3]. A (wireless) body sensor network as defined in the IEEE 802.15.6 standard is a specific wireless network realization that combines body-worn sensors of a single (human or animal) being into a dedicated network. By definition, this makes the most important difference to wearables in the conventional sense or single body-worn sensors, such as a Holter ECG. BSNs, which in general are not strictly limited to wireless communication, have originally been introduced for data acquisition in personalized healthcare. Portable measurement systems can facilitate patient monitoring and alert healthcare professionals in emergencies such as heart attacks or falls. The use of these systems in everyday life could provide access to data that were previously difficult to be obtained and could offer new therapeutic approaches [4]. These inconspicuous interventions could lead to immense financial long-term savings and a considerable improvement in the quality of life. To increase the acceptance of the devices, the sensors must not become a burden for their wearers. In this sense, unobtrusiveness applies to many different aspects, such as wearing location, battery charging, handling for both patients and medical staff, and many others.

The choice of the radio–communication interface has to be carefully considered during the development process since a variety of different communication standards is eligible for the data transfer within a BSN. The carrier-frequency is directly connected to the respective communication standard (Wi-Fi: 2.4 GHz/5 GHz, Bluetooth: 2.4 GHz, ZigBee: 900 MHz/2.4 GHz). As stated in previous studies and theoretical considerations [5–7], lower frequency bands have lower signal attenuation as compared to the 2.4 GHz band. The Industrial, Scientific, and Medical (ISM) band and the Short Range Device (SRD) band provide additional frequency bands at 433 MHz and 868/915 MHz, respectively. These frequency bands allow for the implementation of an application-specific proprietary protocol under certain restrictions. Proprietary communication standards thus can avoid commonly known complications such as collisions with similar communication standards in the 2.4 GHz band, data rate limitations, high power consumption, or shadowing effects of human tissue towards higher frequencies. Especially, the latter has to be considered for a BSN application for gait analysis.

Many research projects have addressed mobile sensorized monitoring systems for varying health and fitness applications. Specifically, in the area of gait analysis, three sensor modalities have proven high usability for wearable monitoring systems: inertial measurement units (IMUs) including magnetic field sensors to monitor posture and motion, insole-integrated force sensing resistors to monitor ground reaction forces (GRFs), and surface-mounted electromyographic (sEMG) sensors to monitor muscular activity [8–12]. The latter is not considered for long-term monitoring due to the use of electrodes that have to be attached to the patient's skin. In contrast, the IMU and GRF sensors are extensively integrated into scientific research platforms in long-term monitoring applications for gait analysis [8,12].

IMUs, which include accelerometers and gyroscopes, were first proposed as a functional tool for gait analysis half a century ago [13] and has gained importance with the development of low-cost piezo-resistive and micro-electromechanical inertial sensors [14]. In recent years, IMUs were integrated into different mobile motion and activity monitoring devices [8,15–23]. These devices usually consist of one single node [23,24] or multiple sensing nodes that are organized in a BSN [8,18,20]. For both single and multiple node applications, different sensor locations were proposed and investigated for different types of targeting diseases.

GRF sensing devices usually consist of different kinds of pressure sensors that are attached to the insole or the shoe of the subject. Pressure sensors are implemented by commercially available sensors, such as force sensing resistors (FSRs) [25–30], flex sensors [29] and barometric pressure sensors [31], or customized solutions, such as electronic textiles-based pressure sensors [32,33], light-sensing chambers with pressure-actuated optical barriers [34] or carbon-embedded piezo-resistive materials [35].

In this work, a modular BSN for long-term motion monitoring is designed and tested. It follows the basic ideas of the Integrated Posture and Activity Network by MedIT Aachen (IPANEMA). The first version of IPANEMA was developed in 2007 [36]. In 2009 and 2011, the BSN was extended by version 2.0 and 2.5, respectively [37]. The third version was recently finalized with a completely revised design. A comparison of the IPANEMA v2.5 mainboard to the mainboard of IPANEMA v3.0 is shown in Figure 1. Table 1 additionally compares the main features of both versions. As an application scenario of the recently developed platform, we present a sensor configuration for unobtrusive gait analysis within this contribution. The sensor network consists of three identical IMU sensors and a single FSR-based insole for recording of GRFs. Since the IMU-based evaluation for this study was already presented in previous publications [38,39], this specific contribution focuses on the evaluation of the technical reliability of the whole system and the extension by an FSR-based GRF sensor.

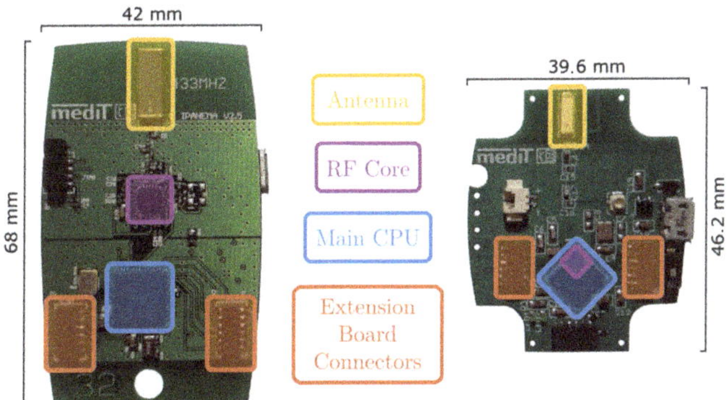


**Figure 1.** Comparison of the board layouts of the Integrated Posture and Activity Network by MedIT Aachen (IPANEMA) v2.5 (**left**) and v3.0 mainboards (**right**). The most important components are color-coded for each board.

**Table 1.** Comparison of main characteristics of the IPANEMA v2.5 and v3.0 mainboard.

| Feature | IPANEMA v2.5 (2011) | IPANEMA v3.0 (2019) |
| --- | --- | --- |
| Controller | MSP430F1611 | CC1310 |
| Architecture | 16-Bit RISC | ARM Cortex-M3 |
| RF Transceiver | CC1101 | integrated |
| Frequency Band | 433 MHz | 868 MHz/915 MHz |
| max. TX Power | 10 dBm | 14 dBm |
| TX Current * | 29.2 mA | 13.4 mA |
| RX Current | 15.7 mA | 5.5 mA |
| Battery | 330 mAh | 230 mAh |
| Operating Time ** | ca. 4 h | ca. 12 h |
| max. Data Rate | 500 kBit/s | 4 MBit/s |
| Sensitivity | −116 dBm | −124 dBm |

* At transmission output power of 10 dBm, ** mainboard equipped with an inertial measurement sensor.

## 2. Methods and Material

### 2.1. The IPANEMA Mainboard

A single network node of a BSN basically contains a CPU, a radio unit (i.e., transceiver and antenna), an energy management system, and specific measurement circuits for the acquisition of physiological or biomechanical parameters. In the case of IPANEMA, each node is equipped with a mainboard, a sensor board, and a 230 mAh lithium-ion battery. The mainboard contains the CPU, the RF transceiver and an antenna, the power supply unit, and a sensor controller (SC) that directly connects to the sensor boards via two socket-connectors (compare (cf.) Figure 1). The key component of the IPANEMA v3.0 mainboard is the CC1310 sub-1 GHz wireless micro-controller (Texas Instruments). It contains an ARM Cortex-M3 main CPU, an RF controller, an integrated sensor controller, and several peripherals such as a Real-Time Clock (RTC) module, an AES-128 encryption module, a Direct Memory Access (DMA) controller, different communication interfaces, and timer modules. The RF controller is based on an ARM Cortex-M0 architecture. It supports several sub-1 GHz ISM and SRD bands such as the 433 MHz, 868 MHz, and 915 MHz. In contrast to IPANEMA v2.5, the physical layer of v3.0 is based on the 868/915 MHz band. To overcome communication conflicts, European regulations prescribe channel sensing for duty cycle-free occupancy of these sub-bands [40]. In the IPANEMA v3.0 protocol stack, Polite Spectrum Access (PSA) is implemented by a Listen-Before-Talk (LBT) approach, where the current Receive Signal Strength Indicator (RSSI) is used to determine the channel occupancy.

The basic idea of the IPANEMA BSN is to serve as a flexible scientific communication platform for different sensor modalities. Therefore, the mainboard contains expansion ports to connect to different types of sensor boards. The connectors contain the 3.3 V supply and battery voltage, and additional pins that directly connect to the multi-functional general-purpose input/output (GPIO) pins of the integrated SC. The SC operates completely independent from the main CPU and organizes the communication with the respective sensor board, i.e., the initialization and configuration process, reading data from analog or digital sensors, and triggering the CPU.

For communicating with the sensor boards, the SC provides different standard interfaces such as the Inter-Integrated Circuit ($I^2C$) bus, the Inter-IC Sound ($I^2S$) bus, the Serial Peripheral Interface (SPI), and the Universal Asynchronous Receiver Transmitter (UART) bus. It also contains an integrated 12-bit Analog-to-Digital Converter (ADC), several interrupt functions, comparators, and counting modules to interact with measurement circuits. The sampling rate can be adapted individually for each different sensor modality within the SC configuration. The SC and the main CPU share a dedicated memory space to exchange data to be further processed and sent over the network. The main CPU organizes the communication following the proprietary network protocol. Sensor data are retrieved from the shared memory and transmitted in network packets. The data are received by the master node and either forwarded to a PC system via a USB port or stored on an integrated micro SD card for long-term monitoring purposes. For telehealth applications, the data are additionally transferred to large-scale memory servers for further processing the obtained data. A graphical overview of the system is given in Figure 2.

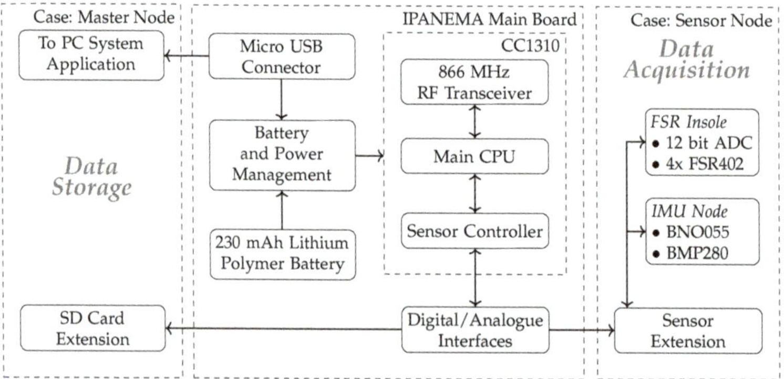


**Figure 2.** The IPANEMA mainboard can be equipped with either the master node extension board or a specific sensor board. The master board contains a USB port that establishes the data connection to the corresponding PC application. The sensor board is equipped with different types of sensors depending on the target modality.

## 2.2. Proprietary Network Protocol

The IPANEMA BSN consists of a master node (MN) and several sensor nodes (SNs) that are organized in a star-shaped network. The proprietary internal network protocol is based on this topology. The MN initializes the network structure for a predefined maximum number of SNs connecting to the MN. The communication is controlled by a time division multiple access (TDMA) process that assigns dedicated time slots to each SN connected. A higher-level communication frame allows for periodical re-synchronization of the SNs and network error handling by the MN. For this purpose, the MN sends re-synchronization beacons each second and eventually reconnects lost SNs. In this case, a careful re-synchronization is not only necessary to comply with the dedicated time slots of the TDMA protocol, but additionally, ensure an accurate time-stamping of a single sampling point for any sensing modality. After each SN has registered and received its individual unique network ID, the MN broadcasts the start beacon initiating the TDMA protocol on all nodes (Figure 3).

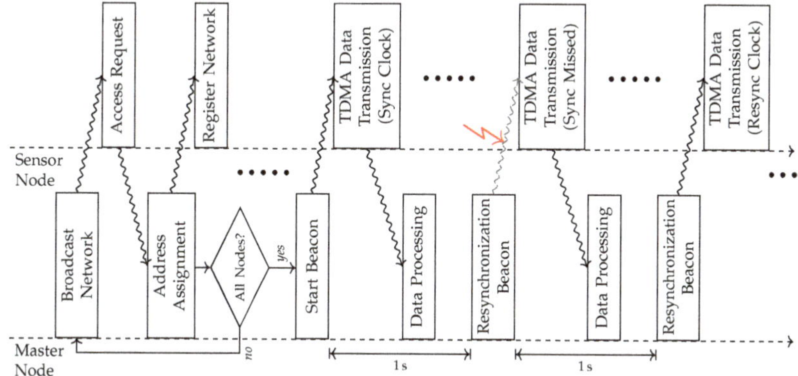


**Figure 3.** Timing diagram according to the communication procedure between the master node (MN) and sensor node (SN) including initialization routine and standard data acquisition procedure.

The core communication and sampling procedures take into account different timing modules to maintain reliably identical clocks. These include the MN core clock module, the SN core clock module, and the RF clock module of the respective SN. The MN clock provides the internal global network time $T_{MN}|_{sync}$ broadcasted within each re-synchronization beacon with a range of 32 bit and a resolution of $\Delta T_{LSB} = 10\,\mu s$. The RF module of each slave node captures the exact receiving time $T_{RF,SN}|_{sync}$ of the re-synchronization packet from the isolated RF clock. When a packet is received by the RF module, it triggers an event for further processing. The temporal delay between receiving and processing the packet is obtained from the RF clock module

$$\Delta T_{proc} = T_{RF,SN}|_{proc} - T_{RF,SN}|_{sync} \tag{1}$$

where the current sensor timestamp $T_{SN}|_{sync}$ and the processing timestamp $T_{RF,SN}|_{proc}$ are obtained simultaneously from the local sensor clock module and the RF clock module, respectively. This way, each SN can calculate the exact sampling time with respect to the unique global network clock $^{MN}T_{SN}$

$$^{MN}T_{SN}[k] = T_{MN}|_{sync} + \left( T_{SN}[k] - \left( T_{SN}|_{sync} - \Delta T_{proc} \right) \right) \tag{2}$$

After receiving a synchronization beacon, the corresponding timestamps, $T_{MN}|_{sync}$ and $T_{SN}|_{sync}$, and the processing time $\Delta T_{proc}$ are updated accordingly. If a synchronization beacon is missed, e.g., due to transmission errors, the previous timestamps are retained for further calculation of the current global network time until the next synchronization frame is received. This way, possible deviations of the oscillator components are compensated and both the TDMA protocol timing and the sampling clock are robust against hardware-dependent variances.

The internal 32-bit timer together with the timer resolution $\Delta T_{\mathrm{LSB}}$ allows for a maximum recording time $T_{\mathrm{max}}$ of

$$T_{\mathrm{max}} = \frac{2^{32} \cdot 10\,\mu s}{3600} = 11.93\,\mathrm{h}, \tag{3}$$

which is in the range of the theoretical maximum operating time due to limitations of the battery capacity (cf. Table 1). Still, the network provides a comparably fine sampling resolution of $10\,\mu s$. The integrated oscillator (TSX-3225 24 MHz, Seiko Epson Corporation, Suwa, Japan) has a frequency tolerance of $f_{\mathrm{tol}} = \pm 10\,\mathrm{ppm}$. This results in a theoretical temporal deviation $\Delta T_{\mathrm{err,max}}$ according to component tolerances of

$$\Delta T_{\mathrm{err,max}} = 2^{32} \cdot 10\,\mu s \cdot (\pm 10\,\mathrm{ppm}) = \pm 0.43\,\mathrm{s}. \tag{4}$$

In the context of the intended application of long-term gait analysis, this means a theoretical deviation of about one stride over 12 h, if only the compensation of the static time offset is considered.

### 2.3. Inertial Measurement Node

The inertial measurement node consists of the integrated inertial measurement unit (IMU) BNO055 (Bosch Sensortec, Reutlingen, Germany) and an absolute barometric pressure sensor BMP280 (Bosch Sensortec, Reutlingen, Germany). The IMU contains an accelerometer, a gyroscope, a geomagnetic field sensor, and a Cortex M0+ microcontroller for on-chip sensor fusion. However, the magnetic field sensor was not used in our approach to reliable indoor orientation estimation. On the one hand, the usage of only accelerometer and gyroscope saves energy with respect to long-term monitoring applications. On the other hand, magnetic field sensors are often distorted by ferrous material or power lines within an indoor environment [41]. The accelerometer and gyroscope data were fused to obtain the current orientation of a sensor node. The barometric pressure sensor (BPS) includes a piezo-resistive sensing element and a temperature sensor to compensate for temperature-depending drifts. The BPS was utilized to estimate changes in absolute altitude for detection of stairs walking, floor elevating, sit-to-stand transitions, or drops. The communication was realized via the $I^2C$ bus for both integrated circuits. An overview of the characteristic parameters for the integrated sensors is given in Table 2.

**Table 2.** Characteristic parameters of the IMU sensor nodes implemented for the BSN.

| Parameter | Accelerometer | Gyroscope | Pressure Sensor |
|---|---|---|---|
| Range | $\pm 4\,g$ | $2000°/s$ | $300\ldots1100\,\mathrm{hPa}$ |
| Resolution | 14 bits | 16 bits | 17 bits |
| Noise | $150\,\mu g/\sqrt{\mathrm{Hz}}$ | $0.014°/s/\sqrt{\mathrm{Hz}}$ | $1.3\,\mathrm{Pa}$ |
| Sampling Rate | $100\,\mathrm{Hz}$ | $100\,\mathrm{Hz}$ | $125\,\mathrm{Hz}$ |

### 2.4. Measurement of Ground Reaction Forces

The measurement of ground reaction forces (GRF) was implemented by integrating four force sensing resistors (FSRs) into an insole. The four FSR402 sensors (Interlink Electronics, Irvine, CA, USA) with a diameter of 13 mm of sensing region were placed below (a) the distal phalanges (i.e., the big toe), (b) the first and (c) the fifth metatarsophalangeal joints, and (d) the heel (Figure 4a). The FSR sensors are subjected to a varying load in the successive phases of one gait cycle, as depicted in Figure 4b. This way, the GRF is sampled spatially along the different regions of the foot that mainly contribute to the overall GRF. The quantitative GRF approximation is determined by calculating the average (arithmetic mean) of the digital FSR signal components

$$\text{FSR}_{\text{sum}}[k] = \frac{1}{4} \sum_{i=1}^{4} \text{FSR}_i[k]. \tag{5}$$

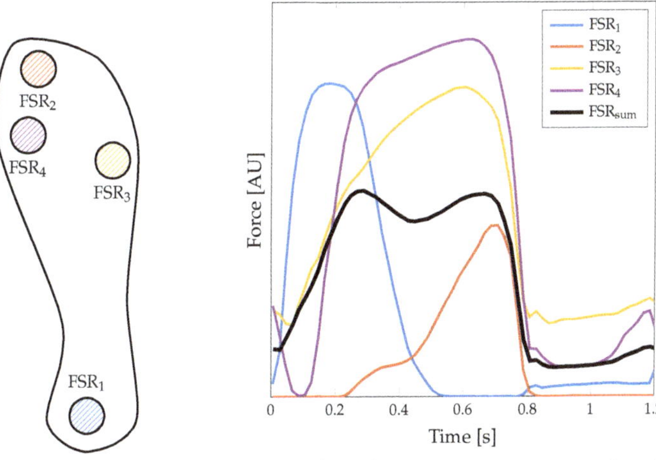


(**a**) Positioning of the FSR sensors.   (**b**) Exemplary FSR measurement over one gait cycle.

**Figure 4.** Implementation of the force sensing resistor (FSR) insole. (**a**) The insole is equipped with four FSR402 sensor pads located at the distal phalanges (FSR$_2$), the first (FSR$_4$) and the fifth metatarsophalangeal joints (FSR$_3$), and the heel (FSR$_1$). (**b**) An exemplary measurement of a step with this sensor setup. Note that the FSR$_{\text{sum}}$ does not reflect the actual force, since the setup is not calibrated to the individual pressure distribution.

Each FSR sensor is part of a voltage divider supplied by the reference voltage $U_{Ref} = 3.3\,\text{V}$ (Figure 5). The resulting signal is filtered by a low-pass circuit with a corner frequency of $f_c = 15.92\,\text{Hz}$ and subsequently sampled by the 12-bit Analog-to-Digital Converter (ADC) MCP3204 (Microchip Technology Inc.). The ADC is connected to the sensor controller via the SPI bus.

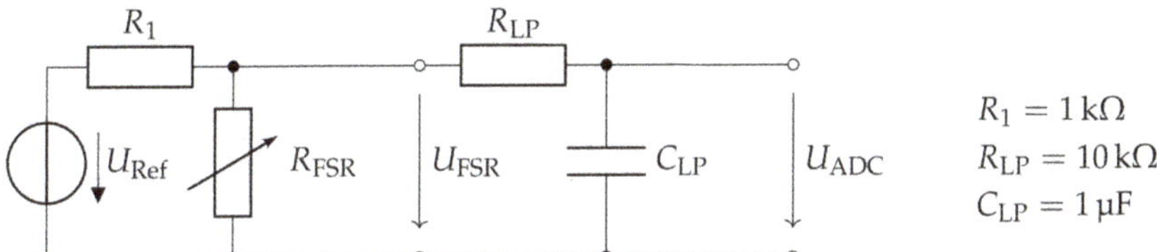


**Figure 5.** Hardware design of the FSR measurement circuit. Each FSR sensor is low-pass filtered to eliminate measurement noise and adapt to the gait dynamics.

### 2.5. Unobtrusive Sensor Setup for Gait Monitoring

While stationary gait analysis systems are traditionally used in movement laboratories, long-term monitoring systems require a maximized level of acceptance and compliance by the patient. Unobtrusive measurement of physiological or biomechanical parameters has become increasingly important in recent decades, as it provides insights into the everyday-life behavior of the patient. An unobtrusive sensor setup in a BSN application for gait analysis is characterized by wearable sensors that are located on the human body so that they do not affect natural motion and the patient ideally does not take note of the additional sensor equipment.

Smart electronic devices, i.e., mobile phones and smart-watches, and their integrated sensor units offer a promising opportunity to unobtrusively collect motion data. Concerning the fundamental idea of an unobtrusive sensor setup for everyday-life monitoring, we chose a setup including three IMU sensors, attached to the thigh, the wrist, and the chest as well as one FSR insole. This setup is motivated by typical wearable devices and their preferable wearing location: A cell phone in the pocket (thigh) and a smart-watch at the wrist supplemented by a pendant sensor for fall detection (cf. Philips Lifeline AutoAlert or similar) and an ordinary insole. With this sensor setup, we ensure on the one hand a broad acceptance for wearing these sensors and on the other hand cover movements of the lower extremities, the trunk, the upper extremities, and the ground contact. The typical applications are twofold. On the one hand, it is possible to conduct long-term measurements with the MN directly attached to the human body, in our case at the second wrist. On the other hand, the BSN can be used to record data in a motion laboratory environment, where the MN is connected to a PC system to enable online tracking of the data streams. The two application scenarios are depicted in Figure 6 showing the individual sensor locations for each case. Please note that in the motion laboratory the distance between MN and the participant walking on the treadmill was approximately 6–8 m due to infrastructural constraints.

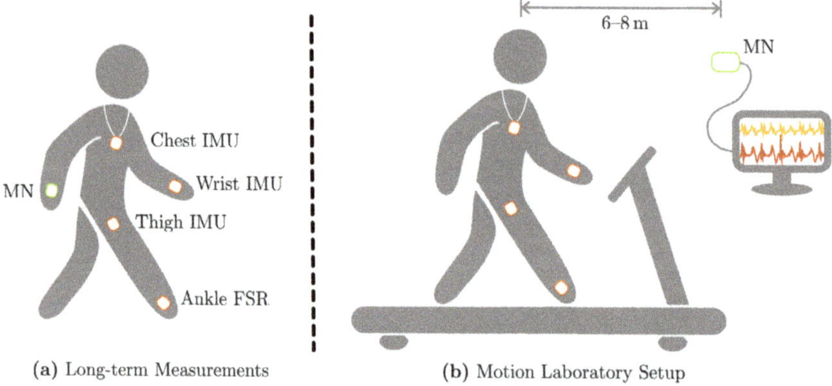


(**a**) Long-term Measurements          (**b**) Motion Laboratory Setup

**Figure 6.** Application scenarios of the proposed sensor system realized in within this contribution. (a) The typical unobtrusive sensor setup for long-term measurements in everyday life. (b) Measurement setup during treadmill walking trials in the motion laboratory conducted for the validation process.

### 2.6. Detection of Physical Activity in Long-Term Measurements

Detection of physical activity, more specifically the extraction of gait phases, during long-term measurements of everyday life activities is a key feature of the proposed sensor system. We identified the sensor node located at the thigh as the most reliable node for determining gait phases in specific. We implemented a simple algorithm to detect phases of physical activity using the signal vector magnitude (SVM) of the three-dimensional thigh acceleration data $\mathbf{a}_{\text{thigh}} = [a_x \, a_y \, a_z]^{\mathsf{T}}$

$$\text{SVM}[k] = ||\,\mathbf{a}[k]\,||_2 \tag{6}$$

$$= \sqrt{a_x^2[k] + a_y^2[k] + a_z^2[k]}. \tag{7}$$

The standard deviation of the SVM, $\sigma_{\text{SVM}}[i]$, was calculated window-based in the first step with a fixed window size of 1 s. The SVM was binary sampled with a threshold of 0.1 g,

$$\sigma_{\text{SVM,bin}}[i] = \begin{cases} 1, & \sigma_{\text{SVM}}[i] \geq 0.1\,\text{g} \\ 0, & else \end{cases}, \tag{8}$$

which suppresses low-energy artifacts, such as low dynamic movements, and the global signal noise of the accelerometer of approximately 1.5 mg. The resulting decision vector was median-filtered with a filter length of five samples, which equals a time section of five seconds. The averaged decision vector was finally filtered with a 1-dimensional opening function to eliminate short periods of physical activity, which for some reason may not be significant for further analysis. Please note that based on the algorithm, it is not possible to differentiate between any kind of physical activity, where the leg or thigh is majorly involved. Since every long-term recording associated with this contribution took place in the office environment and none of the participants reported differently, physical activity in this particular case is only referred to as walking activity.

### 2.7. Algorithm for Gait Segmentation Based on FSR Measurements

The most important characteristic points for wearable-based gait analysis are the heel strike (HS) and toe off (TO) events, respectively. These events indicate the two main phases in human gait, the stance phase, and the swing phase (Figure 7) and usually represent prominent and, therefore, easily detectable landmarks in wearable sensor data. Therefore, they are commonly used as key features in gait segmentation. Based on the specific application, other key features such as mid swing or mid stance can also be detected in wearable-based data. Since the FSR-based gait segmentation is in the main scope of this contribution we focus on the detection of HS and TO events. The summed-up FSR time series serves as the input signal for the segmentation algorithm. Since insole-integrated FSR measurements besides GRF are subject to additional motion artifacts due to contact pressure of the shoe and, therefore, differ from standard GRF measurements (i.e., force plate), we implemented a specific algorithm for HS and TO detection in FSR data.

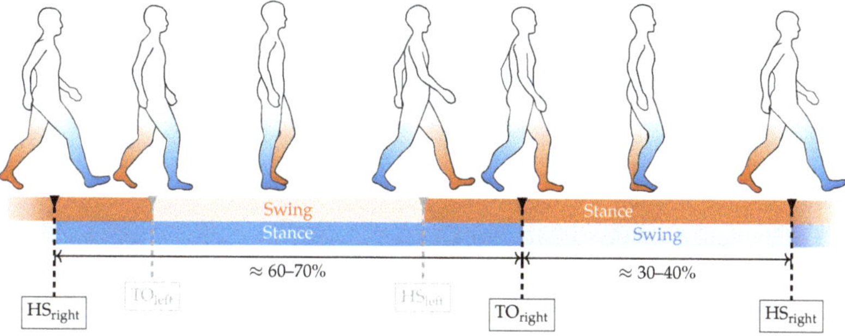


**Figure 7.** The two main components of the basic gait cycle consist of the stance phase and the swing phase. During normal gait, the stance phases of both left and right leg overlap during the double support phase. The stance and swing phases are separated by the heel strike (HS) and toe off (TO) event, which are the key characteristic points to be extracted from the GRF.

Firstly, we calculate the local minimum of the summed-up FSR signal, which corresponds to the lowest pressure during one gait cycle. Since in higher walking speeds the plateau of the maximum foot pressure alters to a double peak due to the swing phase initiation of the opposite leg (cf. Figure 4b), the maximum peak pressure is not a reliable indicator. We assume a minimum distance of 0.7 s between two adjacent peaks for the parameterization of the peak detection. Then, we identify the beginning of the falling (TO) and the rising edge (HS), respectively, starting from the local minimum. Therefore, we find the closest positive and negative value exceeding certain threshold values in the approximated derivative (diff()) of the summed-up FSR signal. These thresholds values were determined based on the FSR measurements conducted during the walking trials related to this contribution. Therefore, we investigated the empirical cumulative distribution function (eCDF) of the single difference signals of the summed-up FSR data. As shown in Figure 8, the eCDF can be separated in three main parts: a flat distribution region of negative signal slope, a relatively small transition region, and a flat region indicating positive slopes. This characteristic is mainly due to the signal shape of a band-limited rectangular signal. We identified the key characteristic points of the eCDFs, which we defined as the turning points from negative slope flat since these points statistically separate falling and rising edges from signal plateaus. For every single normalized eCDF, we identified the turning points by searching the minimum distance between the eCDF curve and $[0, 1]$ for $d^+_{\text{min}}$ and $[0, 0]$ for $d^-_{\text{min}}$, respectively. Subsequently, we calculated the mean value of the normalized amplitudes to define a fixed threshold with respect to the maximum and minimum amplitude values of each eCDF. Due to computational reasons we rounded the values to integer fractions.

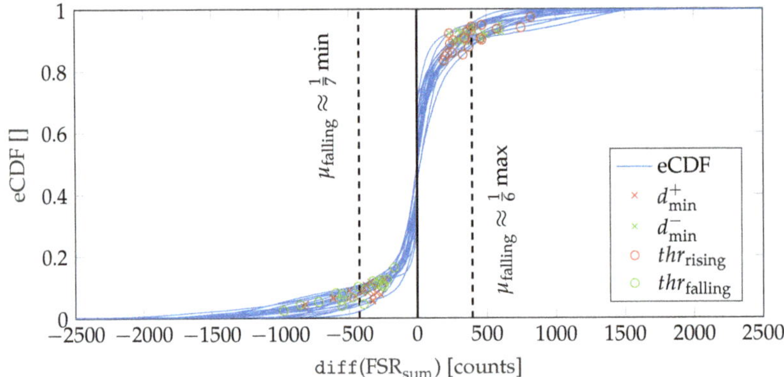


**Figure 8.** Illustration of the procedure for determining the threshold values used for the detection of HS and TO events.

Since the falling edge was found to be steeper as compared to the rising edge, the thresholds were identified as

$$thr_{\text{rising}} = \frac{1}{6} \max_k \text{diff}\left(\text{FSR}_{\text{sum}}\right)[k] \quad \text{and} \tag{9}$$

$$thr_{\text{falling}} = \frac{1}{7} \min_k \text{diff}\left(\text{FSR}_{\text{sum}}\right)[k]. \tag{10}$$

The basic procedure of identifying the threshold values is depicted in Figure 8. In addition, the final threshold values $thr_{\text{rising}}$ and $thr_{\text{falling}}$ according to Equations (9) and (10) are given. Finally, an exemplary illustration of the FSR-based gait segmentation algorithm is shown in Figure 9.

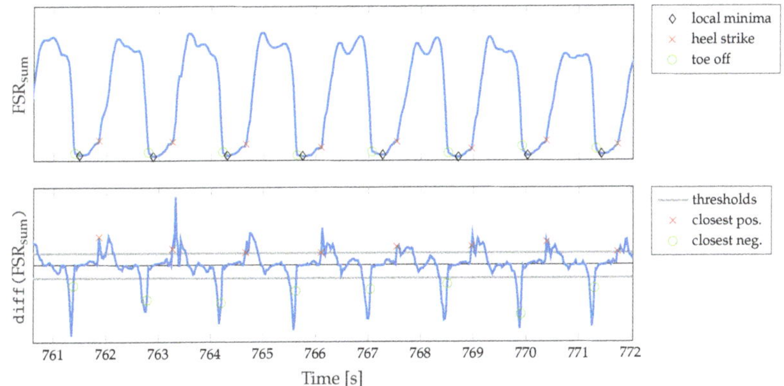


**Figure 9.** Exemplary illustration of the segmentation algorithm. The algorithm utilizes the summed-up FSR signal and its derivative, respectively, to identify TO and HS events during gait phases.

*2.8. Trial Designs for the Validation of the Sensor Setup*

Three different test scenarios were chosen to validate the performance of the proposed sensor setup. Firstly, the functional capability of the system was investigated in a technical validation setup in terms of transmission performance and timings. Secondly, we recorded long-term motion data during everyday life activities to validate the system reliability in a real-life application scenario. Finally, we conducted a smaller clinical study with ten participants to validate the activity detection algorithm and confirm the ability to correctly extract gait parameters from the obtained sensor data, respectively.

- Technical laboratory setup: The validation of the technical functionality in terms of transmission and timing performance was performed in a technical laboratory. The static measurements were conducted inside a laboratory room with as low electromagnetic interference as possible. The spatial signal strength measurements were conducted outside the building to reduce reflection disturbances. The SN was mounted on a turnable robot arm (Mitsubishi Movemaster) to avoid human tissue interference during the measurements and guarantee accurate spatial recordings.
- Long-term measurements: We conducted several long-term measurements with healthy subjects equipped with the proposed system during their everyday office life, while normally continuing their business. The records were used to validate the system reliability in a real-life scenario. Comparable to the technical validation, we focused on the evaluation of the transmission performance of the system.
- Gait parameter study: Since gait analysis is in the main scope of this contribution and the main purpose of the proposed system, we validated the system with respect to the extraction of gait parameters. We conducted a clinical trial in the movement laboratory of the Department of Geriatrics of RWTH Aachen University Hospital to compare the outcome of the system to clinically relevant gait parameters. The study was approved by the ethical committee of the University Hospital at RWTH Aachen University (Ref. No. EK 024-20) and included ten healthy, young participants ($29.8 \pm 4.24$ years, 3 female, 7 male, average body height $175.7 \pm 8.65$ cm). In each trial, the participant was asked to walk on a treadmill (Zebris FDM-T, Zebris Medical GmbH) at different walking speeds ($1$–$4\ \mathrm{km\,h^{-1}}$). The treadmill includes integrated force plates and generates the reference data for the gait parameters.

## 3. Results

### 3.1. Transmission Performance

The transmission performance of a wireless communication system can be quantified by two main characteristics: the Receive Signal Strength Indicator (RSSI) and the Packet Error Rate (PER). Several measurements were conducted recording the RSSI and the number of transmitted packets (NoP) for each SN under different environmental conditions to evaluate the transmission performance of the IPANEMA v3.0 mainboard. The PER can directly be calculated from the NoP. The performance tests were conducted in a technical laboratory setup, as well as during long-term measurements of everyday-life activities and gait trials in a movement laboratory. The RSSI and the NoP are automatically captured by the MN for each packet arriving.

The technical evaluation measurements include recordings of RSSI at an increasing distance with varying transmission power output in a laboratory environment, where the antennas of both sender (SN) and receiver (MN) have been aligned accordingly. As expected according to theoretical considerations following the free-space path loss, the RSSI decreases with increasing distance and decreasing transmission power output. Here, also the transmission output power, matching network loss, and antenna gain are included in the measurement results, which results in an overall offset of approximately $-30$ dBm. In a second experiment, the RSSI was recorded from different spatial angles in a free-space environment, where the sensor node was placed on a turnable end-effector of a robot arm (Figure 10). During the measurement, the sensor node was rotated by $360°$ in the x-y-plane. The receive signal quality was recorded by the master node that was placed at varying distances (Figure 10b).

Additionally, measurements with static and moving human tissue within the direct signal path were performed. An additional signal attenuation of approximately 5–15 dBm occurs with human tissue interference. For moving human tissue, the RSSI pattern shows inconsistent behavior with both increased and reduced signal attenuation compared to the non-moving measurements. For loss-less data transfer, a minimum of $-78$ dBm RSSI was determined. At 10 dBm transmission output power, a maximum of 2.2% PER was found. The proposed sensor system can, therefore, be applied in telehealth sensor systems with human tissue interference.

Finally, we investigated the long-term measurements with regard to the transmission performance of the proposed setup in an everyday-life scenario. Different healthy subjects were asked to wear the measurement setup (sensor nodes at the left wrist, in the right pocket (thigh), at the chest and as an insole (ankle), the master node at the right wrist) in their everyday office life. In addition to the actual sensor data, we additionally recorded the RSSI and the number of packages received by the master node to calculate the PER. The overall results are given in Table 3. The performance evaluation yielded satisfying results with an average PER of 5.45% on average and 7.38% during gait. We performed a correlation analysis between the RSSI and PER values from every single measurement and sensing location, to investigate the relationship between successful packet transmissions and the received signal strength. The correlation analysis yielded an overall correlation coefficient of $\rho_{\mathrm{raw}} = 0.778$. The correlation analysis of the averaged performance parameters from Table 3 even resulted in a correlation coefficient of $\rho_{\mathrm{av}} = 0.993$. Please note, that we used the absolute RSSI values to correlate a decreasing radio signal strength with an increasing PER tendency. We additionally identified the gait phases within the long-term measurements using the algorithm described in Section 2.6 and calculated both RSSI and PER for each sensor location. The correlation analysis yielded correlation coefficients of $\rho_{\mathrm{gait,raw}} = 0.154$ and $\rho_{\mathrm{gait,av}} = 0.986$, respectively.

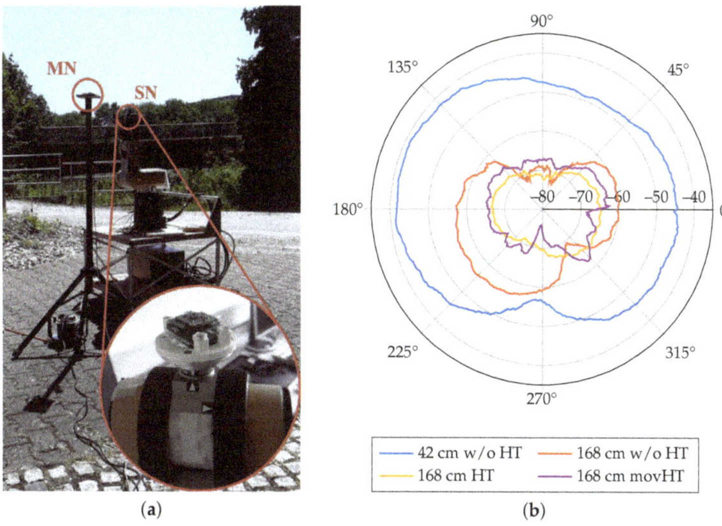



**Figure 10.** Receive Signal Strength Indicator (RSSI) measurements. (**a**) Measurement setup of the spatial RSSI recordings. The sensor node is placed at the end-effector of a turnable robot arm (red circled enlargement). The master node is statically placed at a fixed distance in front of the sensor node. (**b**) Receive Signal Strength over spatial angles for different distances (42 cm and 168 cm) and with additional interference of non-moving and moving human tissue (HT). Antenna (cf. PCB in Figure 1) is aligned with 180°.

**Table 3.** Transmission performance during long-term measurements from unsupervised everyday office life activities. The calculation of RSSI and Packet Error Rate (PER) was performed for the entire data set (all) and gait phases only (gait) to specifically investigate the transmission behavior during gait.

| Data | Parameter | Wrist | Thigh | Chest | Ankle | Averaged |
|---|---|---|---|---|---|---|
| All | RSSI [dBm] | −62.80 | −63.41 | −56.95 | −68.32 | −62.87 |
| | PER [%] | 5.42 | 6.36 | 1.16 | 8.87 | 5.45 |
| Gait | RSSI [dBm] | −65.07 | −60.11 | −57.10 | −67.32 | −62.40 |
| | PER [%] | 9.34 | 4.76 | 3.75 | 11.68 | 7.38 |

### 3.2. Synchronization Performance

Several long-term measurements were conducted to evaluate the timing performance and long-term stability of the RF communication implementation. The devices under test were placed in a standard technical laboratory environment during the long-term recordings. The SNs were placed around the MN with a distance of about 20 cm to ensure optimal transmission performance. We recorded the master timestamp $T_{\mathrm{MN}}|_{\mathrm{sync}}$ received by the SN, the local slave timestamp $T_{\mathrm{SN}}|_{\mathrm{sync}}$, and the processing delay $\Delta T_{\mathrm{proc}}$ that are used to compensate for the static timer offset due to different times of initialization and for linear timer drifts due to component tolerances (cf. Section 2.2).

We recorded data of 20 devices under test during the long-term measurements of approximately 1 h. Two of the devices showed unsatisfying timing performance yielding many missed synchronization beacons and, thus, non-synchronized timestamps. Subsequently, we separated the test measurements into two groups to further analyze the results: passed timing performance test (18 devices) and failed timing performance test (2 devices). We obtained different indicators to quantify the timing performance of every single device based on the time differences of master and sensor clocks. Therefore, we first calculated the deviation between the sensor clock and the master clock

$$\Delta T_{\text{MN,SN}}[k] = T_{\text{MN}}|_{\text{sync}}[k] - T_{\text{SN}}|_{\text{sync}}[k] \tag{11}$$

and approximated this time series by a second-order polynomial fit

$$\Delta T_{\text{MN,SN}}[k] \overset{!}{=} \text{p}_2 \left( \frac{k}{F_s} \right)^2 + \text{p}_1 \frac{k}{F_s} + \text{p}_0. \tag{12}$$

Based on this approximation, we further analyzed the key characteristics of the oscillator deviations. The results are given in Table 4 on average (mean) for the group of devices that passed the test and for the device with the worst result. Please note that the two devices that failed the test were excluded from the averaging. The divergence $\frac{\Delta T}{s}$ is given by the linear slope $\text{p}_1$ from Equation (12). Additionally, we calculated the average of the absolute divergence $|\text{p}_1|$ for all devices. The temporal linearity of the oscillator misalignment is characterized by the ratio of $\left| \frac{\text{p}_2}{\text{p}_1} \right|$. We determined the root mean squares error with regard to the linear fit $\text{RMSE}_{\text{lin}}$ to identify short-time irregularities. Finally, we calculated the divergence of the adjusted timestamps $^{\text{MN}}T_{\text{SN}}$ to prove the correct operation of the adaption formula in Equation (2). As can be seen, the absolute divergence after adjusting the local clock is significantly below 1 µs/s and, therefore, the system can be used for synchronized multi-sensor gait analysis.

**Table 4.** Results of the synchronization evaluation.

| Indicator | Average Passed Test | Worst Case |
|---|---|---|
| Divergence [µs/s] | $-11.69$ | $--$ |
| Abs. Divergence [µs/s] | 18.63 | 46.445.59 |
| Linearity | $2.907 \times 10^{-6}$ | $3.29 \times 10^{-8}$ |
| $\text{RMSE}_{\text{lin}}$ [s] | $1.081 \times 10^{-4}$ | 0.025 |
| Abs. Adj. Divergence [µs/s] | $20.532 \times 10^{-3}$ | $7.539 \times 10^{-3}$ |

### 3.3. Qualitative GRF Measurement

The GRF can be obtained from the FSR sensor data and contains information about the gait stability, gait control strategy, or pathological gait patterns. For example, the position of the Center of Pressure (CoP) and its derivatives are key characteristics in terms of gait stability. It can be deduced from the 2D plantar GRF that is reconstructed by interpolating between the single sensor locations and recalculating the local foot pressure for the specific foot geometry. Further analysis of GRF-related gait quality indicators is not part of this contribution. Instead, we focus on a qualitative validation of the GRF signal morphology. The quality of the signal morphology has a high impact on the detection of HS and TO events using the proposed algorithm. Of course, the morphology individually differs due to foot size, footwear, and contact pressure. Therefore, the insole-based FSR signal is more sensitive to individual gait behavior as compared to the force plates.

To validate the reproducibility of GRF by the FSR insole, we conducted a clinical trial study with 10 healthy, young participants, who volunteered in a treadmill walking trial. The subjects walked on the treadmill for 7 min per trial at increasing speeds from 1–4 km h$^{-1}$. Each speed step was kept constant for one minute and increased by 0.5 km h$^{-1}$ in between by the trial advisor. The treadmill included integrated force sensors underneath the walking plane, which was used to generate the reference data. The force data from the treadmill can be integrated over the 2D foot area to obtain the vertical GRF of the corresponding gait cycle. Gait cycles within the reference force can be easily detected by thresholding the signal with the smallest possible threshold since the force plate yields zero force during the swing phase. The recorded FSR signal is separated using the algorithm presented in Section 2.7. To analyze the qualitative reproducibility, we averaged over all gait cycles at one speed step for both summed-up FSR and reference force signal. A visual comparison of the FSR measurement and the reference force is depicted in Figure 11.

We additionally calculated the coefficient of variation (CoV), which is defined as the standard deviation in relation to the mean value of a quantity, for the individual stride morphologies of both FSR and reference force plate data to give a measure of variability for the corresponding measurement principle. It should be noted that human gait, of course, inherits a certain amount of stride-to-stride variability. Therefore, the CoV in this case is not a measure of precision for the corresponding technical system. The CoV rather needs to be compared to the reference outcome. Secondly, the swing phase does not contribute to the CoV for the reference system since the effective output is zero when the foot does not have ground contact. As can be seen in Table 5, the CoV of the FSR insole data on average is approximately 40% higher compared to the reference force plate data.

**Table 5.** Coefficient of variation [] for the mean FSR morphology compared to the reference morphology obtained from the treadmill force plates.

| Data | Walking Speed [km h$^{-1}$] | | | | | | |
|---|---|---|---|---|---|---|---|
| | **1.0** | **1.5** | **2.0** | **2.5** | **3.0** | **3.5** | **4.0** |
| CoV$_{\text{FSR}}$ | 0.126 | 0.108 | 0.095 | 0.091 | 0.099 | 0.091 | 0.108 |
| CoV$_{\text{Ref}}$ | 0.112 | 0.084 | 0.071 | 0.065 | 0.061 | 0.061 | 0.065 |

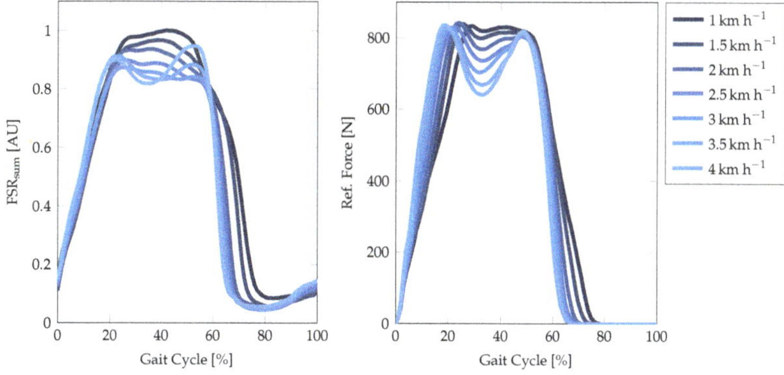


**Figure 11.** Exemplary comparison of the summed-up FSR signal and the force plate measurement during one walking trial of subject 1 from 1–4 km h$^{-1}$. The graphs show the mean morphology over all gait cycles at one particular speed. The gait cycles have been separated by the HS events detected by the proposed algorithm and the gait analysis software of the treadmill, respectively.

### 3.4. Speed-Related Gait Parameters

The capability of the IMU sensor nodes have been intensively investigated before [38,39] and is, therefore, not in the scope of this contribution. Instead, we focus on the speed-related gait parameters, that we can additionally obtain from the GRF measurements. As key characteristic parameters of human gait, we extract the total number of strides, the stride time, and the stance time from the summed-up FSR signal.

The stride time is defined as the temporal distance between two successive ground contacts of the same foot. The exact stride time can, therefore, be extracted by any periodically recurring feature in the FSR signal, such as the HS or TO events. Since HS and TO are the two most dominant featured in the FSR signal, we calculated the stride time from both HS and TO and compared it to the reference treadmill data. The correlation and Bland–Altman plots for the HS-based and TO-based stride time extraction are given in Figure 12a. Since HS-based stride time extraction provided more reliable results, we decided to use the HS events for any further parameter calculation. The stride time is thus calculated by

$$\mathrm{ST}[l] = T|_{\mathrm{HS}_{l+1}} - T|_{\mathrm{HS}_l} \tag{13}$$

and averaged over a period of one minute to calculate the mean stride time.

The number of strides (NoS) is given by the sum of detected HS events. Concerning the conducted gait study trial, we counted the number of strides in each speed interval, which yields a second important gait parameter: the cadence (strides per minute). We calculated the detection accuracy to give a quantitative measure of the overall stride detection

$$\mathrm{Accuracy} = 1 - \frac{|\mathrm{NoS}_{\mathrm{Ref}} - \mathrm{NoS}_{\mathrm{FSR}}|}{\mathrm{NoS}_{\mathrm{Ref}}}. \tag{14}$$

The percentage stance time (PST) indicates the temporal percentage of the stance time in relation to the entire gait cycle (stance time + swing time of one leg) and is calculated by

$$\mathrm{PST}[l] = \frac{T|_{\mathrm{TO}_l} - T|_{\mathrm{HS}_l}}{T|_{\mathrm{HS}_{l+1}} - T|_{\mathrm{HS}_l}} = \frac{T|_{\mathrm{TO}_l} - T|_{\mathrm{HS}_l}}{\mathrm{ST}[l]}. \tag{15}$$

Both the mean stride time and the percentage stance time can be validated using the RMSE

$$\mathrm{RMSE} = \sqrt{\sum_{l=1}^{N} \frac{(y_{\mathrm{FSR}}[l] - y_{\mathrm{Ref}}[l])^2}{N}}, \tag{16}$$

where $y_{\mathrm{FSR}}$ is the respective measure to be validated against $y_{\mathrm{Ref}}$ and $N$ equals the number of observations.

The gait sequences were extracted from the overall recording time by applying the gait phase detection from Section 2.6. All measures were subsequently extracted from the summed-up FSR signal according to Equations (13)–(16). The extracted gait parameters are given in Table 6. Additionally, the results of the validation of the speed-related gait parameters are depicted in Figure 12b.

**Table 6.** Results of the extraction of speed-related gait parameters: the number of strides, the mean stride time, and the percentage of the stance phase. Values are given in mean $\pm$ standard deviation.

| Speed [km h$^{-1}$] | Number of Strides/Cadence | | | Mean Stride Time | | | Percentage Stance Phase | | |
|---|---|---|---|---|---|---|---|---|---|
| | FSR [min$^{-1}$] | Ref. [min$^{-1}$] | Accuracy [%] | FSR [s] | Ref. [s] | RMSE [ms] | FSR [%] | Ref. [%] | RMSE [%] |
| 1.0 | 29.8 $\pm$ 4.25 | 29.1 $\pm$ 4.15 | 97.8 $\pm$ 1.62 | 2.05 $\pm$ 0.31 | 2.05 $\pm$ 0.32 | 22.36 | 77.36 $\pm$ 3.96 | 76.30 $\pm$ 2.99 | 2.76 |
| 1.5 | 35.5 $\pm$ 3.73 | 35.5 $\pm$ 3.66 | 99.7 $\pm$ 0.58 | 1.71 $\pm$ 0.21 | 1.71 $\pm$ 0.21 | 2.80 | 72.53 $\pm$ 3.22 | 72.37 $\pm$ 1.79 | 2.47 |
| 2.0 | 40.5 $\pm$ 3.46 | 40.5 $\pm$ 3.44 | 99.9 $\pm$ 0.38 | 1.49 $\pm$ 0.15 | 1.49 $\pm$ 0.15 | 7.93 | 70.45 $\pm$ 3.02 | 70.45 $\pm$ 1.53 | 2.53 |
| 2.5 | 44.5 $\pm$ 3.55 | 44.4 $\pm$ 3.56 | 99.9 $\pm$ 0.34 | 1.36 $\pm$ 0.12 | 1.36 $\pm$ 0.12 | 0.35 | 68.65 $\pm$ 2.10 | 69.22 $\pm$ 1.35 | 1.70 |
| 3.0 | 47.9 $\pm$ 3.73 | 47.9 $\pm$ 3.77 | 99.9 $\pm$ 0.33 | 1.26 $\pm$ 0.11 | 1.26 $\pm$ 0.11 | 0.40 | 68.18 $\pm$ 2.19 | 68.20 $\pm$ 1.21 | 1.99 |
| 3.5 | 50.1 $\pm$ 3.09 | 50.1 $\pm$ 3.05 | 99.9 $\pm$ 0.31 | 1.20 $\pm$ 0.08 | 1.20 $\pm$ 0.08 | 0.77 | 67.81 $\pm$ 2.70 | 67.40 $\pm$ 1.31 | 2.52 |
| 4.0 | 52.8 $\pm$ 3.38 | 52.9 $\pm$ 3.35 | 99.9 $\pm$ 0.29 | 1.14 $\pm$ 0.08 | 1.14 $\pm$ 0.08 | 0.58 | 67.24 $\pm$ 2.00 | 66.55 $\pm$ 1.27 | 2.10 |
| Average | 43.0 $\pm$ 8.5 | 42.9 $\pm$ 8.6 | 99.6 $\pm$ 1.0 | 1.46 $\pm$ 0.3 | 1.46 $\pm$ 0.3 | 5.027 | 70.32 $\pm$ 4.4 | 70.07 $\pm$ 3.6 | 2.296 |

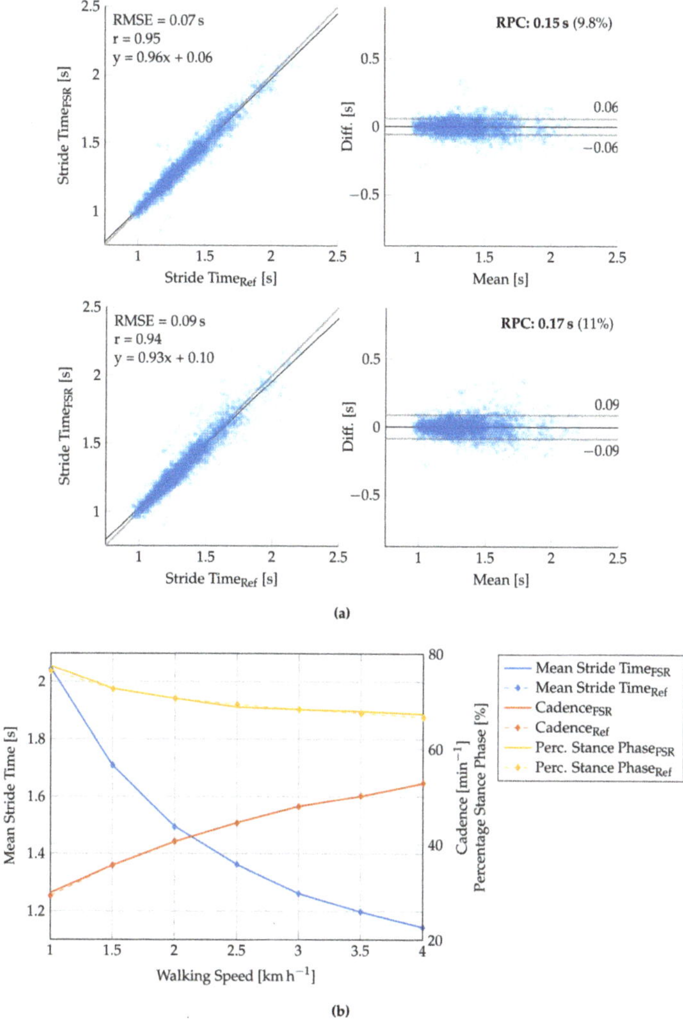

**Figure 12.** Validation results of the speed-related gait parameters. (**a**) Correlation (**left**) and Bland–Altman (**right**) plot for the HS-based stride time extraction (**above**) and the TO-based stride time extraction (**below**) for all strides detected during walking trials. (**b**) Graphical representation of the overall extraction performance with regard to the calculation of the mean stride time, the cadence, and the percentage stance phase (cf. Table 6).

## 4. Discussion

Within this contribution, we introduced the recently redesigned version 3.0 of the IPANEMA BSN based on an 868 MHz communication interface. The current version was implemented for the specific application in gait analysis. Therefore, we integrated two different kind of sensor nodes: an IMU node and a sensorized insole for GRF measurement. We proposed a sensor setup to unobtrusively obtain movement data in long-term measurements.

Our results show that the concept of an 868 MHz communication interface for a BSN is generally applicable in a measurement setup for gait analysis in long-term data. The averaged RSSI values of approximately −62.5 dBm for both the entire data set and gait phases only (cf. Table 3) correspond to the technical evaluation of the transmission performance (approximately $-75$ to $-45\,$dBm, cf. Figure 10b). As expected, the PER increased during the real-life application in the long-term measurements compared to the laboratory measurements (2.2% versus 5.5–7.4%). During gait phases, we also found an increased PER, which can be explained due to additional movement artifacts. Nevertheless, it should be noted that we identified erroneous packets by a simple cyclic redundancy check (CRC) and, thus, the PER equals the packet loss rate. Minor bit errors are currently not corrected, but will be part of future work to reduce the PER. Nevertheless, we showed in Section 3.4 and in [39], respectively, that gait parameter extraction is not drastically affected by missing data points. The correlation between RSSI and PER interestingly decreases during gait phases ($\rho_{\text{gait,raw}} = 0.154$ versus $\rho_{\text{raw}} = 0.778$). This can be explained by highly dynamic changes of the received signal strength during the gait cycles, which can cause a comparatively high PER, while the average RSSI suggests a stable transmission performance. Please note that transmissions with very low signal strengths are not detected by the receiving master node, which means that theoretically low RSSIs can not be considered for the averaged RSSI calculation. Based on these findings, we consider to adaptively adjust the transmitter output power during the gait phases using a model-predictive approach by characterizing the transmission channel of each individual sensor node. This would not only improve the PER, but also the operating time of the system.

According to Table 1, the maximum operating time of the system due to battery limitations was determined to be about 12 h. This corresponds to the currently implemented, maximum timestamp of the local BSN clock (cf. Section 2.2. During the timing evaluation measurements, we found two of the tested devices (10%) that showed large oscillator deviations and, thus, did not function appropriately. Since the deviation of these devices was approximately 50 ms/s, the synchronization beacon will be missed every second time and transmission collisions are not guaranteed to be avoided by the TDMA. Additionally, the power consumption of these devices is excessively higher, since once a synchronization beacon is missed, their RF core remains active until the next re-synchronization beacon is received. For these reasons, devices that failed the timing test (absolute deviation $\Delta T_{\text{MN,SN}} > 1\,$ms/s) were excluded from further consideration. The polynomial fit of the absolute deviation of each individual sensor clock yielded high linearity for all devices during the long-term measurements. Therefore, additional nonlinear time-invariant influences on the system need not be expected. Both the mean deviation and mean of the absolute deviation of the devices that passed the test corresponded to the tolerances of the oscillator. Short-time fluctuations of the clock precision were found to be approximately 0.1 ms, which is below the global sampling period of $T_s = 20\,$ms of the current BSN system. Thus, the setup satisfies the requirements for a gait analysis application in long-term measurements. In case further different sensor modalities are required, the evaluation of the timing performance must be revised with regard to the required sampling rate. As expected, the absolute divergence of the sensor clock is significantly below the global clock tick of 10 µs and is therefore negligible.

A further indicator for validating a system for monitoring and analyzing gait in everyday life is the range of recorded sensor signals with respect to the configured measurement range given in Table 2. During the walking trials in the motion laboratory, we extracted maximum acceleration ranges of approximately $[\pm 3.5\,g]$ for the thigh sensor, $[\pm 1.9\,g]$ for the chest sensor, and $[\pm 2\,g]$ for the wrist sensor. For walking speeds covered by the proposed study, the chosen range is, therefore, sufficient. In the case that higher values are expected, for example from an accelerometer located at the ankle or foot, the range can be extended by software. Similar values were also observed during the long-term measurements. Temporary instances of senor saturation could be attributed to short-term, high-energy events that were not associated with the gait phases. In addition to the IMU, the BPS is also an important component of the mobility sensor nodes at the thigh, the wrist, and the chest. In the current sensor configuration, the BPS provides an effective precision of approximately $\pm 10.9\,$cm at sea level due to the measurement noise. Nevertheless it should be noted that the barometric pressure measurement is highly sensitive to minor pressure fluctuations, such as closing or opening of doors or windows, and changes in temperature. During the long-term measurements, we could observe that significant changes in altitude, such as floor elevating and stairs walking could be simply detected within the BPS data, whereas the detection of minor changes requires additional signal processing. First approaches to detect falls or even sit-to-stand transitions yielded promising results, while taking into account the additionally recorded motion data. This topic will, therefore, be subject of future investigations. In the current sensor setup, the signal noise of the pressure sensor equals a standard deviation of the altitude estimate of approximately 10.9 cm at sea level (cf. Table 2). In addition, the pressure measurement is highly sensitive to temperature changes and minor pressure fluctuations, such as the opening or closing of doors or windows.

The FSR insole is used to calculate important speed-related gait parameters, such as the total number of strides, the cadence, the stride time, and the percentage stance phase. GRF-based analysis is known to provide reliable results for these parameters and therefore often used in movement laboratories. The applicability to a wearable long-term monitoring system is evaluated in this contribution. The results of the validation measurements in the movement laboratory are given in Table 6 and Figure 12, respectively. First, we investigated the stride detection performance of both HS and TO-based by comparing the extracted stride time to the reference stride time (Figure 12a). As the correlation plots and Bland–Altman plots show, the HS-based stride time extraction slightly outperforms the TO-based calculation (RMSE 0.07 s versus 0.09 s). These results may be explained by the foot anatomy and resulting pressure distributions and additional forces due to contact pressure of the shoe. We assume that the heel sensor ($FSR_1$, Figure 4a) yields a more precise and better defined signal contribution as compared to the ball of the foot and the toes ($FSR_2$–$FSR_4$). Therefore, we decided to use the HS-based event detection for further parameter extraction. Furthermore, the reference data of the force plates (cf. Figure 11) suggest the conclusion that besides HS and TO further gait phase parameters such as mid stance can be extracted from the GRF signal. Nevertheless, this is a challenging task with regard to FSR-based measurements since the signal morphology is often corrupted by additional motion artifacts of the insole.

The qualitative validation of the GRF measurements demonstrated that the proposed sensor setup satisfies the requirements of an unobtrusive gait analysis system. Although the signal morphology differs between the implemented insole and the reference force plate data (cf. Figure 11), it can be concluded that four FSR sensors located at the proposed landmarks provide sufficient information for further gait parameter extraction. Similar setups with only FSR sensors integrated also reported the reliability of this arrangement [27,42]. Since the system makes no claim to an exact calculation of the CoP lines or even pressure prints for the detection of foot malpositioning, an extension of the FSR insole is currently not necessary. However, an appropriate calibration routine of the FSR sensors to map the measured resistance to an effective force or pressure, can improve the FSR-based parameter extraction. Nevertheless, we proved the basic concept with the presented FSR sensor arrangement to extract gait parameters. Since the FSR sensors cover the most important points that have ground contact during stance phase, the summed-up FSR signal can be assumed to majorly reflect the GRF morphology (cf. Figure 11). Since the detection of HS and TO events is based on the summed-up FSR signal, no matter which sensor is the main contributor at which time of the gait cycle, it can be assumed that this also applies to pathological gait patterns such as they may occur in equinus foot. Within this contribution, we presented a study with healthy young participants without any diagnosed gait impairments, such that we were not able to prove this hypothesis. A more intensive investigation of the FSR sensor technology regarding the ability to reflect pathological gait patterns, the dynamic repeatability and long-term stability for the development of a calibration routine will therefore be part of the future work.

The number of strides reflect the sum of the detected HS events. The cadence can be calculated from the number of strides per minute. Table 6 shows that the proposed stride detection algorithm is able to detect most strides with an overall accuracy of 99.6 $\pm$ 0.8%, which is comparable to the recently published setup of an FSR insole for step counting purposes [27]. Here, the authors conducted a trial with indoor and outdoor walking at self-selected (approximately $5.2\,\mathrm{km\,h^{-1}}$) and maximal speed (approximately $6.3\,\mathrm{km\,h^{-1}}$) and achieved an accuracy of 99.5% and 99.6%, respectively. It should be noted that we intentionally included slow walking speeds $1$–$4\,\mathrm{km\,h^{-1}}$ in our study due to the target application in the monitoring of movement disorders, which are often accompanied by reduced walking speeds. At very slow walking speeds ($1$–$1.5\,\mathrm{km\,h^{-1}}$) our proposed system provided worst results due to the increased amount of movement artifacts during the transitions of swing and stance phase and the swing phase itself. The increased error in gait parameter extraction at very slow walking speeds is, therefore, due to incorrect gait segmentation. As can be seen in Figure 12a, the error variance of the individual stride times most likely increases with decreasing walking speed. Nevertheless, averaging over a certain period of time can compensate for this effect to a certain extent since deviations in the stride time determination will have the opposite effect on the subsequent gait cycle (cf. Table 6). At higher walking speeds of $2$–$4\,\mathrm{km\,h^{-1}}$, the system yielded satisfying results (99.9%).

The mean stride time can be either calculated by the reciprocal cadence or the mean over the temporal distance between two HS events. In this case, we have calculated the stride-to-stride interval and averaged over the recording period of one minute. In contrast to the direct comparison of the stride times (cf. Figure 12a), the mean stride time yields a very high correlation with the reference data featuring an RMSE of approximately $5\,\mathrm{ms}$. An accurate calculation of the mean stride time is of course directly connected to the accuracy of the stride detection. Therefore, these results correspond to the formerly discussed performance of the stride detection algorithm.

As can be seen in Figure 11, the summed-up FSR signal highly correlates with the force plate measurements regarding the falling and rising edges due to the HS and TO events. However, the determination of the exact time of HS and TO still is a challenging task due to movement artifacts of the shoe contact pressure during the swing phase. The percentage of the stance phase in relation to the total stride time varies by approximately 10% at the given walking speeds of $1$–$4\,\mathrm{km\,h^{-1}}$ (cf. Table 6). This change was equally observed in both FSR and reference data. The FSR-based calculation of the PST resulted in an RMSE of approximately 2.3%, which is acceptable for a wearable gait monitoring system. A significant difference was found in the standard deviation of the PST calculation, which corresponds to the assumption that FSR-based HS and TO detection are less accurate compared to the force plate measurements. Since the FSR-based PST extraction has proven to be reliable, we will expand the system in the future to include a second insole that is worn in the opposite shoe. This will provide the possibility to extract additional gait parameters, such as the single and the double support time, to identify the load response and the pre-swing phase or even to obtain an approximation of the CoP lines.

The cadence, the mean stride time and the percentage stance phase are of course important parameters to observe the progression of the locomotor system or superordinated movement disorders, which initially requires a reliable event detection. Nevertheless, the variability of different gait parameters, such as the stride time variability, contains valuable information about gait stability [39]. The robust stride detection thus fulfills an important requirement of a long-term monitoring system, but also an exact temporal determination of the stride event is of high interest. Therefore, the focus of the future work is to improve the stride-to-stride interval extraction, which not only concerns the algorithmic approaches but also a robust transmission and an accurate time synchronization. On the sensor side, future developments include the addition of an IMU on the insole for a more precise detection of HS and TO by fusing this information with the FSR data. The inclusion of information from the other IMU sensors at the wrist, chest, and thigh is an important part of the future algorithmic work to further improve the accuracy of the gait parameter extraction.

## 5. Conclusions

In this contribution, we presented the recently developed version 3.0 of the IPANEMA BSN based on an $868\,\mathrm{MHz}$ communication interface. We evaluated the transmission performance and the synchronization protocol with regard to an application in a wearable long-term monitoring system for unobtrusive gait analysis. We demonstrated the general functionality of the system within technical laboratory measurements and during long-term measurements in everyday life activities in an office environment. We proposed two different kinds of sensing modalities, IMU motion sensors and FSR-based insoles for GRF measurements, and validated the insoles in a study trial against reference data obtained in a movement laboratory. Therefore, we presented an algorithm to identify the HS and TO events in the summed-up FSR signal. We extracted three different clinically relevant gait parameters, the number of strides, or the cadence, respectively, the mean stride time and the percentage stance time. The system proved an overall satisfying performance for an application in a long-term monitoring setup to gain information of clinically relevant parameters in the patient's everyday life.

**Author Contributions:** M.L. and L.M. were responsible for the development and implementation of the basic communication platform in hardware and firmware. M.L. and M.G.D. developed and evaluated the sensor node extension boards. M.L. and M.G.D. conducted the clinical trial. M.L. was responsible for the trial conceptualization. C.B., S.L., and C.N. supervised the study conduction. C.N. was responsible for the overall supervision and project administration. M.L. prepared the original draft. L.M., M.G.D., C.B., S.L., and C.N. carefully reviewed the draft and gave response for further improvements. All authors have read and agreed to the published version of the manuscript.

**Funding:** This work was financially supported by Philips Research, Eindhoven, The Netherlands. The newly established movement laboratory at the Department of Geriatrics of RWTH Aachen University was funded by Robert Bosch Stiftung (32.5.1140.0009.0).

**Acknowledgments:** The authors would like to thank Heribert Baldus, Warner ten Kate, and Giulio Valenti from Philips Research, who assisted the work in an advisory capacity.

**Conflicts of Interest:** The authors declare no conflict of interest.

## References

1. Centers for Medicare & Medicaid Services. National Health Expenditure Accounts. Available online: https://www.cms.gov/Research-Statistics-Data-and-Systems/Statistics-Trends-and-Reports/NationalHealthExpendData/NationalHealthAccountsHistorical (accessed on 20 October 2020).
2. Yang, G.Z. *Body Sensor Networks*, 2nd ed.; Springer: London, UK, 2014.
3. Edward, S.; Neuman, M.R. *Wearable Sensors*; Elsevier: Amsterdam, The Netherlands, 2014. [CrossRef]
4. Dishman, E. Inventing wellness systems for aging in place. *Computer* **2004**, *37*, 34–41. [CrossRef]
5. Chirwa, L.C.; Hammond, P.A.; Roy, S.; Cumming, D.R.S. Electromagnetic radiation from ingested sources in the human intestine between 150 MHz and 1.2 GHz. *IEEE Trans. Bio-Med. Eng.* **2003**, *50*. [CrossRef] [PubMed]
6. Chan, Y.; Meng, M.Q.; Wu, K.; Wang, X. Experimental study of radiation efficiency from an ingested source inside a human body model*. In Proceedings of the 2005 IEEE Engineering in Medicine and Biology 27th Annual Conference, Shanghai, China, 31 August–3 September 2005; pp. 7754–7757. [CrossRef]
7. Mellios, E.; Goulianos, A.; Dumanli, S.; Hilton, G.; Piechocki, R.; Craddock, I. Off-body channel measurements at 2.4 GHz and 868 MHz in an indoor environment. In Proceedings of the 9th International Conference on Body Area Networks, London, UK, 29 September–1 October 2014; Fortino, G., Suzuki, J., Andreopoulos, Y., Yuce, M., Hao, Y., Gravina, R., Eds.; ICST: South Portland, ME, USA, 2014. [CrossRef]
8. Tao, W.; Liu, T.; Zheng, R.; Feng, H. Gait analysis using wearable sensors. *Sensors* **2012**, *12*, 2255–2283. [CrossRef] [PubMed]
9. Maetzler, W.; Domingos, J.; Srulijes, K.; Ferreira, J.J.; Bloem, B.R. Quantitative wearable sensors for objective assessment of Parkinson's disease. *Mov. Disord. Off. J. Mov. Disord. Soc.* **2013**, *28*, 1628–1637. [CrossRef] [PubMed]
10. Shull, P.B.; Jirattigalachote, W.; Hunt, M.A.; Cutkosky, M.R.; Delp, S.L. Quantified self and human movement: A review on the clinical impact of wearable sensing and feedback for gait analysis and intervention. *Gait Posture* **2014**, *40*, 11–19. [CrossRef] [PubMed]
11. Muro-de-la Herran, A.; Garcia-Zapirain, B.; Mendez-Zorrilla, A. Gait analysis methods: An overview of wearable and non-wearable systems, highlighting clinical applications. *Sensors* **2014**, *14*, 3362–3394. [CrossRef] [PubMed]

12. Chen, S.; Lach, J.; Lo, B.; Yang, G.Z. Toward Pervasive Gait Analysis with Wearable Sensors: A Systematic Review. *IEEE J. Biomed. Health Inform.* **2016**, *20*, 1521–1537. [CrossRef]

13. Morris, J. Accelerometry—A technique for the measurement of human body movements. *J. Biomech.* **1973**, *6*, 729–736. [CrossRef]

14. Evans, A.L.; Duncan, G.; Gilchrist, W. Recording accelerations in body movements. *Med. Biol. Eng. Comput.* **1991**, *29*, 102–104. [CrossRef]

15. Liu, R.; Zhou, J.; Liu, M.; Hou, X. A wearable acceleration sensor system for gait recognition. In Proceedings of the 2nd IEEE Conference on Industrial Electronics and Applications, Harbin, China, 23–25 May 2007; IEEE Operations Center: Piscataway, NJ, USA, 2007; pp. 2654–2659. [CrossRef]

16. Moore, S.T.; MacDougall, H.G.; Gracies, J.M.; Cohen, H.S.; Ondo, W.G. Long-term monitoring of gait in Parkinson's disease. *Gait Posture* **2007**, *26*, 200–207. [CrossRef]

17. Jiang, S.; Zhang, B.; Wei, D. The elderly fall risk assessment and prediction based on gait analysis. In Proceedings of the 2011 IEEE 11th International Conference on Computer and Information Technology, Las Vegas, NV, USA, 11–13 April 2011; Staff, I., Ed.; IEEE: Piscataway, NJ, USA, 2011; pp. 176–180. [CrossRef]

18. Salarian, A.; Burkhard, P.R.; Vingerhoets, F.J.G.; Jolles, B.M.; Aminian, K. A novel approach to reducing number of sensing units for wearable gait analysis systems. *IEEE Trans. Bio-Med. Eng.* **2013**, *60*, 72–77. [CrossRef] [PubMed]

19. Tadano, S.; Takeda, R.; Miyagawa, H. Three dimensional gait analysis using wearable acceleration and gyro sensors based on quaternion calculations. *Sensors* **2013**, *13*, 9321–9343. [CrossRef] [PubMed]

20. Fusca, M.; Negrini, F.; Perego, P.; Magoni, L.; Molteni, F.; Andreoni, G. Validation of a Wearable IMU System for Gait Analysis: Protocol and Application to a New System. *Appl. Sci.* **2018**, *8*, 1167. [CrossRef]

21. Steinmetzer, T.; Bonninger, I.; Priwitzer, B.; Reinhardt, F.; Reckhardt, M.C.; Erk, D.; Travieso, C.M. Clustering of human gait with Parkinson's disease by using dynamic time warping. In Proceedings of the 2018 IEEE International Work Conference on Bioinspired Intelligence (IWOBI), Alajuela, Costa Rica, 18–20 July 2018; pp. 1–6. [CrossRef]

22. Qiu, S.; Liu, L.; Zhao, H.; Wang, Z.; Jiang, Y. MEMS Inertial Sensors Based Gait Analysis for Rehabilitation Assessment via Multi-Sensor Fusion. *Micromachines* **2018**, *9*, 442. [CrossRef] [PubMed]

23. Bongartz, M.; Kiss, R.; Lacroix, A.; Eckert, T.; Ullrich, P.; Jansen, C.P.; Feißt, M.; Mellone, S.; Chiari, L.; Becker, C.; et al. Validity, reliability, and feasibility of the uSense activity monitor to register physical activity and gait performance in habitual settings of geriatric patients. *Physiol. Meas.* **2019**, *40*, 095005. [CrossRef] [PubMed]

24. Yogev, G.; Giladi, N.; Peretz, C.; Springer, S.; Simon, E.S.; Hausdorff, J.M. Dual tasking, gait rhythmicity, and Parkinson's disease: Which aspects of gait are attention demanding? *Eur. J. Neurosci.* **2005**, *22*, 1248–1256. [CrossRef] [PubMed]

25. Aggarwal, A.; Gupta, R.; Agarwal, R. Design and development of integrated insole system for gait analysis. In Proceedings of the 2018 11th International Conference on Contemporary Computing (IC3), Noida, India, 2–4 August 2018; pp. 1–5. [CrossRef]

26. Cho, H. Design and implementation of a lightweight smart insole for gait analysis. In Proceedings of the 16th IEEE International Conference on Trust, Security and Privacy in Computing and Communications, Sydney, Australia, 1–4 August 2017; IEEE: Piscataway, NJ, USA, 2017; pp. 792–797. [CrossRef]

27. Ngueleu, A.M.; Blanchette, A.K.; Bouyer, L.; Maltais, D.; McFadyen, B.J.; Moffet, H.; Batcho, C.S. Design and Accuracy of an Instrumented Insole Using Pressure Sensors for Step Count. *Sensors* **2019**, *19*, 984. [CrossRef]

28. Qin, L.Y.; Ma, H.; Liao, W.H. Insole plantar pressure systems in the gait analysis of post-stroke rehabilitation. In Proceedings of the 2015 IEEE International Conference on Information and Automation, Lijiang, China, 8–10 August 2015; pp. 1784–1789. [CrossRef]

29. Bamberg, S.J.M.; Benbasat, A.Y.; Scarborough, D.M.; Krebs, D.E.; Paradiso, J.A. Gait analysis using a shoe-integrated wireless sensor system. *IEEE Trans. Inf. Technol. Biomed.* **2008**, *12*, 413–423. [CrossRef]

30. Hegde, N.; Sazonov, E. SmartStep: A Fully Integrated, Low-Power Insole Monitor. *Electronics* **2014**, *3*, 381–397. [CrossRef]

31. Jacobs, D.A.; Ferris, D.P. Estimation of ground reaction forces and ankle moment with multiple, low-cost sensors. *J. NeuroEng. Rehabil.* **2015**, *12*, 1. [CrossRef]

32. Lin, F.; Wang, A.; Zhuang, Y.; Tomita, M.R.; Xu, W. Smart Insole: A Wearable Sensor Device for Unobtrusive Gait Monitoring in Daily Life. *IEEE Trans. Ind. Inform.* **2016**, *12*, 2281–2291. [CrossRef]

33. Wu, Y.; Xu, W.; Liu, J.J.; Huang, M.C.; Luan, S.; Lee, Y. An Energy-Efficient Adaptive Sensing Framework for Gait Monitoring Using Smart Insole. *IEEE Sens. J.* **2015**, *15*, 2335–2343. [CrossRef]
34. Crea, S.; Donati, M.; de Rossi, S.M.M.; Oddo, C.M.; Vitiello, N. A wireless flexible sensorized insole for gait analysis. *Sensors* **2014**, *14*, 1073–1093. [CrossRef] [PubMed]
35. Tan, A.M.; Fuss, F.K.; Weizman, Y.; Woudstra, Y.; Troynikov, O. Design of Low Cost Smart Insole for Real Time Measurement of Plantar Pressure. *Procedia Technol.* **2015**, *20*, 117–122. [CrossRef]
36. Kim, S.; Beckmann, L.; Pistor, M.; Cousin, L.; Walter, M.; Leonhardt, S. A versatile Body Sensor Network for health care applications. In Proceedings of the 2009 International Conference on Intelligent Sensors, Sensor Networks and Information Processing (ISSNIP), Melbourne, VIC, Australia, 7–10 December 2009; pp. 175–180. [CrossRef]
37. Kim, S.; Brendle, C.; Lee, H.Y.; Walter, M.; Gloeggler, S.; Krueger, S.; Leonhardt, S. Evaluation of a 433 MHz band body sensor network for biomedical applications. *Sensors* **2013**, *13*, 898–917. [CrossRef] [PubMed]
38. Lueken, M.; ten Kate, W.; Batista, J.P.; Ngo, C.; Bollheimer, C.; Leonhardt, S. Peak detection algorithm for gait segmentation in long-term monitoring for stride time estimation using inertial measurement sensors. In Proceedings of the 2019 IEEE EMBS International Conference on Biomedical & Health Informatics (BHI), Chicago, IL, USA, 19–22 May 2019; pp. 1–4. [CrossRef]
39. Lueken, M.; Kate, W.T.; Valenti, G.; Batista, J.P.; Bollheimer, C.; Leonhardt, S.; Ngo, C. Estimation of Stride Time Variability in Unobtrusive Long-Term Monitoring Using Inertial Measurement Sensors. *IEEE J. Biomed. Health Inform.* **2020**, *24*. [CrossRef] [PubMed]
40. European Telecommunications Standards Institut. ETSI EN 300 220–1 V2. 4.1 (2012-05) Electromagnetic Compatibility and Radio Spectrum Matters (ERM); Short Range Devices (SRD). Available online: https://www.etsi.org/deliver/etsi_en/300200_300299/30022001/02.04.01_60/en_30022001v020401p.pdf (accessed on 19 December 2020).
41. Yadav, N.; Bleakley, C. Accurate orientation estimation using AHRS under conditions of magnetic distortion. *Sensors* **2014**, *14*, 20008–20024. [CrossRef]
42. Fulk, G.D.; Edgar, S.R.; Bierwirth, R.; Hart, P.; Lopez-Meyer, P.; Sazonov, E. Identifying activity levels and steps of people with stroke using a novel shoe-based sensor. *J. Neurol. Phys. Ther. JNPT* **2012**, *36*, 100–107. [CrossRef]

**Publisher's Note:** MDPI stays neutral with regard to jurisdictional claims in published maps and institutional affiliations.

*Article*

# Machine-Learning Based Determination of Gait Events from Foot-Mounted Inertial Units

Matteo Zago [1], Marco Tarabini [2], Martina Delfino Spiga [1], Cristina Ferrario [2], Filippo Bertozzi [3], Chiarella Sforza [3,*] and Manuela Galli [1]

[1] Dipartimento di Elettronica, Informazione e Bioingegneria, Politecnico di Milano, 20133 Milano, Italy; matteo2.zago@polimi.it (M.Z.); martina.delfino@mail.polimi.it (M.D.S.); manuela.galli@polimi.it (M.G.)
[2] Dipartimento di Meccanica, Politecnico di Milano, 20133 Milano, Italy; marco.tarabini@polimi.it (M.T.); cristina.ferrario@polimi.it (C.F.)
[3] Dipartimento di Scienze Biomediche per la Salute, Università degli Studi di Milano, 20133 Milano, Italy; filippo.bertozzi@unimi.it
* Correspondence: chiarella.sforza@unimi.it; Tel.: +39-02-503-15385

**Abstract:** A promising but still scarcely explored strategy for the estimation of gait parameters based on inertial sensors involves the adoption of machine learning techniques. However, existing approaches are reliable only for specific conditions, inertial measurements unit (IMU) placement on the body, protocols, or when combined with additional devices. In this paper, we tested an alternative gait-events estimation approach which is fully data-driven and does not rely on a priori models or assumptions. High-frequency (512 Hz) data from a commercial inertial unit were recorded during 500 steps performed by 40 healthy participants. Sensors' readings were synchronized with a reference ground reaction force system to determine initial/terminal contacts. Then, we extracted a set of features from windowed data labeled according to the reference. Two gray-box approaches were evaluated: (1) classifiers (decision trees) returning the presence of a gait event in each time window and (2) a classifier discriminating between stance and swing phases. Both outputs were submitted to a deterministic algorithm correcting spurious clusters of predictions. The stance vs. swing approach estimated the stride time duration with an average error lower than 20 ms and confidence bounds between ±50 ms. These figures are suitable to detect clinically meaningful differences across different populations.

**Keywords:** gait analysis; spatio-temporal parameters; wearable sensors; decision trees

**Citation:** Zago, M.; Tarabini, M.; Delfino Spiga, M.; Ferrario, C.; Bertozzi, F.; Sforza, C.; Galli, M. Machine-Learning Based Determination of Gait Events from Foot-Mounted Inertial Units. *Sensors* **2021**, *21*, 839. https://doi.org/10.3390/s21030839

Academic Editor: Giuseppe Vannozzi
Received: 12 December 2020
Accepted: 22 January 2021
Published: 27 January 2021

**Publisher's Note:** MDPI stays neutral with regard to jurisdictional claims in published maps and institutional affiliations.

## 1. Introduction

Clinical gait analysis is routinely performed by medical operators to assess ambulatory functional limitations in people with musculoskeletal or cognitive impairments [1,2], as well as to evaluate an individual's quality of life, morbidity and/or mortality [3]. Spatio-temporal gait parameters (i.e., gait speed, stride duration, step length, step width, etc.) provide an immediate picture of an individual's gait profile [1]. They can be used to predict fall risk [4–6] and/or to quantify rehabilitation outcomes [1,2,7–10].

In the last decade, considerable effort was devoted to provide valid and practical alternatives to overcome the limitations of traditional laboratory testing, namely, expensive equipment and time-consuming setup [11,12]. A clear research and development trend appeared towards systems able to capture people's motion without expensive equipment and with limited expert knowledge required [13]. In this landscape, inertial measurements units (IMUs) emerged as a promising family of devices to enable daily-life, affordable, unobtrusive diagnosis and rehabilitation of gait in a wide plethora of cohorts, ranging from neurological diseases to stroke patients [12,14–17].

Among the most appealing advantages of IMU-based gait evaluation in daily routine activities is the opportunity to capture walking adaptations in response to environmental changes or perturbations [16], thus, substantially enhancing the ecological validity of testing [18]. This poses severe challenges to the development of algorithms able to provide accurate measurements despite the natural gait variability [16].

Estimating spatio-temporal parameters from IMU data is not a trivial task due to inherent sensors noise and drift problems [11]. This partly explains why these systems (despite the recent flowering of commercial products [11]) have achieved moderate and sometimes inconsistent performances to date [17], that in turn limited their widespread use for pervasive healthcare [19]. Drifts, in particular, are susceptible to produce large deviations in the calculated results when a double-integration-based sensor fusion approach is adopted [11,20,21]. This approach heavily depends on raw data quality even when an error state Kalman filter is applied to correct sensors' data [5,14]. Most studies achieved gait events detection by sorting peaks, valleys, and zero-crossing in the signals [22,23]. Other algorithms exploited a combination of continuous wavelet transform to detect initial/final contacts (heel-strike, toe-off) and the inverted (double) pendulum model to extract spatio-temporal parameters from sensors' readings. Examples exist where the sensing device was placed on the pelvis, typically in correspondence of the L4–L5 vertebrae [24,25] (assumed to approximate body center of mass location [26]), or alternatively on the foot, exploiting information on the angular velocity of leg swing and size to obtain stride and step lengths [10]. Solutions were also proposed where the temporal detection of gait cycles was based on the norm of the angular velocity of the foot relative to an empirical threshold [27].

A promising but still scarcely explored strategy for the estimation of gait parameters is based on machine learning techniques. In this context, spatio-temporal gait parameters are predicted using a set of features extracted from the IMU signals [11]. Zhang and collaborators used support vector regression models to estimate fundamental gait parameters from an IMU-equipped insole [19]. Hannink et al. used deep convolutional neural networks to map stride-specific IMU data to the resulting stride length, training the model on a publicly available database of 101 geriatric patients [5]. Stride length was estimated with an error of $0.01 \pm 5.37$ cm [5]. They did not, however, focus on temporal parameters.

The achieved accuracy claimed by most of the existing researches is in principle potentially feasible to enable clinical comparisons in terms of temporal parameters [28]. In particular, stride duration was mostly determined with an error (typically measured against a reference system such as optical motion capture, force platforms or instrumented walkways) lower than 50–60 ms [11,29]. However, the main weakness of these approaches is that they are reliable only for particular subjects' conditions [19], for specific IMU placement on the body, for few protocols [22], or when combined with additional devices [19]. In this paper, we tested an alternative gait-events estimation approach based on machine learning algorithms which is fully data-driven and does not rely on a priori models or assumptions.

Section 2 will describe the experimental design, the equipment involved, the processing steps, and the statistical computations performed. Results will be shown in Section 3 and separately discussed in Section 4, where limitations and possible practical implications of the study are also reported; Section 5 contains concluding remarks and perspectives.

## 2. Materials and Methods

### 2.1. Experimental Design and Participants

This study involved the simultaneous collection of data with a commercial IMU device and a force platform during walking. Ground reaction force data were used as a reference. Forty healthy participants (19 women, 21 men) aged between 22 and 55 years voluntarily took part in the experimental sessions; inclusion criteria were: (i) no diagnosed gait-related impairments and (ii) ability to walk comfortably at different speeds. All of the subjects were able to complete the test and coped with the provided instructions. Only anonymized signals were analyzed and no clinical nor personal information was requested.

## 2.2. Experimental Setup

A wearable IMU device (Physilog 5, GaitUp Ltd., Lausanne, Switzerland) was clipped to the participants' right shoe in correspondence to the navicular bone (Figure 1). This IMU is a six-axes stand-alone unit integrating a three-axis accelerometer and a three-axis gyroscope. Device settings were set to a sampling frequency of 512 Hz, and a dynamic range of $\pm 16$ g for the accelerometer and $2000°/s$ for the gyroscope. The unit is sized 47.5 mm $\times$ 26.5 mm $\times$ 10 mm and weights 36 g. The sensors' inertial right-handed coordinate system is oriented as displayed in Figure 1 (it is worth mentioning that GaitUp Ltd. also provides a commercial gait analysis solution, which has been validated and already used in several studies as described in [11]. As our goal was not to re-evaluate the GaitUp algorithms, only the raw data from the IMU development platform were processed in this study).

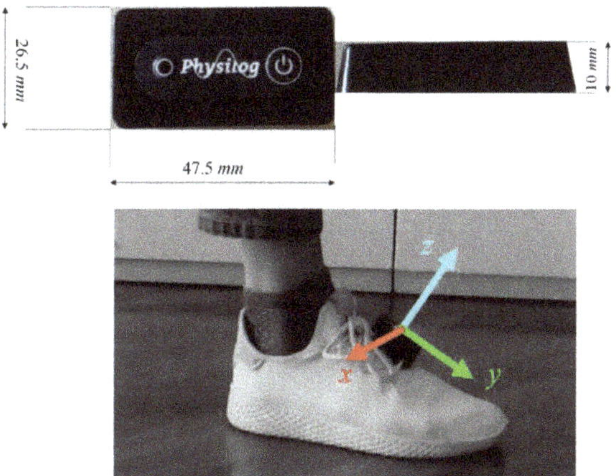


**Figure 1.** GaitUp Physilog size (**top**) and placement on the foot (**bottom**). Direction of the local right-handed reference frame axes is also reported.

A schematic representation of the laboratory setup is depicted in Figure 2. The laboratory was equipped with two $46.5 \times 51.8$ cm$^2$ AMTI OR6-7 force platforms (Advanced Mechanical Technology, Inc., Watertown, MA, USA) sampling at 200 Hz, used to collect ground reaction forces (GRF) data. AMTI OR6-7 are strain gages-based force platforms designed for biomechanics applications. In the adopted configuration (10-V bridges excitation), full scale output in the medial and anteroposterior directions was 4450 N, while in the vertical direction it was 8900 N.

To provide a visual representation of the foot position during gait, the 3D position of four passive reflective markers was recorded with an optoelectronic motion capture system (SMART DX400, BTS Bioengineering, Milano, Italy) with a sampling frequency of 100 Hz. Markers were placed in correspondence to the lateral aspect of the foot at the fifth metatarsal head, on the heel, on the lateral malleolus and on the knee in correspondence to the lateral femoral epicondyle. Optical and ground reaction forces data are automatically synchronized. Global laboratory reference frame was oriented with the $x$ axis horizontal and directed along the walking direction, the $y$ axis pointing upwards, and the $z$ axis mediolateral and pointing to the participant's right (Figure 2).

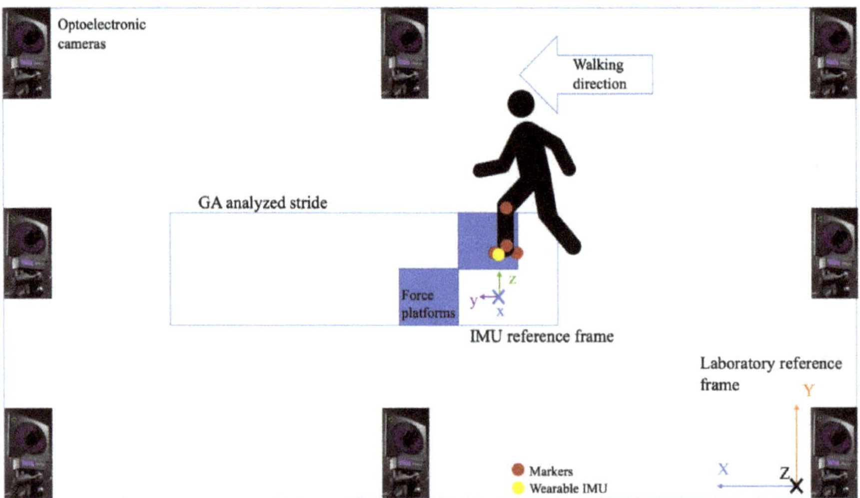



**Figure 2.** Experimental setup—tests were conducted on the middle laboratory lane with force platform embedded on the floor. Motion capture cameras are fixed on the wall in a standard gait analysis configuration.

### 2.3. Procedures

After starting the recording of both motion capture and IMU data, subjects were asked to hit a force platform three times with their right foot. This enabled to synchronize the two measurement systems (motion capture system and IMU) [30], as explained in the following paragraph.

Subsequently, all subjects performed a sequence of short straight-line level walking tests, from three to ten steps each, at self-selected comfortable speed, which were measured by means of the optical system.

### 2.4. Data Processing and Features Engineering

Data were processed by means of custom Matlab (v. 2019b, The Mathworks Inc., Natick, MA, USA) routines. GRF and inertial readings were time-aligned prior to each recording (maximum 30 s each) by determining the delay corresponding to the peak of the cross-correlation function among the vertical force and vertical acceleration signals at the beginning of the recording (Figure 3). The maximum synchronization error when dealing with cross correlation algorithms is typically lower than 1 sampling period [31]. In our case this is 10 ms, thus entailing a standard uncertainty lower than 3 ms. This value is trivial in comparison with the intrinsic variability of the observed phenomenon, as detailed later. Reference time events were obtained from foot–ground contact information, setting a binary GRF threshold of 10 N [32]. Stride time was computed as the time interval between two consecutive initial contacts of the right foot.

The classification of gait phases and events followed a gray-box approach. First, sensor readings related to each step were trimmed in order to contain two consecutive heel strikes. The data stream was subsequently segmented in 64-sample (0.125 s) windows. A fixed-width, 64-sample moving window on the 6 IMU channels (acceleration and angular velocity) with step equal to 1 sample was used to compute 48 statistical features in the time and frequency domains, similarly to what previously done in [33]. In the time domain, the root mean square, variance, kurtosis, skewness, and correlation between each pair of accelerometer and gyroscope (angular velocity) axes were computed. In the frequency domain, the dominant frequency and the power at the dominant frequency were obtained (Table 1). Data processing flow is depicted in Figure 4.

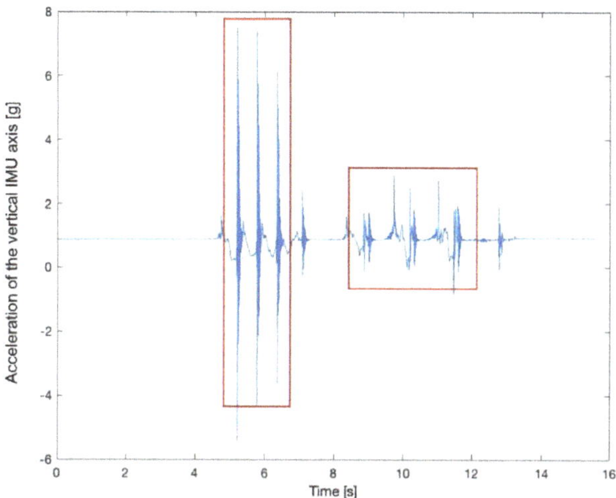

**Figure 3.** Sample vertical (z-axis) raw accelerometer readings during a test. The first three spikes correspond to the synchronization signal and the following data refer to gait events.

**Table 1.** Statistical and frequency-domain features.

| Signal | Time Domain | Frequency Domain |
| --- | --- | --- |
| Acceleration (3 channels) | Root mean squared<br>Variance<br>Kurtosis<br>Skewness<br>Linear Correlation ($x$-$y$, $x$-$z$, $y$-$z$) | Dominant frequency<br>Power at dominant frequency |
| Angular velocity (3 channels) | Root mean squared<br>Variance<br>Kurtosis<br>Skewness<br>Linear Correlation ($x$-$y$, $x$-$z$, $y$-$z$) | Dominant frequency<br>Power at dominant frequency |

*2.5. Gait Events and Phases Classification*

Three binary classifiers, supporting two alternative approaches, were implemented to identify:

- 1a: windows containing a heel strike vs. any other event.
- 1b: windows containing a toe off vs. any other event.
- 2: windows corresponding to stance vs. swing phases.

The results of gait event classifiers (1a) and (1b) were subsequently combined. Each window was labelled based on the reference output: in the case of gait events classifiers, label "1" was attributed to the windows containing a heel strike (1a) or a toe-off (1b), and label "0" elsewhere. In the case of classifier 2, label "0" was assigned to the swing phase and label "1" to the stance (ground contact) phase, evaluated at the time of each window's first sample.

The whole sample of labelled features was split into a training (70% strides) and a test (30% strides) set. Of note, as the sequence of events is crucial to the problem of classifying gait phases, we randomly allocated strides (collection of consecutive observations), and no single observations.

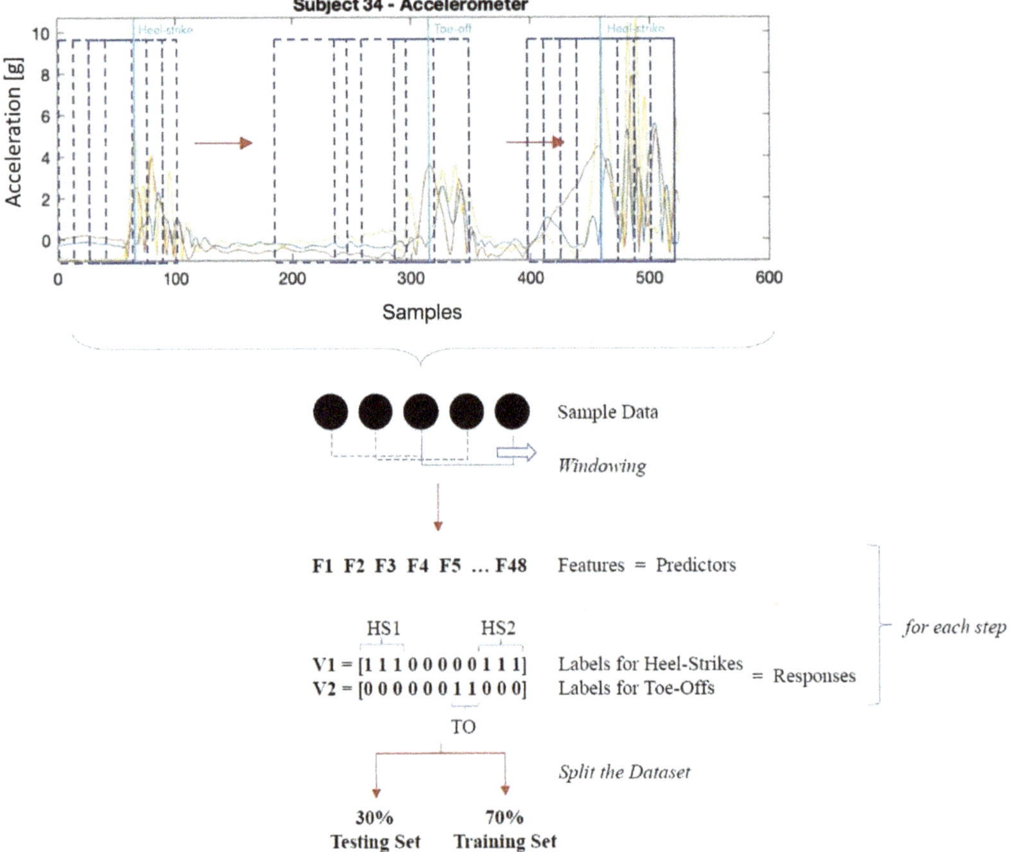



**Figure 4.** Data processing flow: sensors' readings (in the top panel, sample acceleration signal) were windowed (dashed blue boxes represent the window moving across the signal); subsequently a set of features were obtained for each window, which was labelled according to the reference ground reaction force output. The whole set of collected strides (each one containing a collection of features) were randomly split into a training and a test set. HS: heel-strike, TO: toe-off.

The Matlab Machine Learning Toolbox identified decision trees as the most accurate binary classifiers relative to the concurrent application. Decision trees also allow for good classification speed and for features importance evaluation [33]. For these reasons, it was decided to base the following analysis on this classifier method.

The output stream returned by each classification approach was subsequently submitted to a "correction algorithm" aimed at detecting and removing isolated short clusters embedded in larger areas belonging to the opposed class. A cluster was considered "isolated" if it was shorter than half a window (i.e., 32 samples or 0.063 s), preceded and followed by a differently classified array of larger size (Figure 5). The correction algorithm works sequentially and "prefers" the current class. That is, if we are within a "stance" phase, short clusters labelled as "swing" are reversed into "stance". The transitions between classes on the obtained output stream identified the correspondent gait event: for instance, the transition between the stance and the swing phase denoted a toe-off event.

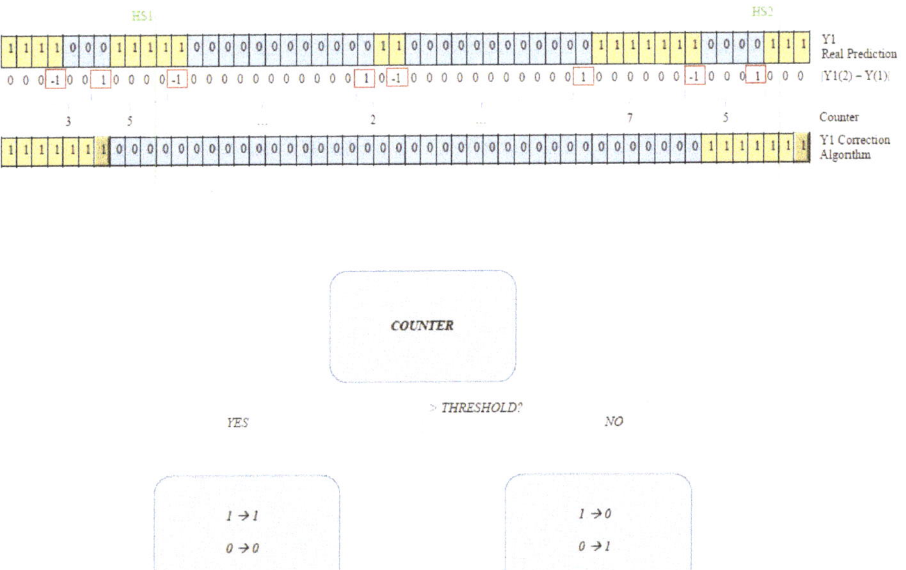



**Figure 5.** Correction algorithm: isolated short (counter < threshold) clusters were corrected according to the surroundings, as schematically reported in the bottom diagram.

### 2.6. Statistical Analysis

The error's bias and random component in the identification of heel-strikes, toe-offs, stride, stance, and swing times were determined on the test set. To do so, the mean and standard deviation of the difference between the reference and the estimated value were computed, as well as the corresponding root-mean-square (RMS) error and 95% confidence intervals (95%CI). The standard error of the mean was computed for the whole test set (U) and for 10 strides ($U_{10}$), the latter considering a common number of repetitions per subject in clinical gait applications [1].

Paired Students' *t*-tests were performed between estimated and reference stride, stance and swing times, and associated with the corresponding Cohen's d effect size (ES): values of $d \leq 0.5$, $0.5 < d \leq 0.8$, and $d > 0.8$ were considered low, moderate, and large effects, respectively. The coefficient of determination ($R^2$) was obtained between estimated and reference stride time, and between the related error and the corresponding gait speed. A statistical significance threshold of 0.05 was implemented throughout.

### 3. Results

Overall, participants performed 10–15 strides each, for a total of 500 recorded strides. Of them, 75 were discarded due to inconsistent or incomplete data. Therefore, the training set included 298 strides and the test set 127 strides. The global number of collected observations (64-sample windows) was 311,802.

Figure 6 and Table 2 report the output of the three binary classifiers: the gait events classifiers (1a and 1b) returned an accuracy of 91–93%, while the stance vs. swing classifier (2) reached 95.6% before being submitted to the correction algorithm. Some features were remarkably more predictive of the correct class. In particular, vertical acceleration played a substantial role in all the models, while angular velocity was less important to discriminate between stance and swing phases.

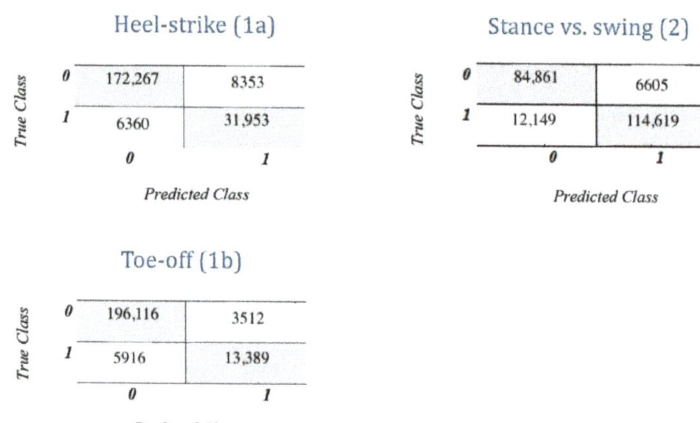



**Figure 6.** Classification performance of the three tested approaches. Left: heel-strike (**1a**) and toe-off (**1b**) events detection; Right: stance vs. swing moving windows classification (**2**).

**Table 2.** Classifiers (decision trees) technical figures. The reported accuracy refers to the classification outcome, not to the final gait event estimation result. The outputs of Methods 1a and 1b were subsequently combined.

| Item | Heel-Strike vs. Other (1a) | Toe-Off vs. Other (1b) | Stance vs. Swing (2) |
|---|---|---|---|
| Prediction accuracy | 93.3% | 91.4% | 95.6% |
| Observations | 218,933 | 218,933 | 218,933 |
| Misclassification cost | 14,713 | 18,754 | 9428 |
| Prediction speed | $7 \times 10^6$ observations/s | $7 \times 10^6$ observations/s | $6.9 \times 10^6$ observations/s |
| Training time | 52.145 s | 18.137 s | 49.694 s |
| Size of training data | 87 MB | 85 MB | 87 MB |
| Validation | Hold-out | Hold-out | Hold-out |
| Features whose importance was greater than 5% | Mediolateral mean $\omega$<br>Vertical acc. RMS | Vertical acc. Mean<br>AP $\omega$ mean<br>AP $\omega$ var | Vertical acc. RMS<br>AP acc. Mean |

Acc.: acceleration, AP: anteroposterior, $\omega$: angular velocity, RMS: root-mean-square; var: variance.

Globally, the stance vs. swing approach returned lower errors in determining all the considered parameters (Table 3). In particular, events identification returned an average error between $-11$ and 5 ms (95%CI, heel-strike) and between $-13$ and 50 ms (toe-off). Conversely, the approach involving methods 1a and 1b returned larger uncertainty values, reaching 35 ms (heel-strike) and 74 ms (toe-off) when considered over 10 strides. Gait phases estimation were therefore significantly different with small-to-medium effect sizes for the approach 1a–1b, while the stance vs. swing method showed no statistical differences for stance and swing phases duration ($p = 0.098$ and $p = 0.782$, low effect); the same approach underestimated the stride time by ($p = 0.004$, 95%CI between $-37$ and $-7$ ms) with a low effect size. However, as displayed in Figure 7a, the coefficients of determination were 0.796 and 0.808 for the gait-events classifiers (1a and 1b) and stance vs. swing (2) approaches, respectively ($p < 0.001$ for both).

**Table 3.** Descriptive statistics (in ms) of the heel-strike and toe-off identification, as well as swing/stance/stride time estimation, with respect to the reference system (force platform).

| Method | Mean | SD | RMSE | U | $U_{10}$ | 95%CI | $p$ | ES |
|---|---|---|---|---|---|---|---|---|
| *Heel-strike identification* | | | | | | | | |
| Heel-strike and toe-off (1a, 1b) | −20 | 111 | 113 | 7 | 35 | −33, −6 | - | - |
| Stance vs. swing (2) | −3 | 59 | 59 | 4 | 19 | −11, 5 | - | - |
| *Toe-off identification* | | | | | | | | |
| Heel-strike and toe-off (1a, 1b) | 95 | 233 | 251 | 21 | 74 | 54, 136 | - | - |
| Stance vs. swing (2) | 19 | 165 | 166 | 15 | 52 | −13, 50 | - | - |
| *Stance phase estimation* | | | | | | | | |
| Heel-strike and toe-off (1a, 1b) | −113 | 214 | 241 | 19 | 68 | −150, −75 | <0.001 | 0.514 |
| Stance vs. swing (2) | −26 | 168 | 169 | 16 | 53 | −50, 5 | 0.098 | 0.158 |
| *Swing phase estimation* | | | | | | | | |
| Heel-strike and toe-off (1a, 1b) | 39 | 221 | 224 | 20 | 70 | 0, 78 | 0.048 | 0.205 |
| Stance vs. swing (2) | 5 | 179 | 178 | 17 | 56 | −29, 38 | 0.782 | 0.034 |
| *Stride time* | | | | | | | | |
| Heel-strike and toe-off (1a, 1b) | −74 | 140 | 158 | 12 | 44 | −98, −49 | <0.001 | 0.258 |
| Stance vs. swing (2) | −22 | 79 | 81 | 7 | 25 | −36, −7 | 0.004 | 0.122 |

ES: Cohen's d effect size; RMSE: root mean square error; U: standard error of the mean; $U_{10}$: standard error of the mean on 10 strides; p: paired *t*-tests.

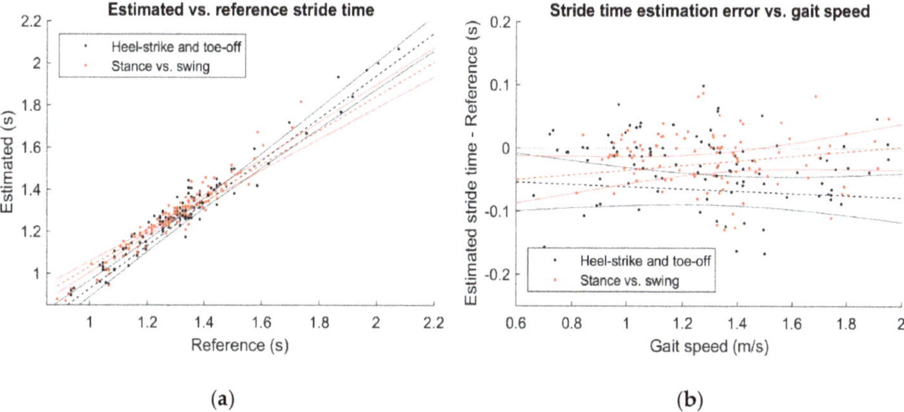


**Figure 7.** Regression plots comparing estimated and reference stride time (**a**) and the measurement error as a function of gait speed (**b**). Regression lines (dashed) and 95% confidence bounds (solid lines) were reported.

In the best case (stance vs. swing approach), we found a single totally erroneous heel-strike (2% of the test set, difference from the reference of about 0.7 s) and seven toe-off events whose error was higher than 0.2 s (6% of the test set). There was not a significant correlation between the estimation error and the walking speed (Figure 7b), being R2 equal to 0.004 ($p = 0.482$, approach 1a and 1b) and 0.021 ($p = 0.135$, approach 2).

## 4. Discussion

By exploiting uncorrected, high-frequency acceleration and angular velocity data readings combined with a grey-box machine-learning approach, in this paper, we showed that it was possible to estimate gait events (heel-strikes and toe-off) with an average error lower than 20 ms and confidence bounds between ±50 ms. This led to determine stride time with a root-mean-squared error of about 80 ms.

### 4.1. A Data-Driven Approach

The discrimination of stance vs. swing phases (approach 2) outperformed the first approach (1a and 1b), intended to detect time windows where the initial/final contacts occur. This was partly expected, as gait events per se are distinct instants in time: a moving window, even if relatively short (64 samples at 512 Hz equals 125 ms) turned out to be prone to systematically anticipate (heel-strike, negative bounds of 95% CI) or delay (toe-off, positive bounds of 95%CI) the events detection. This inevitably led to significant systematic differences between the estimated temporal gait parameters (low-to-moderate effect sizes) and the reference system (force platform).

Conversely, the stance vs. swing approach almost halved the measurement error. Stride time resulted still significantly lower than that detected by the reference system, but with a low effect size and a reduced uncertainty ($U_{10}$ = 25 ms). The RMSE of about 80 ms were slightly higher than a more sophisticated experimental setup based on an instrumented insole [19]. Confidence interval was limited to 29 ms and between the limits of agreement obtained by Yeo and colleagues [29] (60–100 ms), who also reported a very similar error of 20 ms. The correlation coefficient with the reference measure was r = 0.89, substantially in line with [11] and slightly lower than in Zhou and collaborators [16] (r = 0.95). Swing and stance time confidence intervals (−29, 38 ms and −50, 5 ms) were comparable with those provided by Godfrey and collaborators, who obtained (−35, 49 ms and −39, 49 ms) with a pelvis-mounted IMU and event detection based on Gaussian continuous wavelet transform [25].

Notably, these outcomes were purely data-driven. In other words, they were obtained without a priori assumptions, neither concerning subjects' anthropometrics or speed nor regarding raw signal conditioning (i.e., filtering, sensors' bias compensation), which was deliberately avoided to show the potential of the approach. Not relying on any deterministic model, the algorithms provided are potentially able to capture unconventional gait patterns, for instance with long stance phases and shorter steps than normal, as in the case of Parkinson's disease [14,17,28].

Discrepancies from previous research performances should also be read in the light of the numerosity of the experimental cohort. Instead of gathering a large number of strides from a relatively reduced sample of participants, we decided to extend the survey to a wide (*n* = 40) range of subjects, higher than those observed in similar studies (e.g., five subjects in [11], 14 subjects in [19], and 30 subjects in [29]). This was done to enhance the generalizability of results, despite it could have introduced interindividual variability and reduced the estimation accuracy.

Besides their limited computational cost, an advantage of the regression trees used as the main classification algorithm is the opportunity to easily examine features' relative contribution. Vertical acceleration [23] and anteroposterior angular velocity were the most revealing among a reduced set of discriminant statistical features. In that, the grade of variability of a signal throughout a time window (accounted for by the signal's variance) was probably key to capture quick variations due to initial contact and/or toe-off. Additionally, limiting the number of relevant features adds to the feasibility of a real-time implementation. This reduces the computational burden often associated with end-to-end complex machine learning models (as deep convolutional neural networks [19]) or double integration approach (exploiting zero-velocity update technique), which requires sensors fusion algorithms [11]. Lastly, the proposed correction algorithm is simply enabling a sort of data cleaning prior to the determination of gait events, and its execution requires very simple and inexpensive operations.

### 4.2. Effect of Measurement Uncertainty in the Real Instrument Usage Context

Normative data in healthy adults report an interindividual variability (standard deviation) of stride time of about 80–120 ms [34]. This parameter tends to increase with age [34–36] and when a perturbation (physical, pathological, or cognitive as dual-task) arises, reaching values up to 180 ms [17,37]. Other investigations on orthopaedic patients reported a variability of 90 ms in the stride times measured in a single participant at self-selected speed across over ten gait cycles [8]. In this context, the obtained $U_{10}$ value of 25 ms appears tolerable to properly characterize temporal gait parameters, even when the goal is detecting mean differences across populations: as a reference, Beauchet et al. reported between-sexes differences in stride time of about 50 ms [34], while Hollmann and collaborators denoted an increase in stride time higher than 60 ms between 70–74 years and 85+ years old women [35]. Significant differences of about 20–30 ms were also found between healthy controls and Down Syndrome patients at lower walking speed [38].

*4.3. Limitations*

The purpose of this study was to show the feasibility of classifying gait phases and determining the related events during walking in healthy adults. While we claimed that this method could be easily extended to other forms of locomotion like running or hopping, the proposed approach could not necessarily directly apply to patients with locomotor impairments. The collection of new training data in these conditions would be probably required. Likewise, in order to correctly capture the specific walking variability of pathological or pediatric populations it is advised that additional training data would be collected and combined. The real-time implementation of the algorithm was beyond the scope of this paper, and will be addressed in upcoming research.

A second limitation is that the original dataset was randomly split into a training and test set, without explicitly separate subjects. This was done to ensure the highest generalizability of the training set, but it also means that in principle, steps from the same participants could have be assigned to both splits.

The sampling frequency of the reference laboratory equipment (force platforms) was 200 Hz. This brought in an inherent uncertainty of $5 \text{ ms}/2 \cdot \sqrt{3} = 1.4$ ms. In this sense, referring to instrumented walkways (e.g., Microgate's Optogait [39]) the higher time resolution (up to 1 ms) would lead to a minor reduction in the overall measurement uncertainty. However, previous investigations relied on optical systems with a sampling frequency of 100 Hz to perform the same comparison [16]. Moreover, as previously discussed, differences in gait temporal parameters of this magnitude (<10 ms) are not clinically meaningful.

Lastly, we did not face the issue of multiple units' synchronization, that, in principle, could be an additional source of uncertainty [40].

## 5. Conclusions

IMU-based solutions for the assessment of the gait function in real-world settings are continuously improved to provide personalized and pervasive healthcare [16]. In this study, we proposed a novel data-driven approach for the determination of temporal gait parameters based on inertial sensors and simple machine learning algorithms. Measurement errors were comparable to existing IMU-based methods, still, the proposed approach did not rely on any particular raw data processing constraint, and it was robust to inter-subject variability, thus, making it unnecessary to collect patient- or condition-specific training data [19]. Further, the proposed approach opens to further advancements on this path, which offers reduced computational burden and the potential to detect gait phases even when unconventional ambulatory modes are evaluated outside from restricted laboratory settings.

This study also reinforces the use of a movement analysis laboratory as a required reference when testing the measurement uncertainty of new devices. In the current market situation, in which we are witnessing the rapid spread of different systems for measuring a wealth of parameters related to human movement, it is necessary to adopt methods aimed at rigorously verifying the quality of the data produced.

**Author Contributions:** Conceptualization, M.Z., M.D.S., and M.T.; methodology, M.Z. and M.T.; software, M.D.S. and C.F.; validation, M.Z., C.F.; formal analysis, M.Z. and M.D.S.; investigation, C.F., M.D.S., and F.B.; resources, C.S. and M.G.; data curation, M.Z., F.B., and M.D.S.; writing—original draft preparation, M.Z.; writing—review and editing, C.S., and M.G.; supervision, M.T., M.G., and C.S.; project administration, M.G. and C.S. All authors have read and agreed to the published version of the manuscript.

**Funding:** University of Milan (Piano di sostegno alla ricerca 2015–17).

**Institutional Review Board Statement:** Institutional Review Board approval was waived due to collection of anonymized, non-invasive and non-confidential data as mentioned in the text.

**Informed Consent Statement:** Patient consent was waived due to collection of anonymized, non-invasive and non-confidential data as mentioned in the text.

**Data Availability Statement:** The data presented in this study are available on request from the corresponding author.

**Conflicts of Interest:** The authors declare no conflict of interest.

# References


1. Whittle, M.W. Clinical gait analysis: A review. *Hum. Mov. Sci.* **1996**, *15*, 369–387. [CrossRef]
2. Ferber, R.; Osis, S.T.; Hicks, J.L.; Delp, S.L. Gait biomechanics in the era of data science. *J. Biomech.* **2016**, *49*, 3759–3761. [CrossRef]
3. Studenski, S.; Perera, S.; Patel, K.; Rosano, C.; Faulkner, K. Gait Speed and Survival in Older Adults. *JAMA* **2011**, *305*, 50–58. [CrossRef]
4. Allseits, E.; Agrawal, V.; Lučarević, J.; Gailey, R.; Gaunaurd, I.; Bennett, C. A practical step length algorithm using lower limb angular velocities. *J. Biomech.* **2018**, *66*, 137–144. [CrossRef] [PubMed]
5. Hannink, J.; Kautz, T.; Pasluosta, C.F.; Barth, J.; Schulein, S.; Gabmann, K.G.; Klucken, J.; Eskofier, B.M. Mobile Stride Length Estimation with Deep Convolutional Neural Networks. *IEEE J. Biomed. Health Inform.* **2018**, *22*, 354–362. [CrossRef] [PubMed]
6. Valkanova, V.; Ebmeier, K.P. What can gait tell us about dementia? Review of epidemiological and neuropsychological evidence. *Gait Posture* **2017**, *53*, 215–223. [CrossRef]
7. Temporiti, F.; Zanotti, G.; Furone, R.; Molinari, S.; Zago, M.; Loppini, M.; Galli, M.; Grappiolo, G.; Gatti, R. Gait analysis in patients after bilateral versus unilateral total hip arthroplasty. *Gait Posture* **2019**, *72*, 46–50. [CrossRef]
8. Lovecchio, N.; Sciumè, L.; Zago, M.; Panella, L.; Lopresti, M.; Sforza, C. Lower limbs kinematic assessment of the effect of a gym and hydrotherapy rehabilitation protocol after knee megaprosthesis: A case report. *J. Phys. Ther. Sci.* **2016**, *28*, 1064–1070. [CrossRef]
9. Perry, J.; Burnfield, J.N. *Gait Analysis: Normal and Pathological Function*, 2nd ed.; SLACK Incorporated: Thorofare, NJ, USA, 2010; ISBN 1-55642-192-3.
10. Salarian, A.; Russmann, H.; Vingerhoets, F.J.G.; Dehollain, C.; Blanc, Y.; Burkhard, P.R.; Aminian, K. Gait assessment in Parkinson's disease: Toward an ambulatory system for long-term monitoring. *IEEE Trans. Biomed. Eng.* **2004**, *51*, 1434–1443. [CrossRef]
11. Zhou, L.; Fischer, E.; Tunca, C.; Brahms, C.M.; Ersoy, C.; Granacher, U.; Arnrich, B. How we found our imu: Guidelines to IMU selection and a comparison of seven IMUs for pervasive healthcare applications. *Sensors* **2020**, *20*, 4090. [CrossRef]
12. Iosa, M.; Picerno, P.; Paolucci, S.; Morone, G. Wearable inertial sensors for human movement analysis. *Expert Rev. Med. Devices* **2016**, *13*, 641–659. [CrossRef]
13. Mundt, M.; Koeppe, A.; David, S.; Witter, T.; Bamer, F.; Potthast, W.; Markert, B. Estimation of Gait Mechanics Based on Simulated and Measured IMU Data Using an Artificial Neural Network. *Front. Bioeng. Biotechnol.* **2020**, *8*, 1–16. [CrossRef] [PubMed]
14. Tunca, C.; Pehlivan, N.; Ak, N.; Arnrich, B.; Salur, G.; Ersoy, C. Inertial sensor-based robust gait analysis in non-hospital settings for neurological disorders. *Sensors* **2017**, *17*, 825. [CrossRef] [PubMed]
15. Wouda, F.J.; Giuberti, M.; Bellusci, G.; Veltink, P.H. Estimation of Full-Body Poses Using Only Five Inertial Sensors: An Eager or Lazy Learning Approach? *Sensors* **2016**, *16*, 2138. [CrossRef] [PubMed]
16. Zhou, L.; Tunca, C.; Fischer, E.; Brahms, C.M.; Ersoy, C.; Granacher, U.; Arnrich, B. Validation of an IMU Gait Analysis Algorithm for Gait Monitoring in Daily Life Situations. In Proceedings of the Annual International Conference of the IEEE Engineering in Medicine & Biology Society (EMBC), Montreal, QC, Canada, 20–24 July 2020; pp. 4229–4232.
17. Zago, M.; Sforza, C.; Pacifici, I.; Cimolin, V.; Camerota, F.; Celletti, C.; Condoluci, C.; De Pandis, M.F.M.F.; Galli, M. Gait evaluation using inertial measurement units in subjects with Parkinson's disease. *J. Electromyogr. Kinesiol.* **2018**, *42*, 44–48. [CrossRef]
18. Lopez-Nava, I.H.; Muñoz-Meléndez, A. Wearable Inertial Sensors for Human Motion Analysis: A review. *IEEE Sens. J.* **2016**, *16*, 7821–7834. [CrossRef]
19. Zhang, H.; Guo, Y.; Zanotto, D. Accurate Ambulatory Gait Analysis in Walking and Running Using Machine Learning Models. *IEEE Trans. Neural Syst. Rehabil. Eng.* **2020**, *28*, 191–202. [CrossRef]
20. Ferrari, A.; Ginis, P.; Hardegger, M.; Casamassima, F.; Rocchi, L.; Chiari, L. A mobile Kalman-filter based solution for the real-time estimation of spatio-temporal gait parameters. *IEEE Trans. Neural Syst. Rehabil. Eng.* **2016**, *24*, 764–773. [CrossRef]
21. Nhat Hung, T.; Soo Suh, Y. Inertial sensor-based two feet motion tracking for gait analysis. *Sensors* **2013**, *13*, 5614–5629. [CrossRef]
22. Caldas, R.; Mundt, M.; Potthast, W.; Buarque, F.; Neto, D.L. Gait & Posture A systematic review of gait analysis methods based on inertial sensors and adaptive algorithms. *Gait Posture* **2017**, *57*, 204–210.

23. Bugané, F.; Benedetti, M.G.; Casadio, G.; Attala, S.; Biagi, F.; Manca, M.; Leardini, A. Estimation of spatial-temporal gait parameters in level walking based on a single accelerometer: Validation on normal subjects by standard gait analysis. *Comput. Methods Programs Biomed.* **2012**, *108*, 129–137. [CrossRef] [PubMed]
24. Del Din, S.; Hickey, A.; Hurwitz, N.; Mathers, J.C.; Rochester, L.; Godfrey, A. Measuring gait with an accelerometer-based wearable: Influence of device location, testing protocol and age. *Physiol. Meas.* **2016**, *37*, 1785–1797. [CrossRef] [PubMed]
25. Godfrey, A.; Del Din, S.; Barry, G.; Mathers, J.C.; Rochester, L. Instrumenting gait with an accelerometer: A system and algorithm examination. *Med. Eng. Phys.* **2015**, *47*, 400–407. [CrossRef] [PubMed]
26. Mapelli, A.; Zago, M.; Fusini, L.; Galante, D.; Colombo, A.; Sforza, C. Validation of a protocol for the estimation of three-dimensional body center of mass kinematics in sport. *Gait Posture* **2014**, *39*, 460–465. [CrossRef] [PubMed]
27. Mariani, B.; Hoskovec, C.; Rochat, S.; Büla, C.; Penders, J.; Aminian, K. 3D gait assessment in young and elderly subjects using foot-worn inertial sensors. *J. Biomech.* **2010**, *43*, 2999–3006. [CrossRef] [PubMed]
28. Keloth, S.M.; Viswanathan, R.; Jelfs, B.; Arjunan, S.; Raghav, S.; Kumar, D. Which gait parameters and walking patterns show the significant differences between Parkinson's disease and healthy participants? *Biosensors* **2019**, *9*, 59. [CrossRef]
29. Yeo, S.S.; Park, G.Y. Accuracy Verification of Spatio-Temporal and. *Sensors* **2020**, *20*, 1343. [CrossRef]
30. Rhudy, M. Time Alignment Techniques for Experimental Sensor Data. *Int. J. Comput. Sci. Eng. Surv.* **2014**, *5*, 1–14. [CrossRef]
31. Costa-Júnior, J.F.S.; Cortela, G.A.; Maggi, L.E.; Rocha, T.F.D.; Pereira, W.C.A.; Costa-Felix, R.P.B.; Alvarenga, A.V. Measuring uncertainty of ultrasonic longitudinal phase velocity estimation using different time-delay estimation methods based on cross-correlation: Computational simulation and experiments. *Meas. J. Int. Meas. Confed.* **2018**, *122*, 45–56. [CrossRef]
32. Tirosh, O.; Sparrow, W.A. Identifying heel contact and toe-off using forceplate thresholds with a range of digital-filter cutoff frequencies. *J. Appl. Biomech.* **2003**, *19*, 178–184. [CrossRef]
33. Zago, M.; Sforza, C.; Dolci, C.; Tarabini, M.; Galli, M. Use of machine learning and wearable sensors to predict energetics and kinematics of cutting maneuvers. *Sensors* **2019**, *19*, 3094. [CrossRef] [PubMed]
34. Beauchet, O.; Allali, G.; Sekhon, H.; Verghese, J.; Guilain, S.; Steinmetz, J.P.; Kressig, R.W.; Barden, J.M.; Szturm, T.; Launay, C.P.; et al. Guidelines for assessment of gait and reference values for spatiotemporal gait parameters in older adults: The biomathics and canadian gait consortiums initiative. *Front. Hum. Neurosci.* **2017**, *11*, 353. [CrossRef] [PubMed]
35. Hollman, J.H.; McDade, E.M.; Petersen, R.C. Normative spatiotemporal gait parameters in older adults. *Gait Posture* **2011**, *34*, 111–118. [CrossRef] [PubMed]
36. Ciprandi, D.; Bertozzi, F.; Zago, M.; Ferreira, C.L.P.; Boari, G.; Sforza, C.; Galvani, C. Study of the association between gait variability and physical activity. *Eur. Rev. Aging Phys. Act.* **2017**, *14*, 19. [CrossRef] [PubMed]
37. Beauchet, O.; Freiberger, E.; Annweiler, C.; Kressig, R.; Herrmann, F.; Allali, G. Test-retest reliability of stride time variability while dual tasking in healthy and demented adults with frontotemporal degeneration. *J. Neuroeng. Rehabil.* **2011**, *8*, 37. [CrossRef] [PubMed]
38. Agiovlasitis, S.; McCubbin, J.A.; Yun, J.; Mpitsos, G.; Pavol, M.J. Effects of Down syndrome on three-dimensional motion during walking at different speeds. *Gait Posture* **2009**, *30*, 345–350. [CrossRef]
39. Carbajales-Lopez, J.; Becerro-de-Bengoa-Vallejo, R.; Losa-Iglesias, M.E.; Casado-Hernández, I.; Benito-De Pedro, M.; Rodríguez-Sanz, D.; Calvo-Lobo, C.; Antolín, M.S. The optogait motion analysis system for clinical assessment of 2D spatio-temporal gait parameters in young adults: A reliability and repeatability observational study. *Appl. Sci.* **2020**, *10*, 3726. [CrossRef]
40. Coviello, G.; Avitabile, G.; Florio, A. A synchronized multi-unit wireless platform for long-term activity monitoring. *Electronics* **2020**, *9*, 1118. [CrossRef]

*sensors*

*Article*

# A Comparative Analysis of Shoes Designed for Subjects with Obesity Using a Single Inertial Sensor: Preliminary Results

Veronica Cimolin [1], Michele Gobbi [2], Camillo Buratto [3], Samuele Ferraro [3], Andrea Fumagalli [2], Manuela Galli [1] and Paolo Capodaglio [2,4,*]

1    Department of Electronics, Information and Bioengineering, Politecnico di Milan, Piazza Leonardo da Vinci 32, 20133 Milan, Italy; veronica.cimolin@polimi.it (V.C.); manuela.galli@polimi.it (M.G.)
2    Orthopaedic Rehabilitation Unit and Research Lab for Biomechanics, Rehabilitation and Ergonomics, Ospedale San Giuseppe, Istituto Auxologico Italiano, IRCCS, via Cadorna 90, 28824 Piancavallo di Oggebbio, Italy; m.gobbi@auxologico.it (M.G.); a.fumagalli@auxologico.it (A.F.)
3    Podartis SRL, via Erizzo 123/c, 31035 Piancavallo, Italy; camillo.buratto@podartis.it (C.B.); samuele.ferraro@podartis.it (S.F.)
4    Department Surgical Sciences, Physical and Rehabilitation Medicine, University of Torino, 10126 Torino, Italy
*    Correspondence: p.capodaglio@auxologico.it

**Abstract:** Walking remains a highly recommended form of exercise for the management of obesity. Thus, comfortable and adequate shoes represent, together with the prescription of a safe adapted physical activity, an important means to achieve the recommended physical activity target volume. However, the literature on shoes specific for obese individuals is inadequate. The aim of the present study was to compare the performance of shoes specifically designed for subjects with obesity with everyday sneakers during instrumented 6-min walking test and outdoor 30-min ambulation in a group of subjects with obesity using a single wearable device. Twenty-three obese individuals (mean age 58.96 years) were recruited and classified into two groups: deconditioned (n = 13) and non-deconditioned patients (n = 10). Each participant was evaluated with his/her daily sneakers and the day after with shoes specifically designed for people with obesity by means of a questionnaire related to the comfort related to each model of shoes and instrumentally during the i6MWT and an outdoor walking test. The results showed that the specifically designed shoes displayed the higher score as for comfort, in particular in the deconditioned group. During the i6MWT, the distance walked, and step length significantly increased in the deconditioned group when specifically designed shoes were worn; no significant changes were observed in the non-deconditioned individuals. The deconditioned group displayed longer step length during the outdoor 30-min ambulation test. In the non-deconditioned group, the use of specific shoes correlated to better performance in terms of gait speed and cadence. These data, although preliminary, seem to support the hypothesis that shoes specifically conceived and designed for counteracting some of the known functional limitations in subjects with obesity allow for a smoother, more stable and possibly less fatiguing gait schema over time.

**Keywords:** obesity; walking; 6-min walking test; wearable system; inertial sensor; rehabilitation

**Citation:** Cimolin, V.; Gobbi, M.; Buratto, C.; Ferraro, S.; Fumagalli, A.; Galli, M.; Capodaglio, P. A Comparative Analysis of Shoes Designed for Subjects with Obesity Using a Single Inertial Sensor: Preliminary Results. *Sensors* **2022**, *22*, 782. https://doi.org/10.3390/s22030782

Academic Editor: Chi Hwan Lee

Received: 22 December 2021
Accepted: 15 January 2022
Published: 20 January 2022

**Publisher's Note:** MDPI stays neutral with regard to jurisdictional claims in published maps and institutional affiliations.



## 1. Introduction

Footwear conditions influence gait variables. This is of particular importance in subjects experiencing difficulties in deambulation and balance, such as with frail, elderly, and obese subjects. For the latter, finding appropriate footwear that meet comfort and safety expectations is sometimes difficult. Their options rely often on sports footwear capable of accommodating a wider foot, or on orthopaedic shoes for specific foot pain conditions, but choice in-between for everyday life shoes appears limited. Gait in subjects with obesity has been studied with 3D motion capture and it is known that subjects with obesity walk slower than lean individuals, with reduced step length and frequency, wider stance, reduced hip and knee flexion and increased ankle plantarflexion [1–4]. Joint loading is increased during walking and obesity elevates the risk for musculoskeletal disorders such as osteoarthritis, low back pain, soft tissue injury, tendinitis and plantar fasciitis [5,6]. Musculoskeletal injuries represent a frequent cause of dropping out of physical activity programs. However, despite being a critical source of biomechanical loading, walking remains a highly recommended form of exercise for the management of obesity. Comfortable and adequate shoes represent, together with the prescription of a safe adapted physical activity, an important means to achieve the recommended physical activity target volume.

Literature on shoes specific for obese individuals is limited. Peduzzi de Castro et al. [7] tested the effects of two pressure relief insoles developed for people carrying an extra-weight, like backpackers and subjects with obesity, based on the ground reaction forces and plantar pressure peaks during gait. They demonstrated that insoles showed positive effects for either the plantar pressure distribution or the ground reaction forces parameters. Russel et al. [8] showed by means of gait analysis positive effects of a prophylactic wedged insole for reducing the magnitude of the load on the knee's medial compartment in obese women who are at risk for knee osteoarthritis development. Griffon et al. [9] demonstrated that the orthosis intervention at foot level significantly improved the ambulatory performances of obese individuals during the 6-min walking test (6MWT), reduced the perception of fatigue and the postural changes occurring after the 6MWT.

Despite the widespread use of 6MWT in clinical studies, the mere output of the test is the distance walked. Additional subject-specific factors such as stride length, velocity or body posture, granularity of overall gait patterns or body segment kinematics, and gait parameters are not measured during the test. In spite of the increasing body of evidence for the use of wearable sensors during gait, their use during the 6MWT (i6MWT: instrumented 6MWT) is not common [10]. i6MWT provides a comprehensive assessment of gait without adding further burden to the patient and represents an alternative to traditional gait analysis and postural control assessment. Those, in fact, require expensive equipment, are time-consuming and provide detailed information only for a very limited number of consecutive gait cycles [11]. Considering the hundreds of steps taken during the 6MWT, the information that can be extracted from this clinical test is potentially very valuable.

Thus, the aim of the present study was to compare the performance of shoes specifically designed for subjects with obesity with everyday sneakers during i6MWT and an outdoor 30-min ambulation in a group of subjects with obesity using an existing single wearable device.

## 2. Materials and Methods

### 2.1. Participants

A sample of 23 obese individuals (OG, 6 males, 17 females, mean age 58.96 years, BMI > 30 kg/m$^2$) admitted for a comprehensive multidisciplinary rehabilitation program at the Istituto Auxologico Italiano, Piancavallo (VB, Italy), were recruited for the study on a voluntary basis. At the time of the experimental tests, all of them were free from any acute musculoskeletal, neuromuscular, psychological and/or cardiopulmonary conditions able to significantly affect their walking abilities and postural control.

According to the value of 6MWt distance, participants were classified into two groups: deconditioned (n = 13) and non-deconditioned patients (n = 10). The patient is classified as deconditioned if the 6MWT walked distance is lower than 400 m; in this case patient is considered sarcopenic or potentially frail [12]. If the distance walked during 6MWT is over 400 m, the patient is classified as non-deconditioned. Their anthropometric features are reported in Table 1.

**Table 1.** Participant characteristics.

|  | Deconditioned | Non-Deconditioned | *p*-Value |
|---|---|---|---|
|  | (n = 13) | (n = 10) |  |
| **Gender, n (%)** |  |  |  |
| Male | 3 (23.1%) | 4 (40%) |  |
| Age (years) | 63.78 ± 8.67 | 58.55 ± 8.61 | 0.989 |
| Height (m) | 1.59 ± 0.07 | 1.64 ± 0.07 | 0.092 |
| Body mass Index (kg/m$^2$) | 41.34 ± 3.69 | 39.82 ± 3.17 | 0.118 |

Each participant was evaluated by means of a questionnaire related to the comfort related to each model of shoes and instrumentally during the i6MWT and an outdoor walking test. The tests were repeated in different sessions: in the first session, with his/her daily sneakers and in the second session with shoes specifically designed for people with obesity. The second session was performed the day after the first one, asking the individuals to wear the specifically designed shoes during their normal daily activities.

The questionnaire related to the comfort was a 7 items self-assessment questionnaire:

- Rearfoot pain
- Midfoot pain
- Forefoot pain
- Comfort at rest
- Comfort during walking
- Stability
- Safety

The score of each item is from 0 (absence) and 3 (presence).

All participants were required to sign a written informed consent form, in which the details of the experimental tests were reported. The study was carried out in compliance with the World Medical Association Declaration of Helsinki and its later amendments.

### 2.2. Data Collection and Processing

For the i6MWT and the outdoor 30-min outdoor ambulation test, a single miniaturized inertial sensor (G-Sensor®, BTS Bioengineering, Milan, Italy), previously validated for investigations on gait in unaffected individuals and people with several pathological conditions [13–17] was placed on participants' lower backs, approximately at the L4-L5 vertebrae position. The sensor, which is sized 70 mm × 40 mm × 18 mm and weighs 37 g, was built with a triaxial accelerometer 16 bit/axes with multiple sensitivity (±2, ±4, ±8, ±16 g), a triaxial magnetometer 13 bit (±1200 mT) and a triaxial gyroscope 16 bit/axes with multiple sensitivity (±250, ±500, ±1000, ±2000 °/s).

The i6MWT was performed indoors, along a long, flat, undisturbed 30-m hospital corridor with the length marked every 5 m; turnaround points were marked with a cone. Patients were instructed to walk as fast as they could. Encouragement was given every minute during the test until subject exhaustion using only standardized phrases as specified in the "ATS Statement: Guidelines for the Six-minute Walk Test" [18]. Chest pain, severe dyspnoea, physical exhaustion, muscle cramps, sudden gait instability or other signs of severe distress were additional criteria for stopping the test [18].

After a 15-min recovery, participants performed a supervised outdoor 30-min ambulation test in the hospital park. Participants were asked to refrain from speaking throughout the test.

In each test, acceleration data, acquired at 100 Hz frequency, were transmitted via Bluetooth to a PC and processed using dedicated software (BTS® G-Studio, BTS Bioengineering S.p.A.; ver. 3.2.20) which performs data acquisition, elaboration, reporting and storage. The software used is BTS G-Studio has a specific protocol capable of analysing the i6MWT and gait test, which automatically generates a report for each trial. The first 5 s of acquisition (during which the subject was asked to stand still) were used to verify the orientation of the sensor and then use such information to correct the acceleration vectors data during the acquired trials.

Based on the raw acceleration data, the main spatio-temporal parameters were calculated following the approaches described by literature for each stride [15,19,20] in each session. As for the i6MWT the values of walked distance, gait speed, step length and cadence were computed during the entire test; as for the outdoor 30-min gait test, the values of gait speed, step length and cadence were computed during the early (1st minute) and last (30th minute) segment of the test.

### 2.3. Obesity-Specific Shoes

The shoes specifically designed for obese individuals (Figure 1) were developed in order to decrease the foot hyper-pronation with valgus of the knees, to compensate the lack of the natural shock absorber given by the arch of the foot (collapsed due to excess weight) and by the heel (the tissues have an excess of fat) and to accommodate the foot extra-volume (usually, subjects with obesity wear 1–2 size larger shoes).

(a)

| ID | PART | FEATURES | GOAL |
|---|---|---|---|
| 1 | UPPER | SELF MODELING AND STRECHABLE MATERIAL | ABLE TO SHAPED ITSELF TO THE ANATOMY OF THE FOOT AND ANY DEFORMITIES, ENSURING COMFORT |
| 2 | TIP | BUTTRESS | CUSHIONS PAINS CAUSED BY ACCIDENTAL BUMPS |
| 3 | LAST | HIGH VOLUME | ALLOW TO LODGE OBESITY FOOT AND THICK CUSTOME INSOLE |
| 4 | HEEL | BUTTRESS | HELP STABILITY THE GAIT |
| 5 | LINING | SEAMLESS | PROVIDE COMFORT AND PREVENT PAINFUL FRICTION |
| 6 | OUTSOLE STRUCTURE | ELASTIC RETURN | GIVE DYNAMIC HELP IN PROPULSION PHASE |
|  |  | REINFORCED INSERT | PROVIDE CONTROL OF PRONATION AND SUPINATIONS AND PREVENT THE SOLE FROM COLLAPSE IN CASE OF OVERWEIGHT |
| 7 | OUTSOLE OUTLINE | WIDTH IN FOREFOOT | GUARANTEE GREATER SUPPORT SURFACE AND STABILITY |

(b)

**Figure 1.** Image related to the obesity-specific shoes (**a**) and description of key elements of obesity-specific shoes with the role of each part (**b**).

Their sole is composed of rubber tread with high resistance to abrasion, an EVA midsole (light material with shock absorber properties able to cushion the impact between foot and ground) and an insert in composite fibers with high resistance and elasticity. In the medial and lateral part of the heel, the sole insert has 2 piers of 16 mm that prevent excessive pronation and supination and at the same time the collapsing of the sole. This elastic insert in composite fibers also has the ability to deform itself during the stance phase, accumulating energy, and to release it during the push off phase. The sole is displayed in detail in Figure 2.

**Figure 2.** Details of the sole of the obesity-specific shoes used in the study.

These shoes are able to provide propulsive boost, support and stabilization to the foot.

### 2.4. Statistical Analysis

All the previously defined parameters were computed for each participant in the two sessions and used for analysis.

The Kolmogorov–Smirnov test was necessary to verify if the parameters were normally distributed. Because assumptions of normality were fulfilled, media and standard deviations relating to all indices were calculated for the obese groups (deconditioned and non-deconditioned group) of participants.

A repeated measure ANOVA was performed on the data of i6MWT with the within-subject factor of shoes (daily sneakers vs. specific shoes) and the between-subject factor of group (deconditioned vs. non-deconditioned). A repeated measure ANOVA was performed on the data of the outdoor 30-min ambulation test with the within-subject factor of shoes (daily sneakers vs. specific shoes) and the between-subject factor of time (1st minute vs. 30th minute). Post-hoc tests were performed where appropriate by applying Fisher's correction for the significance threshold. A significance level of 0.05 was implemented throughout. The statistical analysis was performed using Minitab® (version 18.1, State College, PA, USA).

### 3. Results

Out of the total sample of 23 subjects, 7 were men and 16 were women. The age and anthropometric characteristics of the two study groups are shown in Table 1; their characteristics were not significantly different between groups.

In Table 2 the median and range values of the score for each item of the questionnaire related to comfort for the two groups are reported for the two sessions (daily sneakers vs. specific shoes). It is possible to observe that in the deconditioned group no differences appeared in terms of pain between the two sessions (daily sneakers vs. specific shoes), while as for the other items (comfort stability and safety) the specific shoes seemed to show the higher score. In the non-deconditioned group only, stability and safety appeared to have the higher score in the session when the specific shoes were worn.

**Table 2.** Median (minimum and maximum) values of the score for the questionnaire related to comfort for the two groups are reported for the two sessions (daily sneakers vs. specific shoes). * = $p < 0.05$, statistically significant in the post hoc comparison daily sneakers vs. specific-obesity shoes.

|  | Deconditioned | | Non-Deconditioned | |
|---|---|---|---|---|
|  | **Daily Sneakers** | **Specific Shoes** | **Daily Sneakers** | **Specific Shoes** |
| Rearfoot pain | 0 (0–3) | 0 (0–2) | 0 (0–1) | 0 (0–1) |
| Midfoot pain | 0 (0–3) | 0 (0–3) | 0 (0–1) | 0 (0–1) |
| Forefoot pain | 0 (0–3) | 0 (0–1) | 0 (0–1) | 0 (0–1) |
| Comfort at rest | 2 (1–3) * | 3 (2–3) | 2 (0–3) | 3 (2–3) |
| Comfort during walking | 2.5 (0–3) * | 3 (2–3) | 3 (0–3) | 3 (2–3) |
| Stability | 2 (0–3) * | 3 (2–3) | 2 (0–3) | 3 (3–3) * |
| Safety | 2 (0–3) * | 3 (2–3) | 2 (0–3) | 3 (2–3) * |

In Table 3 the mean and the standard deviations of i6MWT parameters for the entire test of the two groups are reported for the two sessions (daily sneakers vs. specific shoes).

**Table 3.** Mean and standard deviation of i6MWT parameters of the two groups are reported for the two sessions (daily sneakers vs. specific shoes). * = $p < 0.05$, statistically significant in the post hoc comparison daily sneakers vs. specific-obesity shoes; + = $p < 0.05$, statistically significant in the post hoc comparison deconditioned vs. non-deconditioned group.

|  | Deconditioned | | Non-Deconditioned | |
|---|---|---|---|---|
|  | **Daily Sneakers** | **Specific Shoes** | **Daily Sneakers** | **Specific Shoes** |
| Walked distance (m) | 332.15 (125.82) + | 358.32 (122.04) *+ | 539.14 (59.73) | 533.53 (62.36) |
| Gait speed (m/s) | 1.08 (0.28) + | 1.12 (0.28) + | 1.62 (0.18) | 1.61 (0.19) |
| Step length (m) | 0.58 (0.11) + | 0.61 (0.11) *+ | 0.78 (0.09) | 0.78 (0.08) |
| Cadence (step/min) | 109.69 (17.53) + | 115.91 (12.82) + | 124.03 (5.69) | 122.64 (6.69) |

*Walked distance* (Figure 3a): The main effect of Group [$F(1, 38) = 38.84$; $p < 0.001$; partial $\eta^2 = 0.5054$] was significant, with lower value for the deconditioned patients and also the main effect of Shoes was significant [$F(1, 38) = 0.11$; $p = 0.039$; partial $\eta^2 = 0.1030$], with lower value with the personal shoes; the Group × Shoes interaction was confirmed as significant [$F(1, 38) = 1.19$; $p = 0.032$; partial $\eta^2 = 0.1070$].

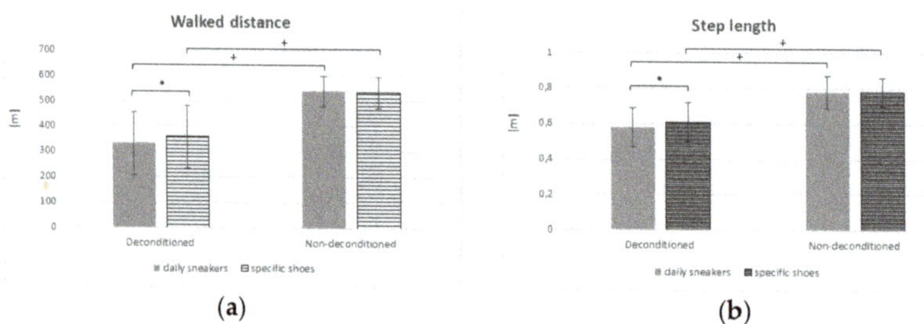


**Figure 3.** Mean and standard deviation of walked distance (**a**) and step length (**b**) during the i6MWT of the two groups are reported for the two sessions (daily sneakers vs. specific shoes). * = $p < 0.05$, statistically significant in the post hoc comparison daily sneakers vs. specific-obesity shoes; + = $p < 0.05$, statistically significant in the post hoc comparison deconditioned vs. non-deconditioned group.

*Gait speed*: The main effect of Group [$F(1, 38) = 46.60$; $p < 0.001$; partial $\eta^2 = 0.5508$] was significant, with lower value for the deconditioned patients, while the main effect of Shoes was not significant [$F(1, 38) = 0.05$; $p = 0.821$; partial $\eta^2 = 0.0014$]; the Group $\times$ Shoes interaction revealed to be not significant [$F(1, 38) = 0.09$; $p = 0.770$; partial $\eta^2 = 0.0023$].

*Step length* (Figure 3b): The main effect of Group [$F(1, 38) = 39.66$; $p < 0.001$; partial $\eta^2 = 0.5107$] was significant, with lower value for the deconditioned patients and also the main effect of Shoes was significant [$F(1, 38) = 0.03$; $p = 0.043$; partial $\eta^2 = 0.1034$], with lower value with the personal shoes; the Group $\times$ Shoes interaction was confirmed as significant [$F(1, 38) = 1.09$; $p = 0.041$; partial $\eta^2 = 0.1008$].

*Cadence*: The main effect of Group [$F(1, 38) = 8.20$; $p = 0.007$; partial $\eta^2 = 0.1775$] was significant, with lower value for the deconditioned patients, while the main effect of Shoes was not significant [$F(1, 38) = 0.44$; $p = 0.513$; partial $\eta^2 = 0.0114$]; the Group $\times$ Shoes interaction revealed to be not significant [$F(1, 38) = 1.08$; $p = 0.306$; partial $\eta^2 = 0.0276$].

Values of parameter in the 1st and 6th segment of 6MWT, where no statistical results were obtained, are not reported.

In Tables 4 and 5 the mean and the standard deviations of the parameters during the outdoor 30-min gait test were reported during the early (1st minute) and last (30th minute) segment of the test during the two sessions (daily sneakers vs. specific shoes) for the deconditioned (Table 3) and for the non-deconditioned group (Table 4).

**Table 4.** Mean and standard deviation of parameters during the outdoor 30-min gait test for the deconditioned group during the early (1st minute) and last (30th minute) segment of the test during the two sessions (daily sneakers vs. specific shoes). * = $p < 0.05$, statistically significant in the post hoc comparison daily sneakers vs. specific shoes; § = $p < 0.05$, statistically significant in the post hoc comparison 1st minute vs. 30th minute.

| | Deconditioned | | | |
| | Daily Sneakers | | Specific Shoes | |
| | 1st Minute | 30th Minute | 1st Minute | 30th Minute |
|---|---|---|---|---|
| Gait speed (m/s) | 0.92 (0.28) | 0.90 (0.18) | 0.96 (0.28) | 0.99 (0.32) |
| Step length (m) | 0.68 (0.61) | 0.60 (0.08) §* | 0.67 (0.08) | 0.68 (0.07) |
| Cadence (step/min) | 87.04 (31.10) | 90.96 (14.09) | 89.40 (26.91) | 93.52 (30.75) |

**Table 5.** Mean and standard deviation of parameters during the outdoor 30-min gait test for the non-deconditioned group during the early (1st minute) and last (30th minute) segment of the test during the two sessions (daily sneakers vs. specific-obesity shoes). * = $p < 0.05$, statistically significant in the post hoc comparison daily sneakers vs. specific-obesity shoes.

| | Non-Deconditioned | | | |
| | Daily Sneakers | | Specific Shoes | |
| | 1st Minute | 30th Minute | 1st Minute | 30th Minute |
|---|---|---|---|---|
| Gait speed (m/s) | 1.13 (0.36) * | 1.14 (0.22) * | 1.28 (0.28) | 1.23 (0.32) |
| Step length (m) | 0.75 (0.10) | 0.76 (0.10) | 0.76 (0.25) | 0.76 (0.25) |
| Cadence (step/min) | 92.63 (27.38) * | 95.85 (16.78) * | 104.98 (24.76) | 99.91 (19.91) |

Here, the results for the deconditioned group are reported (Table 4):

*Gait speed*: The main effects of shoes [$F(1, 48) = 0.07$; $p = 0.798$; partial $\eta^2 = 0.0014$] was not significant, the main effect of Time [$F(1, 48) = 6.15$; $p = 0.017$; partial $\eta^2 = 0.1135$] was significant; the Time $\times$ Shoes interaction revealed to be not significant [$F(1, 48) = 1.06$; $p = 0.309$; partial $\eta^2 = 0.0216$]

*Step length* (Figure 4): The main effect of shoes [$F(1, 48) = 0.27$; $p = 0.043$; partial $\eta^2 = 0.1056$] was significant, with lower value with the daily sneakers, the main effect of Time was significant [$F(1, 48) = 0.15$; $p = 0.046$; partial $\eta^2 = 0.1032$], with lower values as for the 30th minute; the Group $\times$ Shoes interaction was confirmed as significant

[F(1, 48) = 0.39; $p$ = 0.041; partial $\eta^2$ = 0.1081], with significant difference in the comparison between the 1st and the 30th minute with the daily sneaker and between the daily sneaker and the specific-obesity shoes at the 30th minute.

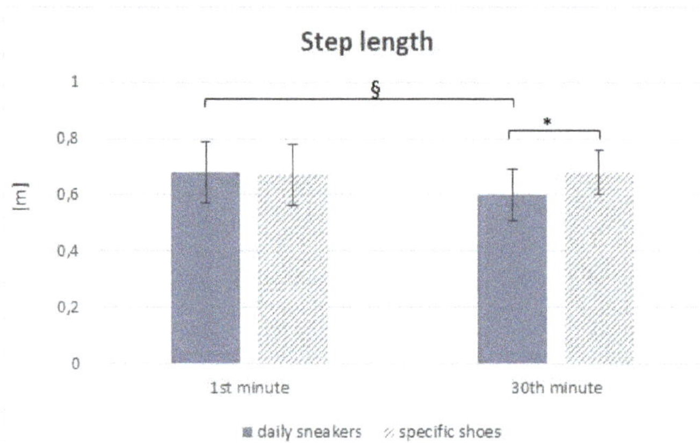



**Figure 4.** Mean and standard deviation of step length during the outdoor 30-min gait test for the deconditioned group during the early (1st minute) and last (30th minute) segment of the test during the two sessions (daily sneakers vs. specific shoes). * = $p < 0.05$, statistically significant in the post hoc comparison daily sneakers vs. specific shoes; § = $p < 0.05$, statistically significant in the post hoc comparison 1st minute vs. 30th minute.

*Cadence*: The main effects of shoes [F(1, 48) = 0.02; $p$ = 0.888; partial $\eta^2$ = 0.0004] and Time [F(1, 48) = 2.84; $p$ = 0.099; partial $\eta^2$ = 0.0558] were not significant; the Time × Shoes [F(1, 48) = 0.05; $p$ = 0.819; partial $\eta^2$ = 0.0011] interaction was not significant.

Below, the results for the non-deconditioned group are reported (Table 5):

*Gait speed*: The main effect of shoes [F(1, 36) = 0.10; $p$ = 0.043; partial $\eta^2$ = 0.1028] was significant, with lower values with daily sneakers, the main effect of Time [F(1, 36) = 1.85; $p$ = 0.182; partial $\eta^2$ = 0.0489] was not significant and the Time × Shoes [F(1, 36) = 3.21; $p$ = 0.082; partial $\eta^2$ = 0.0819] interaction was not significant.

*Step length*: The main effects of shoes [F(1, 36) = 0.02; $p$ = 0.902; partial $\eta^2$ = 0.0004] and Time [F(1, 36) = 0.01; $p$ = 0.910; partial $\eta^2$ = 0.0004] were not significant; the Time × Shoes interaction was confirmed to be not significant [F(1, 36) = 0.001; $p$ = 0.972; partial $\eta^2$ = 0.0001].

*Cadence*: The main effects of shoes [F(1, 36) = 0.15; $p$ = 0.046; partial $\eta^2$ = 0.1060] was significant, with lower values with daily sneakers, the main effect of Time [F(1, 36) = 1.89; $p$ = 0.178; partial $\eta^2$ = 0.0498] was not significant and the Time × Shoes [F(1, 36) = 4.27; $p$ = 0.705; partial $\eta^2$ = 0.0040] interaction was not significant.

## 4. Discussion

The influence of obesity on gait biomechanics has been extensively investigated with quantitative gait analysis, but far less attention has been devoted to the effects of footwear on walking capacity in individuals with obesity.

The purpose of this study was to compare the comfort and the performance of shoes specifically designed for subjects with obesity with everyday sneakers during i6MWT and outdoor 30-min ambulation test in a group of obese subjects using a single wearable device. We wanted to test whether subjects with obesity could improve distance walked during the 6MWT and reduce fatigue using shoes specifically designed for them as compared to habitual shoes. Whereas the technology of new shoes for subjects with obesity remains the focus of this paper, its originality relates to the simple, indirect method used to assess their performance by means of commonly used clinical and ecological tests instrumented with a single wearable device.

Our results showed that in terms of comfort the specifically designed shoes that were worn displayed the higher score, in particular in the deconditioned group; no differences appeared in terms of pain reduction.

As for the instrumental tests, during the i6MWT, the distance walked and step length significantly increased in the deconditioned group when specifically designed shoes were worn. No differences were found in terms of gait speed and cadence, even if a trend towards an increase could be observed. No significant changes were observed in the non-deconditioned individuals. As for the results related to the outdoor 30-min ambulation test, the deconditioned group displayed better performance when using specific shoes, with longer step length as compared to that obtained using daily sneakers. While no statistically significant differences were found in the comparison between the beginning (1st minute) and the end (30th minute) of the outdoor ambulation test when using specific shoes, a significant reduced step length was observed at the 30th minute of walking outdoor when using daily sneakers. In the non-deconditioned group, the use of specific shoes correlated to better performance in terms of gait speed and cadence, with no differences in terms of time (1st vs. 30th minute). Those data seem to support the hypothesis that shoes specifically conceived and designed for counteracting some of the known functional limitations in subjects with obesity allow a smoother, more stable and possibly less fatiguing gait schema over time.

These preliminary results show encouraging data about the use of the specific shoes tested as compared to daily sneakers in a group of individuals with obesity. In particular, the positive effects are evident in the deconditioned group, characterised by more limited parameters as compared to the non-deconditioned group.

Our study presents with some limitations. Firstly, the small number of participants for each group (deconditioned and non-deconditioned) with a wide age range (from 35 to 83 years), which result in limited strength of the statistical findings. A larger and more homogeneous group of patients in terms of age could strengthen the results. Secondly, both males and females were recruited for this study in order to improve the generalizability of the results. Combining male and female subjects, though, introduces a source of potential variability, as obesity modifies the body geometry by adding mass to different regions and dissimilar fat distribution in males and females could produce gender-related effects. However, with our sample it was not possible to consider them separately. Thus, potential differences between males and females who are obese will need further study. Another limitation is related to the absence of a stratification of the participants in terms of the severity of obesity. Future larger studies for each obesity class are needed for a deeper understanding of the effects of specific shoes, as effects may be related to the degree of severity of obesity.

**Author Contributions:** V.C.: conceptualization, methodology, formal analysis, writing—original draft; M.G. (Michele Gobbi): clinical assessment of the participants, review and editing; C.B. design and production of the specific shoes, provision of the shoes for experimental study, technical support to the study; S.F. technical support to the study; A.F.: review and editing; M.G. (Manuela Galli): review and editing; P.C.: conceptualization, methodology, review. All authors have read and agreed to the published version of the manuscript.

**Funding:** This research received no external funding. Podartis provided the specific shoes for the study participants.

**Institutional Review Board Statement:** The study was conducted according to the guidelines of the Declaration of Helsinki, and approved by the Institutional Ethics Committee of Istituto Auxologico Italiano IRCCS (2013_04_23_03).

**Informed Consent Statement:** Informed consent was obtained from all subjects involved in the study.

**Data Availability Statement:** Data available on request due to restrictions e.g., privacy or ethical.

**Acknowledgments:** The authors would like to acknowledge Eng. Marina Indalizio for her valuable contribution in data processing.

**Conflicts of Interest:** The authors declare no conflict of interest.

# References

1. Gilleard, W. Functional Task Limitations in Obese Adults. *Curr. Obes. Rep.* **2012**, *1*, 174–180. [CrossRef]
2. Lai, P.P.K.; Leung, A.K.L.; Li, A.N.M.; Zhang, M. Three-Dimensional Gait Analysis of Obese Adults. *Clin. Biomech. (Bristol, Avon)* **2008**, *23* (Suppl. S1), S2–S6. [CrossRef] [PubMed]
3. Runhaar, J.; Koes, B.W.; Clockaerts, S.; Bierma-Zeinstra, S.M.A. A Systematic Review on Changed Biomechanics of Lower Extremities in Obese Individuals: A Possible Role in Development of Osteoarthritis. *Obes. Rev.* **2011**, *12*, 1071–1082. [CrossRef] [PubMed]
4. Capodaglio, P.; Gobbi, M.; Donno, L.; Fumagalli, A.; Buratto, C.; Galli, M.; Cimolin, V. Effect of Obesity on Knee and Ankle Biomechanics during Walking. *Sensors* **2021**, *21*, 7114. [CrossRef] [PubMed]
5. Orpana, H.M.; Tremblay, M.S.; Finès, P. Trends in Weight Change among Canadian Adults. *Health Rep.* **2007**, *18*, 9–16. [PubMed]
6. De Souza, S.A.F.; Faintuch, J.; Valezi, A.C.; Sant' Anna, A.F.; Gama-Rodrigues, J.J.; de Batista Fonseca, I.C.; Souza, R.B.; Senhorini, R.C. Gait Cinematic Analysis in Morbidly Obese Patients. *Obes. Surg.* **2005**, *15*, 1238–1242. [CrossRef] [PubMed]
7. Peduzzi de Castro, M.; Abreu, S.; Pinto, V.; Santos, R.; Machado, L.; Vaz, M.; Vilas-Boas, J.P. Influence of Pressure-Relief Insoles Developed for Loaded Gait (Backpackers and Obese People) on Plantar Pressure Distribution and Ground Reaction Forces. *Appl. Ergon.* **2014**, *45*, 1028–1034. [CrossRef] [PubMed]
8. Russell, E.M.; Miller, R.H.; Umberger, B.R.; Hamill, J. Lateral Wedges Alter Mediolateral Load Distributions at the Knee Joint in Obese Individuals: KNEE JOINT IN OBESE INDIVIDUALS. *J. Orthop. Res.* **2013**, *31*, 665–671. [CrossRef] [PubMed]
9. Griffon, P.; Vie, B.; Weber, J.P.; Jammes, Y. Effect of 4 Weeks of Foot Orthosis Intervention on Ambulatory Capacities and Posture in Normal-Weight and Obese Patients. *J. Am. Podiatr. Med. Assoc.* **2020**, *110*, 2. [CrossRef] [PubMed]
10. Storm, F.A.; Cesareo, A.; Reni, G.; Biffi, E. Wearable Inertial Sensors to Assess Gait during the 6-Minute Walk Test: A Systematic Review. *Sensors* **2020**, *20*, 2660. [CrossRef] [PubMed]
11. Muro-de-la-Herran, A.; Garcia-Zapirain, B.; Mendez-Zorrilla, A. Gait Analysis Methods: An Overview of Wearable and Non-Wearable Systems, Highlighting Clinical Applications. *Sensors* **2014**, *14*, 3362–3394. [CrossRef] [PubMed]
12. Cruz-Jentoft, A.J.; Bahat, G.; Bauer, J.; Boirie, Y.; Bruyère, O.; Cederholm, T.; Cooper, C.; Landi, F.; Rolland, Y.; Sayer, A.A.; et al. Sarcopenia: Revised European Consensus on Definition and Diagnosis. *Age Ageing* **2019**, *48*, 16–31. [CrossRef] [PubMed]
13. Pau, M.; Caggiari, S.; Mura, A.; Corona, F.; Leban, B.; Coghe, G.; Lorefice, L.; Marrosu, M.G.; Cocco, E. Clinical Assessment of Gait in Individuals with Multiple Sclerosis Using Wearable Inertial Sensors: Comparison with Patient-Based Measure. *Mult. Scler. Relat. Disord* **2016**, *10*, 187–191. [CrossRef] [PubMed]
14. Zago, M.; Sforza, C.; Pacifici, I.; Cimolin, V.; Camerota, F.; Celletti, C.; Condoluci, C.; De Pandis, M.F.; Galli, M. Gait Evaluation Using Inertial Measurement Units in Subjects with Parkinson's Disease. *J. Electromyogr. Kinesiol.* **2018**, *42*, 44–48. [CrossRef] [PubMed]
15. Cimolin, V.; Cau, N.; Sartorio, A.; Capodaglio, P.; Galli, M.; Tringali, G.; Leban, B.; Porta, M.; Pau, M. Symmetry of Gait in Underweight, Normal and Overweight Children and Adolescents. *Sensors* **2019**, *19*, 2054. [CrossRef]
16. Schifino, G.; Cimolin, V.; Pau, M.; da Cunha, M.J.; Leban, B.; Porta, M.; Galli, M.; Souza Pagnussat, A. Functional Electrical Stimulation for Foot Drop in Post-Stroke People: Quantitative Effects on Step-to-Step Symmetry of Gait Using a Wearable Inertial Sensor. *Sensors* **2021**, *21*, 921. [CrossRef] [PubMed]
17. Pau, M.; Mulas, I.; Putzu, V.; Asoni, G.; Viale, D.; Mameli, I.; Leban, B.; Allali, G. Smoothness of Gait in Healthy and Cognitively Impaired Individuals: A Study on Italian Elderly Using Wearable Inertial Sensor. *Sensors* **2020**, *20*, 3577. [CrossRef] [PubMed]
18. ATS Committee on Proficiency Standards for Clinical Pulmonary Function Laboratories. ATS Statement: Guidelines for the Six-Minute Walk Test. *Am. J. Respir. Crit. Care Med.* **2002**, *166*, 111–117. [CrossRef] [PubMed]
19. Brandes, M.; Zijlstra, W.; Heikens, S.; van Lummel, R.; Rosenbaum, D. Accelerometry Based Assessment of Gait Parameters in Children. *Gait Posture* **2006**, *24*, 482–486. [CrossRef] [PubMed]
20. Bugané, F.; Benedetti, M.G.; Casadio, G.; Attala, S.; Biagi, F.; Manca, M.; Leardini, A. Estimation of Spatial-Temporal Gait Parameters in Level Walking Based on a Single Accelerometer: Validation on Normal Subjects by Standard Gait Analysis. *Comput. Methods Programs Biomed.* **2012**, *108*, 129–137. [CrossRef] [PubMed]

*Article*

# Use of a Single Wearable Sensor to Evaluate the Effects of Gait and Pelvis Asymmetries on the Components of the Timed Up and Go Test, in Persons with Unilateral Lower Limb Amputation

Maria Stella Valle [1,2,*,†], Antonino Casabona [1,2,3,†], Ilenia Sapienza [3,4], Luca Laudani [5], Alessandro Vagnini [6], Sara Lanza [4] and Matteo Cioni [1,7]

1. Laboratory of Neuro-Biomechanics, Department of Biomedical and Biotechnological Sciences, University of Catania, 95123 Catania, Italy; casabona@unict.it (A.C.); mcioni@unict.it (M.C.)
2. Section of Physiology, Department of Biomedical and Biotechnological Sciences, University of Catania, 95123 Catania, Italy
3. Residency Program of Physical Medicine and Rehabilitation, Department of Biomedical and Biotechnological Sciences, University of Catania, 95123 Catania, Italy; ile.sapienz@gmail.com
4. Department of Physical Medicine and Rehabilitation, Regina Margherita Hospital, 97013 Comiso, Italy; dott.saralanza@gmail.com
5. Faculty of Sports, Campus de los Jerónimos, UCAM Catholic University of Murcia, 30107 Guadalupe, Spain; llaudani@ucam.edu
6. BTS Bioengineering, 20024 Garbagnate Milanese, Italy; alessandro.vagnini@btsbioengineering.com
7. Gait and Posture Analysis Laboratory, Policlinico Vittorio Emanuele-San Marco, University Hospital, 95123 Catania, Italy
* Correspondence: m.valle@unict.it
† These authors contributed equally to this work.

**Citation:** Valle, M.S.; Casabona, A.; Sapienza, I.; Laudani, L.; Vagnini, A.; Lanza, S.; Cioni, M. Use of a Single Wearable Sensor to Evaluate the Effects of Gait and Pelvis Asymmetries on the Components of the Timed Up and Go Test, in Persons with Unilateral Lower Limb Amputation. *Sensors* **2022**, *22*, 95. https://doi.org/10.3390/s22010095

Academic Editors: Paolo Capodaglio and Veronica Cimolin

Received: 29 November 2021
Accepted: 20 December 2021
Published: 24 December 2021

**Publisher's Note:** MDPI stays neutral with regard to jurisdictional claims in published maps and institutional affiliations.



**Abstract:** The Timed Up and Go (TUG) test quantifies physical mobility by measuring the total performance time. In this study, we quantified the single TUG subcomponents and, for the first time, explored the effects of gait cycle and pelvis asymmetries on them. Transfemoral (TF) and transtibial (TT) amputees were compared with a control group. A single wearable inertial sensor, applied to the back, captured kinematic data from the body and pelvis during the 10-m walk test and the TUG test. From these data, two categories of symmetry indexes (SI) were computed: One SI captured the differences between the antero-posterior accelerations of the two sides during the gait cycle, while another set of SI quantified the symmetry over the three-dimensional pelvis motions. Moreover, the total time of the TUG test, the time of each subcomponent, and the velocity of the turning subcomponents were measured. Only the TF amputees showed significant reductions in each SI category when compared to the controls. During the TUG test, the TF group showed a longer duration and velocity reduction mainly over the turning subtasks. However, for all the amputees there were significant correlations between the level of asymmetries and the velocity during the turning tasks. Overall, gait cycle and pelvis asymmetries had a specific detrimental effect on the turning performance instead of on linear walking.

**Keywords:** sensory–motor gait disorders; limb prosthesis; spatial–temporal analysis; kinematics; symmetry index

## 1. Introduction

Motor deficit due to unilateral amputation of a lower limb can lead to functional limitations over almost all the activities of daily living. The presence of a prosthesis can partially restore being physically autonomous, also permitting an aesthetic recovery of an amputee's integrity. All these factors are important in terms of personal image and quality of life, as they facilitate the patient's social and work integration.

One of the consequences related to the use of prostheses is the appearance of asymmetries between the prosthetic and intact sides concerning kinematic and/or kinetic parameters of the gait cycle, such as the step length or velocity, or parameters related to the mobility of body segments, such as the pelvis or the trunk [1–4]. Local body asymmetries lead to a shift in the body's center of gravity that can cause further detrimental effects; however, in many cases, asymmetries produce compensatory strategies that improve postural stability [5–8].

The effects of asymmetries in amputees have typically been studied during linear walking and, as reported by Devan et al. [9], most of the movement asymmetry studies in lower limb amputation have focused on local gait features, such as weight transfer during stance, generation of ground reaction forces, step time, and step length. On the contrary, the asymmetries concerning trunk and pelvic segments have received less attention, with the consequences of neglecting the effects of asymmetries on the mobility of daily living motor tasks, such as standing up, sitting down, or walking along a curved path.

In the current study, we chose to use the Timed Up and Go (TUG) test to quantify the physical mobility of a sample of subjects with lower limb amputation and evaluate the effects of both gait cycle and pelvic asymmetries among the subcomponents that are included in the test. In fact, the test consists of a sequence of natural actions that, starting from the sitting position, includes the following steps: standing up from a chair, walking 3 m forward, turning around an obstacle, walking back for 3 m, turn pivoting on one foot, and sitting down again. This test was introduced by Podsiadlo and Richardson [10] in order to permit the evaluation of the global mobility of frail older people adopting a simple and short duration test. In fact, the TUG test is frequently used not only for the assessment of physical mobility in the elderly population [11], but also in pathological contexts, such as Parkinson's disease [12], muscular dystrophy [13], and stroke [14]. The extensive experience gained in the use of the TUG test in subjects with motor difficulties provides a rationale for its use to evaluate mobility also in lower limb amputees, paying specific attention to the different motor tasks included in the test. The TUG test was validated in people with a lower limb amputation [15,16], showing significant increase in the total time required to travel the entire path, when compared with the times measured in non-amputees [15–19].

Typically, the TUG test is performed using a simple stopwatch, but any errors due to operator reaction times can increase the variability of the measurement. A big leap forward for clinical and basic research took place when the use of wearable inertial measurement units (IMU) measured the TUG test (iTUG). The easy use of the wearable sensors and the capacity to provide spatiotemporal parameters make these devices suitable for capturing timing and kinematic data from both the entire path and from each TUG subcomponent [20]. As far as we know, only Clemens et al. [21] implemented a protocol to parameterize each subcomponent of the TUG test in people with a lower limb amputation, by using a mobile iPad application. These authors showed significant differences among the iTUG components, comparing transtibial (TT) with transfemoral (TF) amputees, especially for the mid-turning task. With respect to the descriptive information from the study of these authors, in the current work we used the TUG test, instrumented by a single wearable sensor, to add insights into the effects of the gait and pelvis asymmetries on the physical performance of the subcomponents of the TUG test, in TF and TT amputees.

## 2. Methods

### 2.1. Ethical Statement

This study was conducted in accordance with the Declaration of Helsinki Ethical Principles and Good Clinical Practices and was approved by the local ethics committee of Catania University Hospital "Policlinico Vittorio Emanuele-San Marco" (BIOART Project, n° 72/2019/PO). Participation was voluntary and all participants read and signed an informed written consent before starting the study.

## 2.2. Participants

Fourteen people with unilateral lower amputation were contacted and invited by the Physical Medicine and Rehabilitation Section of Regina Margherita Hospital (Comiso, Italy), to participate in the study. The inclusion criteria were: $\geq$18 years old, community-dwelling, unilateral leg or thigh amputation at any anatomical level, daily use of a prosthesis, and independent walking. The exclusion criteria were: bilateral amputation, fitting problems of the prosthesis, uncontrolled risk factors for cardiovascular disease, and presence of cognitive disorders. Among the selected subjects, one despondent and apparently inconsolable patient refused to participate as he did not perceive the study as relevant and 3 were unable to attend the health center. Therefore, 10 male amputees, classified by the evaluation practitioners at K3 or K4 level, according to the Medicare Functional Classification [22], were enrolled in the study.

Based on the level of amputation, the participants were divided into 2 groups, 5 belonging to the transfemoral (TF) group and 5 to the transtibial (TT) group. A third group of 5 healthy volunteers (CTRL) was recruited as the control. All the participants included in the three groups were male, with non-significant statistical differences for age, weight, and height (Table 1). The anthropometric and clinical data for each person with a lower limb amputation are reported in Table 2.

**Table 1.** Descriptive statistics (mean $\pm$ standard deviation and range) of the anthropometric data of all participants.

| | Groups | | | *t* Test (*p*) | | |
|---|---|---|---|---|---|---|
| | **TT** | **TF** | **CTRL** | **CTRL vs. TT** | **CTRL vs. TF** | **TT vs. TF** |
| Age (yrs) | 50.6 $\pm$ 9.1 (41–64) | 53.6 $\pm$ 14.9 (34–71) | 49.6 $\pm$ 8.5 (43–63) | 0.862 | 0.617 | 0.712 |
| W (kg) | 84.8 $\pm$ 15.6 (73–110) | 81.2 $\pm$ 14.8 (71–106) | 78.6 $\pm$ 7.2 (72–90) | 0.444 | 0.734 | 0.719 |
| H (cm) | 172.8 $\pm$ 4.9 (165–178) | 171.8 $\pm$ 8.8 (165–187) | 174.4 $\pm$ 6.3 (164–181) | 0.664 | 0.605 | 0.829 |

CTRL, control group; TF, transfemoral group; TT, transtibial group; W, weight; H, height; *p*, level of significance.

**Table 2.** Anthropometric and clinical data of the persons with lower limb amputation.

| Subjects | Level | Age (yrs) | W (kg) | H (cm) | BMI | Onset (yrs) | Side | Cause | Falls (n) | BBS Score | Knee | Foot |
|---|---|---|---|---|---|---|---|---|---|---|---|---|
| #1 | TF | 50 | 106 | 187 | 30.3 | 3 | Left | T | 4 | 43 | RK3 | VF-LP |
| #2 | TF | 66 | 74 | 165 | 27.2 | 4 | Right | V | 0 | 52 | PABHD | AM |
| #3 | TF | 71 | 71 | 171 | 24.3 | 5 | Right | V | 1 | 41 | PABHD | AM |
| #4 | TF | 47 | 71 | 167 | 25.4 | 32 | Right | T | 2 | 50 | P | VF |
| #5 | TF | 34 | 84 | 169 | 29.4 | 4 | Right | T | 2 | 52 | PTK | SACH |
| #6 | TT | 41 | 73 | 165 | 26.8 | 3 | Left | T | 0 | 54 | | VF |
| #7 | TT | 55 | 90 | 172 | 30.4 | 20 | Left | T | 0 | 55 | | PF-XC |
| #8 | TT | 48 | 74 | 174 | 24.6 | 4 | Left | T | 3 | 56 | | VF-HS |
| #9 | TT | 45 | 110 | 178 | 34.7 | 30 | Right | T | 0 | 54 | | PF-XL |
| #10 | TT | 64 | 77 | 175 | 25.1 | 1 | Left | V | 0 | 40 | | 1D1 |

TF, transfemoral; TT, transtibial; T, trauma; V, vascular; RK3, Rheo Knee 3 microprocessor; PABHD, polyfunctional automatic brake hydraulic device; P, polycentric; PTK, polycentric total knee 1900; VF-LP, Vari-Flex-LP; AM, articulated multi-axis: VF, Vari-Flex; VF-HS, Vari-Flex-Harmony; PF-XL, Pro-Flex XL; PF-XC, Pro-Flex XC; 1D1 = 1D1-Dynamic.

## 2.3. Experimental Procedures

All participants, wearing comfortable clothes in order to ensure adequate mobility, received standardized verbal and visual instructions and explanations about the experimental set-up and protocols so that equipment and rules were the same for everyone. As the motor tasks belonged to routine activities of daily life, the training regimen before recording was not considered necessary. The amputees wore their prostheses during the entire recording period of the procedures.

To assess balance ability, gait cycle, and pelvis symmetries, as well as physical mobility, the following procedures were used:

Test 1: Balance ability and risk of falling (Italian Version of Berg Balance Scale (BBS-it) [23]). The static and dynamic balance abilities and possible risk of falling for the amputees were assessed using the BBS-it. This is among the most used scales for balance evaluation in the rehabilitation field and its validity and reliability have also been well documented in persons with lower limb amputation [24]. Participants were required to perform a series of predetermined 14 tasks (14) of varying difficulty, scored from 0 (unable) to 4 (independent), so that the total score ranged from 0 to 56. This procedure took approximately 25 min to complete.

Test 2: 10-m walk test. The path was identified with a line marked with adhesive tape at the beginning and at the end of the 10 m. All participants were instructed to walk the path three times at a self-selected comfortable velocity. This protocol took approximately 10 min to complete, with an inter-trial rest interval of 2 min. We collected at least 24 gait cycles for each participant. From the total number of trials, we selected those that presented comparable measures of the walking velocity over the three groups of participants. For this purpose, over the total trials (3 for each subject), we identified the ranges of velocity values measured in the participant belonging to each group. The range of the TF amputees showed the lowest velocity values over the groups (0.7–1.0 m/s), the CTRL group showed a highest range of velocity values (0.9–1.2 m/s), while the TT amputees group exhibited an intermediate range of velocity values (0.8–1.1 m/s). For the CTRL and TT groups, we eliminated those trials with values out of the TF group range and verified that not significant differences occurred between each pair of groups. To evaluate the asymmetries of the gait cycle and pelvis, the following kinematic parameters were extrapolated for each cycle: body acceleration when the gait cycle was executed by left or right lower limb and pelvis angular displacements over the three planes of space.

Test 3: TUG test (Figure 1). A 3-m-long rubber mat covered the floor and a chair without armrests was correctly placed at the beginning of the path. All participants executed the following sequence of simple tasks: stand up from the chair (sit to stand), walk along a straight line for 3 m (walking forward), turn 180° around a pin 18-cm high (m id-turning), walk back to the chair (walking back), turn for sitting (final turning), and sit down again (stand to sit). All participants were asked to walk at their own normal pace. The test was repeated six times, three times walking clockwise (Figure 1A,B) and the remaining three times counterclockwise (Figure 1C,D), to catch turning with internal and external prosthetic limb, respectively. The median values over the three clockwise and the three counterclockwise turns were used for the subsequent analyses. This protocol took approximately 15 min to complete, with an inter-trial rest interval of 2 min.

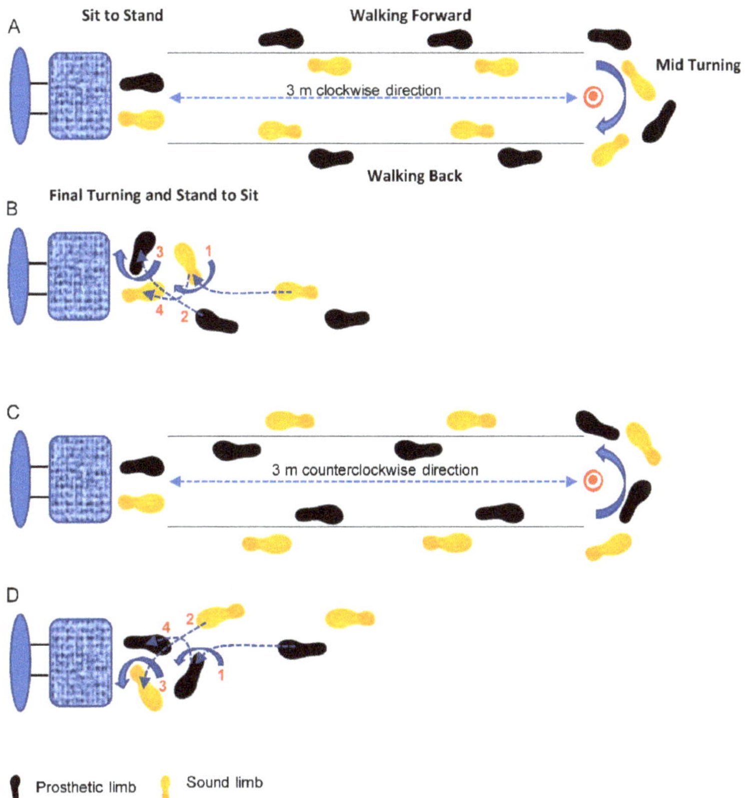



**Figure 1.** Top view of typical path of 3-m long for the TUG test. The test was performed in clockwise (**A**,**B**) and counterclockwise (**C**,**D**) directions. The single subcomponents are indicated in the panels (**A**,**B**). Four basic phases are performed to accomplish the final turning, before sitting down (**B**,**D**): (1) the body rotated of about 90° around the inner limb, (2) allowing the swing of the outer limb; (3) the body rotated of another 90° around the outer limb, (4) allowing the swing of the inner limb and the completion of 180° body rotation.

### 2.4. Data Collection and Processing

For tests 2 and 3, we used a commercial wearable inertial sensor (G-Sensor, BTS Bioengineering, Garbagnate Milanese, Italy) with dimensions of $70 \times 40 \times 18$ mm, applied over the skin of the second lumbar vertebra. The sensor is composed of a triaxial accelerometer 16 bit/axes (sensor range, $\pm 2$ g), a triaxial magnetometer 13 bit ($\pm 1200$ µT), and a triaxial gyroscope 16 bit/axes (sensor range, $\pm 2000°$/s). The signals were sampled with a frequency of 100 Hz and transmitted via Bluetooth to a laptop computer for acquisition and processing using a dedicated software package (BTS® G-Studio, BTS Bioengineering, Garbagnate Milanese, Italy). For test 3, the subcomponents were identified using the criteria described by Negrini et al. [25]. Then, the following movement parameters were considered: total time duration of the iTUG test, duration of the single subcomponents, average velocities of mid- and final turning.

We used the 10-m walk test (test 2) to estimate gait cycle and pelvis symmetries, instead of the TUG test, because the former test provided less variability in the waveform average than the TUG test, which involved the alternation of short linear with curved walking.

Gait cycle symmetry estimation can be obtained by several mathematical tools on the basis of the differences in single measures between the two body sides, such as velocity, joint torque, or based on a more global estimation of the symmetry using correlation analysis on data captured over the entire left and right gait cycle (for a comprehensive review see [26]). In this study the correlation method was used to compute two symmetry measures: a symmetry index (SI) for total gait cycle ($SI_{gait}$) and a SI for pelvis movements ($SI_{pelvis}$).

To estimate the $SI_{gait}$, we correlated anterior–posterior (AP) body acceleration signals detected when the gait cycle was performed by the left or right lower limb (Figure 2). The $SI_{gait}$ was obtained starting from the AP acceleration signal provided by the sensor as follows:

1. AP acceleration signal relative to the left gait cycles has been extracted from the whole acceleration signals;
2. AP acceleration signal relative to the right gait cycles has been extracted from the whole acceleration signals;
3. Mean normalized AP acceleration signal of the left gait cycles has been computed (acceleration signals on the left panels in Figure 2);
4. Mean normalized AP acceleration signal of the right gait cycles has been computed (acceleration signals on the right panels in Figure 2);
5. Compute Pearson's correlation coefficients (r) between 3 and 4;
6. The $SI_{gait}$ is obtained remapping the values of r, ranging from $-1$ to 1, between 0 and 100 with the following formula: $SI_{gait} = (r + 1) \times 100/2$.

Instead, the $SI_{pelvis}$ was obtained from the correlation between the measures of the pelvic angles, when the gait cycle occurred on the left and right sides (Figure 3). Pelvic angular displacements have been computed by the software G-Studio (BTS Bioengineering, Milano) processing the rotational angles provided by the sensor (i.e., roll, pitch, and yaw) to obtain Cardan angles referred to a global reference system. Each $SI_{pelvis}$ was obtained starting from the pelvic angle movement in the plane as follows:

1. Pelvic angle signal relative to the left gait cycles has been extracted from the whole pelvic angle signals;
2. Pelvic angle signal relative to the right gait cycles has been extracted from the whole pelvic angle signals;
3. Mean normalized pelvic angles signal of the left gait cycles has been computed (changes in amplitude of pelvic angle on the left panels in Figure 3);
4. Mean normalized pelvic angle signal of the right gait cycles has been computed (changes in amplitude of pelvic angle on the right panels in Figure 3);
5. Compute Pearson's correlation coefficients (r) between 3 and 4
6. The $SI_{pelvis}$ is obtained remapping the values of r, ranging from $-1$ to 1, between 0 and 100 with the following formula: $SI_{pelvis} = (r + 1) \times 100/2$

In the case of the pelvis, a set of $SI_{pelvis}$ was computed for the movements in the sagittal, frontal and transverse planes, obtaining tilt, obliquity and rotation $SI_{pelvis}$, respectively.

On these bases, the larger value of symmetry index, the more similar body acceleration during gait cycle ($SI_{gait}$) or pelvis angular displacements ($SI_{pelvis}$) between the two sides will be.

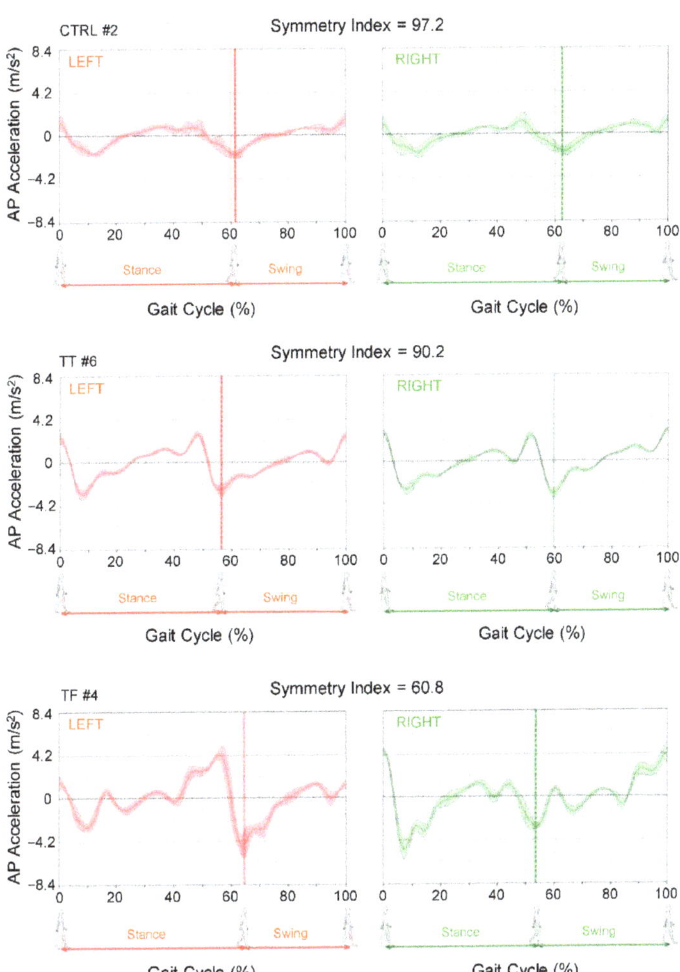


**Figure 2.** Representative examples of anterior–posterior (AP) body acceleration during gait cycle performed on the left and right sides by one control participant (CTRL), one transtibial amputee (TT), and one transfemoral amputee (TF). The waveforms represent the means and the standard deviations of AP accelerations from the total number of cycles during the 10-m walk test. The linear correlation between the left and the right AP accelerations, estimated by the Pearson correlation coefficient, was the basis for the gait cycle symmetry index (see Section 2 for details).

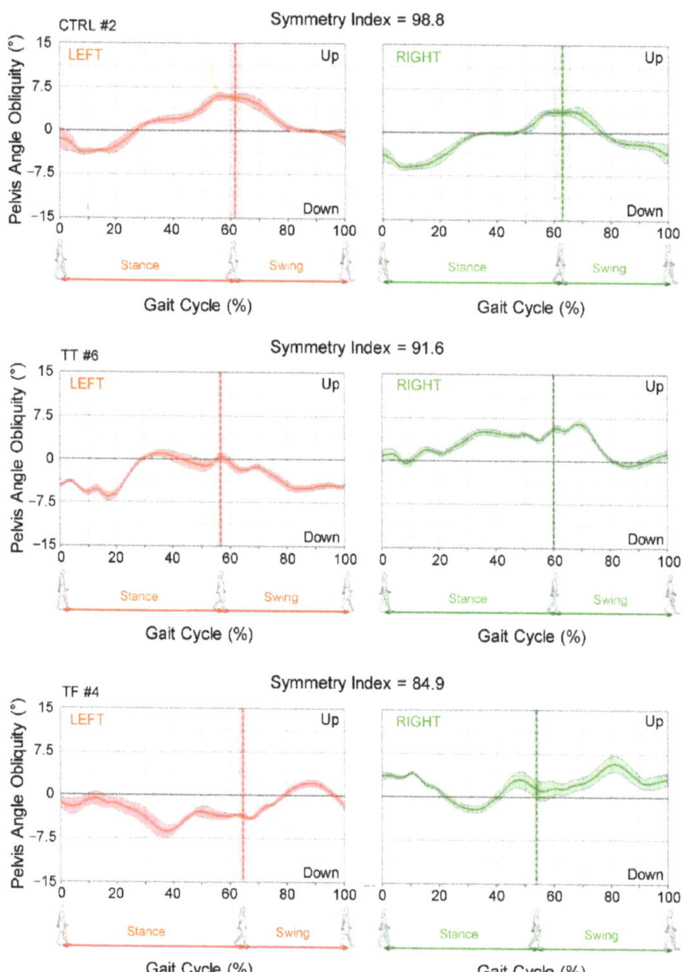

**Figure 3.** Representative examples of pelvis angle obliquity during gait cycle performed on the left and right sides by one control participant (CTRL), one transtibial amputee (TT) and one transfemoral amputee (TF). The waveforms represent the means and the standard deviations of pelvis angle obliquity from the total number of cycles during the 10-m walk test. The linear correlation between the left and the right pelvis angle obliquity, estimated by the Pearson correlation coefficient, was the basis for the gait cycle symmetry index. (See Section 2 for details).

*2.5. Statistical Analysis*

As each group included a small sample, we preliminarily analyzed data for normality distribution using Shapiro–Wilk's test. The assumption of normality was not met for most of the parameters; therefore, nonparametric statistics were applied. Moreover, considering the small and poorly distributed samples, the significance level ($p$ value) was computed based on the exact significance test. Finally, to determine the strength of the results, the magnitude of the effect was evaluated by epsilon-squared ($\varepsilon^2$), as it is considered a less biased effect size estimator compared to others [27].

The outcomes for each group are represented as median, interquartile range and the minimum and maximum values. The mean and the standard deviation are reported when required. The Kruskal–Wallis $H$-statistic was performed to compare the three groups for each SI and TUG parameter, with Dunn's test used for the post hoc analysis. To evaluate the differences between inner vs outer walking in each amputee group, the Wilcoxon Signed Rank test was adopted. Finally, Spearman's test was used to explore the relationship between SI and TUG parameters in the amputees. Separate correlations were computed with respect to the limb prosthetic position during the TUG test. The strength of the linear correlation was assessed by Spearman's Rho coefficient ($r_s$).

For all the statistical comparisons, the alpha level of significance was set at 0.05.

## 3. Results

### 3.1. Gait and Pelvis Symmetry Evaluations

The indices of gait cycle and pelvis symmetries were lower in the participants with amputation with respect to the individuals of the CTRL group for all the evaluations (Figure 4A–D).

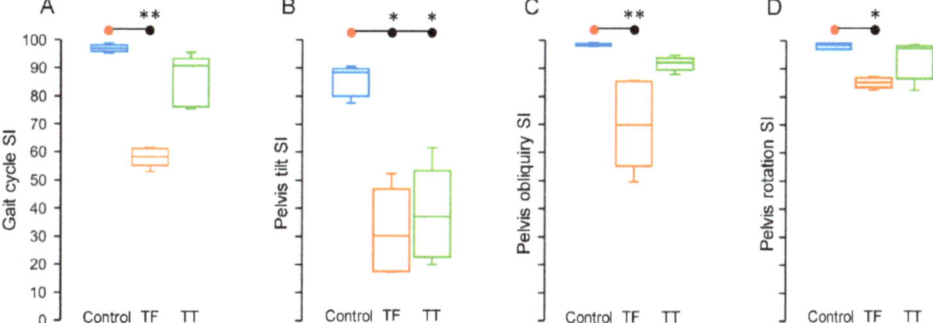

**Figure 4.** Box and whisker plots reporting the median value (horizontal line within the box) and the variability as interquartile range (vertical length of the box) and as the highest and the lowest values (lines above and below the box) of symmetry indexes of gait cycle (**A**), pelvis tilt (**B**), pelvis obliquity (**C**), and pelvis rotation (**D**), for each group. In all subplots, the horizontal lines with asterisks indicate statistically significant differences (* $p < 0.05$; ** $p < 0.01$). Abbreviations and symbols: SI = symmetry index; TF = transfemoral group (orange boxes); TT = transtibial group (green boxes).

Participants with TF amputations showed the lowest values of SI, compared to CTRL and TF groups, for the $SI_{gait}$ (Figure 4A; median [interquartile range]: CTRL, 96.9 [2.2]; TF, 58.3 [6]; TT, 90.6 [17]), the pelvis tilt SI (Figure 4B; CTRL, 88.5 [9.7]; TF, 30.4 [29.4]; TT, 37.2 [30.8]), the pelvis obliquity SI (Figure 4C; CTRL, 98.3 [0.8]; TF, 69.9 [30.3]; TT, 92.1 [4.2]) and pelvis rotation SI (Figure 4D; CTRL, 98.6 [1.9]; TF, 85.2 [3.3]; TT, 97.4 [11.6]).

There were significant differences across the groups for SI evaluated for gait cycle ($H_2 = 12.02$, $p < 0.001$, $\varepsilon^2 = 0.86$), pelvis tilt ($H_2 = 9.62$, $p = 0.002$, $\varepsilon^2 = 0.69$), pelvis obliquity ($H_2 = 12.5$, $p < 0.001$, $\varepsilon^2 = 0.89$), and pelvis rotation ($H_2 = 7.74$, $p = 0.012$, $\varepsilon^2 = 0.55$).

The post hoc comparisons showed that this is the result of a significant difference observed only between the CTRL and TF groups for $SI_{gait}$ (Figure 4A; $p = 0.002$), pelvis obliquity SI (Figure 4C; $p = 0.002$) and pelvis rotation SI (Figure 4D; $p = 0.012$), while in the case of pelvis tilt SI, there were significant differences between the CTRL and TF groups (Figure 4B; $p = 0.011$) and between the CTRL and TT groups (Figure 4B; $p = 0.049$). No significant differences were observed between TF and TT groups for all SI evaluations.

### 3.2. Timed Up and Go Component Results

The TUG total time and the measures obtained in each TUG component among the CTRL, TF, and TT groups were statistically analyzed by two sets of comparisons considering whether the prosthetic limb was in the inner or outer position during mid-turning. Thus, in each box plot of Figure 5, the data regarding the CTRL group (blue box), obtained with the TUG test performed in a clockwise direction, were compared with the results from the

TF (orange box) and TT (green box) groups with separate analysis for the inner or outer prosthetic limb position. To facilitate a comparison between the results of the TUG test obtained in this study and those reported by other studies, in Table 3 we provide numerical data of the TUG test results expressed as mean and standard deviation.

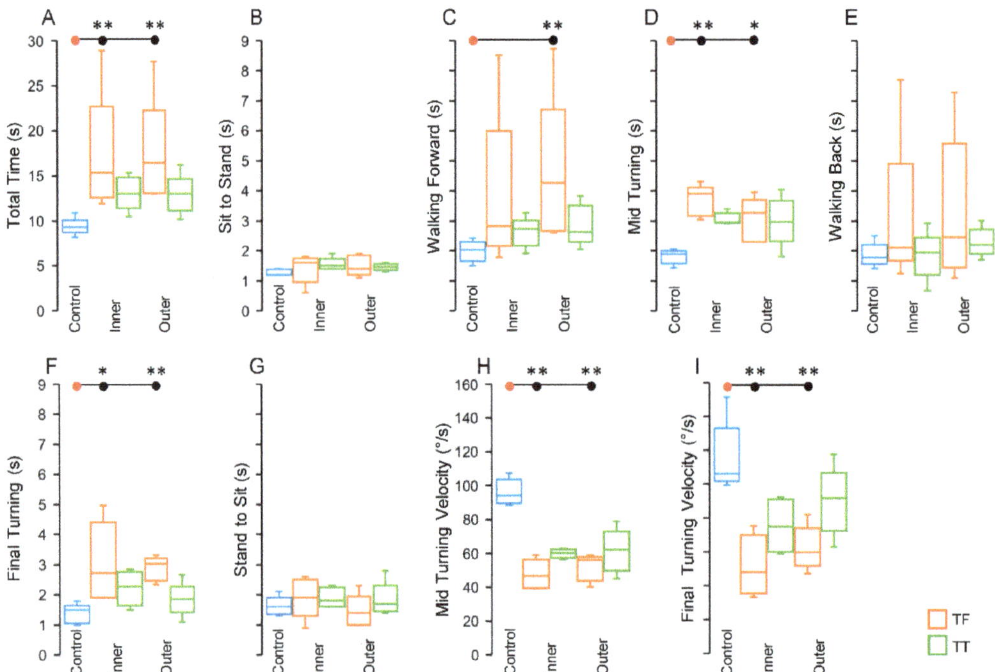

**Figure 5.** Box and whisker plots reporting median and variability (as described in Figure 4) of parameters measured during the Timed Up and Go test: Total time (**A**), sit to stand (**B**), walking forward (**C**), mid-turning (**D**), walking back (**E**), final turning (**F**), (**G**) stand to sit, (**H**) mid-turning velocity, (**I**) final turning velocity. Abbreviations as in Figure 4. Symbols: * $p < 0.05$; ** $p < 0.01$.

**Table 3.** Descriptive statistics (mean ± standard deviation) for temporal and velocity parameters over the TUG subcomponents, in the persons with lower limb amputation and non-amputated.

| Groups | Temporal Parameters (s) | | | | | | | Velocity Parameters (°/s) | |
|---|---|---|---|---|---|---|---|---|---|
| | Total Time | Sit to Stand | Walking Forward | Mid Turning | Walking Back | Final Turning | Stand to Sit | Mid Turning | Final Turning |
| CTRL | 9.4 ± 1.0 | 1.3 ± 0.1 | 2.0 ± 0.4 | 1.8 ± 0.2 | 1.9 ± 0.4 | 1.4 ± 0.3 | 1.6 ± 0.3 | 96.1 ± 7.7 | 115.0 ± 20.9 |
| TF Inner | 17.2 ± 6.8 | 1.4 ± 0.5 | 3.8 ± 2.7 | 3.7 ± 0.5 | 3.1 ± 2.6 | 3.1 ± 1.3 | 1.9 ± 0.7 | 47.7 ± 8.7 | 51.5 ± 17.8 |
| TF Outer | 17.4 ± 6.0 | 1.5 ± 0.3 | 4.6 ± 2.5 | 3.1 ± 0.7 | 3.3 ± 2.5 | 2.9 ± 0.6 | 1.5 ± 0.5 | 51.9 ± 7.8 | 62.1 ± 12.9 |
| TT Inner | 13.1 ± 1.9 | 1.6 ± 0.2 | 2.6 ± 0.5 | 3.1 ± 0.2 | 1.8 ± 0.8 | 2.2 ± 0.6 | 1.9 ± 0.3 | 60.0 ± 2.7 | 75.2 ± 15.4 |
| TT Outer | 12.9 ± 2.2 | 1.5 ± 0.1 | 2.9 ± 0.7 | 3.0 ± 0.8 | 2.3 ± 0.5 | 1.9 ± 0.6 | 1.8 ± 0.6 | 61.4 ± 12.7 | 89.8 ± 20.0 |

CTRL, control group; TF, transfemoral group; TT, transtibial group.

The results of the non-parametric statistical analysis for the TUG test are summarized in Table 4. In the following paragraphs the time measures are expressed in seconds, in cases of linear iTUG subcomponents, and the velocity measures in degree per seconds, in cases of Mid and Final Turning.

**Table 4.** Results of Kruskal–Wallis and Wilcoxon signed rank test analyses.

| TUG Components | Kruskal-Wallis CTRL vs. TF vs. TT | | | Post Hoc (Dunn-Bonferroni) | | | | | | Wilcoxon Signed Rank Test | |
|---|---|---|---|---|---|---|---|---|---|---|---|
| | | | | CTRL vs. TF | | CTRL vs. TT | | TF vs. TT | | | |
| | $H_2$ | $p$ | $\varepsilon^2$ | $z$ | $p$ | $z$ | $p$ | $z$ | $p$ | $z$ | $p$ |
| | Prosthetic limb INNER | | | | | | | | | Transfemoral Inner vs. Outer | |
| Total time | 9.57 | **0.002** | 0.68 | 8.5 | **0.008** | −5.9 | 0.108 | 2.6 | 1 | −0.272 | 0.938 |
| Sit to stand | 4.65 | 0.093 | 0.33 | | | | | | | −0.271 | 0.875 |
| Walking forward | 5.04 | 0.075 | 0.36 | | | | | | | −2.023 | 0.063 |
| Mid-turning time | 11.2 | **<0.001** | 0.80 | 9.4 | **0.003** | −5.6 | 0.143 | 3.8 | 0.536 | −1.483 | 0.188 |
| Walking back | 1.56 | 0.482 | 0.11 | | | | | | | −0.135 | 1.000 |
| Final turning time | 8.31 | **0.007** | 0.59 | 8.0 | **0.013** | −5.2 | 0.194 | 2.8 | 0.96 | −0.135 | 1.000 |
| Stand to sit | 2.06 | 0.376 | 0.15 | | | | | | | −1.219 | 0.313 |
| Mid-turning velocity | 11.58 | **<0.001** | 0.83 | 9.6 | **0.002** | 5.4 | 0.169 | −4.2 | 0.413 | −1.753 | 0.125 |
| Final turning velocity | 10.5 | **<0.001** | 0.75 | 9.0 | **0.004** | 6.0 | 0.102 | −3.0 | 0.867 | −1.753 | 0.125 |
| | Prosthetic limb OUTER | | | | | | | | | Transtibial Inner vs. Outer | |
| Total time | 10.37 | **<0.001** | 0.74 | 9.0 | **0.004** | 5.4 | 0.165 | 3.6 | 0.602 | −0.135 | 1.000 |
| Sit to stand | 3.61 | 0.167 | 0.26 | | | | | | | −0.412 | 0.813 |
| Walking forward | 9.38 | **0.009** | 0.67 | 8.6 | **0.007** | 5.2 | 0.198 | 3.4 | 0.688 | −0.405 | 0.813 |
| Mid-turning time | 7.35 | **0.016** | 0.53 | 7.0 | **0.04** | 6.2 | 0.085 | 0.8 | 1 | −0.135 | 1.000 |
| Walking back | 1.72 | 0.448 | 0.12 | | | | | | | −2.023 | 0.063 |
| Final turning time | 9.59 | **0.002** | 0.68 | 8.7 | **0.006** | 3.6 | 0.607 | 5.1 | 0.213 | −0.674 | 0.625 |
| Stand to sit | 1.83 | 0.433 | 0.13 | | | | | | | −0.404 | 0.812 |
| Mid-turning velocity | 9.98 | **<0.001** | 0.71 | 8.6 | **0.007** | 6.4 | 0.071 | 2.2 | 1 | −0.134 | 1 |
| Final-turning velocity | 9.26 | **0.003** | 0.66 | 8.6 | **0.007** | 4.0 | 0.472 | 4.6 | 0.312 | −0.943 | 0.437 |

CTRL, control group; TF, transfemoral group; TT, transtibial group; $H_2$, Kruskal–Wallis statistic adjusted for ties; $p$, level of significance; $\varepsilon^2$, epsilon squared effect size; $z$, Wilcoxon signed rank test statistic. Statistically significant differences are marked in bold.

Statistical differences among the three groups were detected for the TUG total time (Figure 5A) both when the prosthetic limb was in the inner position and when it was in the outer position (see Table 4 for the results of Kruskal–Wallis test). The TF group showed the highest value of TUG total time (Table 4 and Figure 4A; median [interquartile range]: CTRL, 9.3 [1.35]; TF inner, 15.3 [10.1]; TT inner, 13 [3.4]; TF outer, 16.4 [9.15]; TT outer, 13 [3.5]), and the only significant difference among the post hoc pairwise comparisons was found between the CTRL and the TF groups (inner, $p = 0.008$; outer, $p = 0.004$; see Table 4 for more details).

The values observed for the TUG total time depended mainly on the changes in time and velocity during mid-turning (Figure 5D,H) and final turning (Figure 5F,I). In fact, as reported in Table 4 (Kruskal–Wallis test), the pattern of statistical differences observed for the total time, with the significant differences focused on the comparison between CTRL and TF groups, was replicated by the measures revealed during the mid-turning time (Figure 5D; CTRL, 1.89 [0.42]; TF inner, 3.9 [0.94]; TT inner, 2.95 [0.32]; TF outer, 3.27 [1.41]; TT outer, 2.97 [1.35]), the final turning time (Figure 5F; CTRL, 1.5 [0.6]; TF inner, 2.72 [2.51]; TT inner, 2.27 [1.12]; TF outer, 3.02 [0.76]; TT outer, 1.86 [0.86]), the mid-turning velocity (Figure 5H; CTRL, 94.1 [14.1]; TF inner, 46.7 [16.9]; TT inner, 60.2 [5.3]; TF outer, 56 [13.9]; TT outer, 62.1 [23]) and the final turning velocity (Figure 5I; CTRL, 106 [31.15]; TF inner, 47.9 [34.35]; TT inner, 74.5 [30.75]; TF outer, 59.6 [21.85]; TT outer, 91.7 [34.35]).

No significant differences among groups, for both inner and outer prosthetic limb positions, were observed for the other TUG components, except for walking forward (Figure 5C), which showed significant differences for groups when the prosthetic limb was in the outer position, with pairwise significant differences between CTRL and TF outer ($p = 0.007$; see Table 4 for more details).

The comparison between inner and outer conditions as repeated measures within TF and TT groups was conducted by the Wilcoxon signed rank test and revealed that no statistically significant differences occurred for all the measured TUG parameters (see details in Table 4).

For all the statistically significant results, there was a good magnitude of the effect size, with $\varepsilon^2$ ranging from 0.59 to 0.83 for the data related to the condition with the prosthetic limb in the inner position, and from 0.53 to 0.74 for the data related to the condition with the prosthetic limb in the outer position.

### 3.3. Correlations between Symmetry Indices and TUG Test Measures

The correlations between the changes in SI parameters and the measures of TUG subcomponents are reported in Table 5. As can be seen from Table 5, almost all the statistically significant correlations ($p < 0.05$; in bold in the Table 5) are those where the SI of the gait cycle, pelvis obliquity, and pelvis rotation are related to the mid-turning time and velocity when the prosthetic limb was in the inner position during mid-turning (Figure 6A–F) and to the final turning time and velocity when the prosthetic limb was in the outer position during mid-turning (Figure 6G–L).

**Table 5.** Results of Spearman's correlation analysis.

| TUG Components | Symmetry Index | | | | | | | | BBS-it | |
|---|---|---|---|---|---|---|---|---|---|---|
| | Gait Cycle | | Pelvis Tilt | | Pelvis Obliquity | | Pelvis Rotation | | | |
| | $r_s$ | $p$ | $r_s$ | $p$ | $r_s$ | $p$ | $r_s$ | $p$ | $r_s$ | $p$ |
| Prosthetic limb INNER | | | | | | | | | | |
| Total time | −0.44 | 0.199 | 0.01 | 0.973 | −0.43 | 0.22 | −0.10 | 0.776 | −0.59 | 0.074 |
| Sit to stand | 0.10 | 0.789 | 0.54 | 0.111 | 0.12 | 0.738 | 0.40 | 0.258 | −0.18 | 0.624 |
| Walking forward | −0.47 | 0.174 | −0.25 | 0.489 | −0.38 | 0.276 | −0.16 | 0.651 | −0.45 | 0.197 |
| Mid-turning time | −0.64 | **0.044** | −0.32 | 0.374 | −0.77 | **0.01** | −0.63 | **0.05** | −0.59 | 0.071 |
| Walking back | −0.20 | 0.575 | 0.08 | 0.827 | −0.61 | 0.063 | −0.36 | 0.304 | −0.75 | **0.013** |
| Final turning time | −0.30 | 0.393 | 0.17 | 0.638 | −0.16 | 0.65 | 0.28 | 0.434 | −0.52 | 0.121 |
| Stand to sit | −0.14 | 0.7 | 0.27 | 0.455 | −0.35 | 0.327 | −0.12 | 0.751 | −0.28 | 0.426 |
| Mid-turning velocity | 0.70 | **0.025** | 0.37 | 0.293 | 0.77 | **0.009** | 0.55 | 0.098 | 0.66 | **0.038** |
| Final turning velocity | 0.55 | 0.098 | −0.12 | 0.751 | 0.43 | 0.214 | 0.02 | 0.96 | 0.68 | **0.03** |

**Table 5.** *Cont.*

| TUG Components | Symmetry Index | | | | | | | | BBS-it | |
|---|---|---|---|---|---|---|---|---|---|---|
| | Gait Cycle | | Pelvis Tilt | | Pelvis Obliquity | | Pelvis Rotation | | | |
| | $r_s$ | $p$ | $r_s$ | $p$ | $r_s$ | $p$ | $r_s$ | $p$ | $r_s$ | $p$ |
| | Prosthetic limb OUTER | | | | | | | | | |
| Total time | −0.65 | **0.041** | −0.29 | 0.422 | −0.67 | **0.034** | −0.33 | 0.353 | −0.60 | 0.068 |
| Sit to stand | −0.13 | 0.724 | −0.27 | 0.454 | −0.01 | 0.973 | 0.06 | 0.867 | 0.21 | 0.569 |
| Walking forward | −0.46 | 0.187 | −0.03 | 0.934 | −0.52 | 0.128 | −0.21 | 0.556 | −0.73 | **0.018** |
| Mid-turning time | −0.12 | 0.738 | 0.14 | 0.7 | −0.42 | 0.228 | −0.28 | 0.434 | −0.34 | 0.337 |
| Walking back | −0.08 | 0.829 | −0.48 | 0.162 | −0.01 | 0.987 | 0.21 | 0.556 | −0.45 | 0.197 |
| Final turning time | −0.77 | **0.009** | −0.07 | 0.855 | −0.90 | **<0.001** | −0.71 | **0.022** | −0.70 | **0.026** |
| Stand to sit | 0.33 | 0.353 | 0.32 | 0.362 | 0.16 | 0.662 | −0.04 | 0.907 | 0.41 | 0.238 |
| Mid-turning velocity | 0.36 | 0.31 | 0.24 | 0.511 | 0.58 | 0.082 | 0.18 | 0.627 | 0.67 | **0.034** |
| Final turning velocity | 0.72 | **0.019** | 0.06 | 0.881 | 0.87 | **0.001** | 0.66 | **0.038** | 0.71 | **0.022** |

$r_s$, Spearman's Rho coefficient; $p$, level of significance; BBS-it, Italian Version of Berg Balance Scale. Statistically significant values are shown in bold.

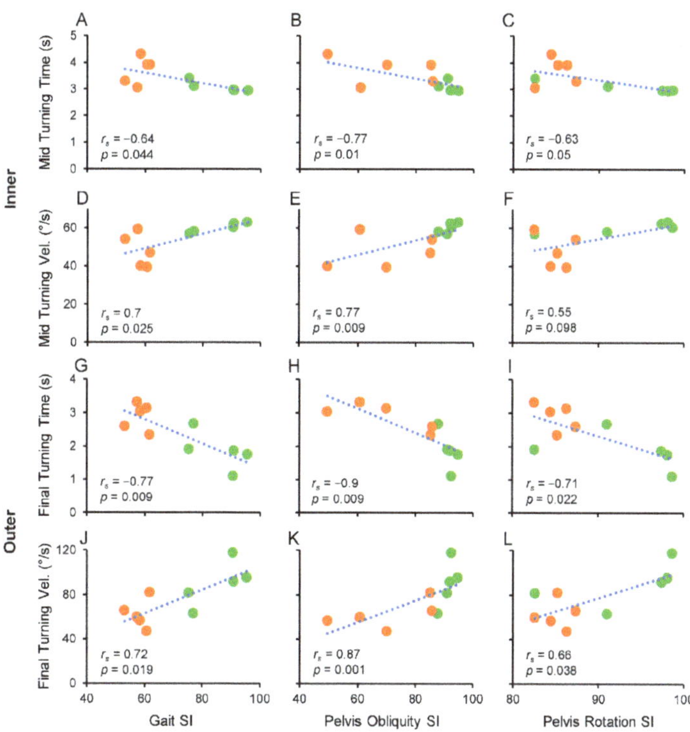



**Figure 6.** Spearman's correlation analysis between symmetry index and mid- and final turning time and velocity. Symmetry indexes of gait cycle, pelvis obliquity, and pelvis rotation were correlated with mid-turning time and velocity, when the amputated limb was in inner position (**A–F**), and with final turning time and velocity, when the amputated limb was in outer position (**G–L**). Abbreviation and symbol: $r_s$, Spearman's Rho coefficient; $p$, level of significance; orange circles, TF; green circles, TT.

The level of correlation, estimated by the Spearman coefficient ($r_s$), ranged from 0.64 to 0.77 for the correlations associated with mid-turning, and from 0.66 to 0.9 for the correlations associated with final turning.

The plots in Figure 6 show that as SI increased, the turning time decreased (Figure 6A–C,G–I) or the turning velocity increased (Figure 6D–F,J–L), indicating that gait cycle and pelvis asymmetries associated with amputation influenced the performance during rotation components of the TUG test more than the performance during the execution of linear TUG components.

Moreover, the two amputee groups tended to cluster into two separate groups, with the TT group (green circle) fitting the linear model better than TF group (orange circle) in the case of the correlations associated with mid-turning time or velocity (Figure 6A–F).

Finally, the BBS-it test showed significant correlations with the measures of TUG subcomponents mainly during mid- and final turning, for both the inner and outer positions of the prosthetic limb (last two columns on the right, in Table 5). The velocity of turning increased as balance increased.

## 4. Discussion

The basic result reported here is that subjects with an amputation took more time to complete the TUG test and showed lower levels of gait and pelvis SI than non-amputee volunteers. However, almost all the statistical differences between amputees and healthy people were focused on the TF group.

The TUG test duration in the case of amputees depended mainly on the slowdown observed during mid- and final turning. Moreover, the velocity reduction during the turning components of the TUG test was well correlated with the reduction in gait cycle and pelvis symmetries observed in the amputees. In particular, the association between reduction in velocity and reduction in gait cycle and pelvis symmetries occurred when the prosthetic limb was in the inner position during mid-turning and in the outer position during final turning.

### 4.1. The Turning Components of the TUG Test Represent the Most Demanding Tasks for the Amputees

The results of the current study provide insights into the possible causes for the specific difficulties showed by the amputees in facing the turning task, suggesting that gait cycle and pelvis asymmetries may contribute to differentiate the level of mobility between linear and curved walking.

To our knowledge, only Clemens et al. [21] have used the TUG test to explore the performance over the single task components. Although the number of participants with lower limb prostheses in the study of Clemens et al. [21] was much larger (a total of 118 participants) than the sample size used in the current work, their main result supports our main finding as these authors indicate that, among the tasks included in the TUG test, the turning components have the greatest impact in the level of mobility of subjects with lower limb amputation. Although the high number of participants in the study of Clemens at al. [21], the mean time duration and the variability observed by these authors between TT and TF amputees, during the mid-turning, were comparable with the scores recorded in the current study: for the TT amputees, 2.73 s $\pm$ 0.65 in Clemens et al. [21], 3.05 s $\pm$ 0.5 in the current study (average between inner and outer limbs); for the TF amputees, 3.53 s $\pm$ 1.2 in Clemens et al. [21], and 3.4 s $\pm$ 0.6 in the current study (average between inner and outer limbs). However, the large difference in the sample size between the two studies determined significant statistical differences between TT and TF amputees for all the TUG subcomponents in the study of Clemens et al. [21], with some subcomponents, especially for walking forward, showing large duration and variability in the current study. This reflects the different time duration in the TUG total time between the two works: for the TT amputees, 10.04 s $\pm$ 2.3 in Clemens et al. [21], 13 s $\pm$ 2.05 in the current study (average between inner and outer limbs); for the TF amputees, 12.77 s $\pm$ 5.04 in Clements et al. [21], 17.3 s $\pm$ 6.4 in the current study (average between inner and outer limbs).

Several authors showed evidence that asymmetries in subjects with lower limb amputations might be the result of motor adaptations to accomplish functional compensations such as, facilitating lower-leg trajectory [8], foot clearance during the swing phase of gait [7], gait stability [6] and functional step length and symmetrical thigh inclinations [5]. Typically, these adaptations are related to walking in a straight line, therefore, the appearance of asymmetry may be functional to a linear path, but it may not be suitable for the path along a curve, impairing the turning task.

### 4.2. Velocity Decrease and Asymmetry Effects When Amputees Face Turning Points: Biomechanical and Neuronal Considerations

In an attempt to provide an explanation for the changes in the kinematics and for the effects of gait and pelvis asymmetries on mid- and final turning, in the following section we consider the basic biomechanics and neuronal processes underlying a successful execution of a turning task.

#### 4.2.1. Biomechanical Considerations

In non-amputees, the process of turning starts before the change of direction, with forward motion deceleration followed by body rotation and the movement toward the new direction [28,29].

The two turning components of the TUG test are performed on the basis of different biomechanical strategies. Mid-turning can be included in the category of step turning, since the body is stepping along an arc of 180°, while final turning can be considered a form of spin turning, as the step is stopped, and the body axis rotates about 180° [28–31]. During step turning, the body center of mass typically moves over the mediolateral axis, shifting in the direction of the turn, at the boundary of the base of support. As the velocity of turning decreases, the center of mass slightly shifts between the two feet, within the base of support [32]. This process provides a straightforward explanation for the reduced velocity observed during mid-turning in the subjects with a lower limb amputation: the increased risk to perturb body stability forced the amputees to reduce their velocity to place the center of mass in a more reliable position to maintain stability during the turning trajectory.

This time compensation could be particularly effective in cases of TF amputees who show more significant reductions in turning velocity than TT amputees when compared with the CTRL group. However, when mid-turning was negotiated with the prosthetic limb in the inner position, both amputee groups showed a reduction in velocity as gait cycle and pelvis obliquity asymmetries increased. This specific influence of pelvis obliquity asymmetry makes sense if we consider that the medial-lateral axis, along which the center of mass moves during mid-turning, is the same axis along which the pelvis oscillates when its obliquity changes. Moreover, the internal shift of the body center of mass produces a greater load on the inner limb, exacerbating the negative effects of pelvis asymmetry over the transverse plane.

Contrary to what we observed in mid- and final turning, there were no significant differences in timing and no influences of asymmetries in the linear components of the TUG test. The different responses observed in linear and turning walks may depend on the intrinsic lower stability characterizing turning with respect to walking in a straight line [33]. In subjects with a lower limb amputation these differences in stability are accentuated, since they are more sensitive to perturbations occurring along the medial-lateral axis, mainly present during turning, than perturbations that affect the anterior-posterior axis, mainly occurring during linear walking [34,35]. The parallel reduction in the BBS-it test and the velocity during turning tasks, observed in our experiments, confirms the tendency of amputees to be less stable in curved than in straight paths.

The reduction in velocity observed during mid-turning for TF amputees and the correlations with the changes in symmetry were also observed in final turning, before sitting down. The large curved path of step turning (mid-turning) offers greater stability compared to the spin turn (final turning) that requires the body to rotate around the vertical axis [28,29,31]. Although we did not observe differences in velocity reduction between the two types of turning, the enhanced instability attributed to final turning may explain the greater negative influences of asymmetries observed for final turning compared to mid-turning. In fact, the levels of correlation detected during final turning where higher than those observed during mid-turning, with significant correlations not only for gait cycle and pelvis obliquity asymmetries but also for the asymmetries associated with pelvis rotation (see correlation analyses illustrated in Figure 6G–L).

Another difference that we observed between the two turning tasks was that the influence of decreased symmetries at final turning occurred when the TUG test was executed with the prosthetic limb in the outer position. The transition from turning and stand to sit, illustrated in Figure 1B,D, provides a possible explanation for the prevalent influence of asymmetries when final turning is performed by the prosthetic limb in the outer position. This transition occurs accomplishing a 180° body rotation through two semi-rotations: the first semi-rotation of the body, of about 90°, occurs around the inner limb (point 1 in Figure 1B,D), allowing the swing of the outer limb (point 2 in the Figure 1B,D); the second semi-rotation, of another 90°, takes place around the outer limb (point 3 in the Figure 1B,D), allowing the swing of the inner limb (point 4 in the Figure 1B,D), and the completion of the 180° rotation of the body. During the first semi-rotation, the static load on the inner limb is reduced due to the residual energy from the dynamics of the last step before starting final turning. Instead, the velocity reduction occurred when the step cycle stopped, producing an increase in static load on the outer limb during the second semi-rotation. Thus, the outer limb receives more load than the inner limb, becoming more susceptible to the effects of pelvis asymmetries.

### 4.2.2. Neuronal Considerations

The best performance observed in amputees during linear compared to curved walking could depend on a specific need to adopt a predictive feedforward strategy in the curved walk, with respect to the straight line walk.

A critical factor for successful turning biomechanics is to plan trunk and pelvis rotations in advance, during the deceleration period preceding the turning phase [28,29,32,36]. A walking direction-dependent neuronal control was shown by Bauby and Kuo [37] and O'Connor and Kuo [38] who observed that reduced visual information destabilized more mediolateral than anteroposterior motions during walking. Since visual sensory feedback is essential for planning in advance the changes in walking direction, these authors concluded that linear walking relies more on the passive mechanical properties of the limbs, while for stabilizing mediolateral motion a significant central active control must be provided. As discussed in the previous section, amputees show greater difficulty in stabilizing perturbations that act along the mediolateral axis, with respect to those that act along the anterior-posterior axis [34,35]. Thus, a deficit in neuronal predictive control could be an explanation for the reduced turning performance in amputees, particularly for those with a TF amputation.

Difficulties in anticipatory strategy were observed in TT amputees when movements of the upper limbs perturbed upright standing [39] or lateral pushes perturbed linear walking [40]. Similar deficits were detected in older subjects who showed limitations in facing mid- and final turning during the TUG test [20] and difficulties in anticipatory adjustments after a later perturbation during gait initiation [41]. Similar to the condition of amputees, it can be assumed that also physical body variations associated with aging make predictive capacity more difficult to implement.

Planning a movement strategy in advance to face future biomechanical changes, the inside shift of the center of mass in the case of mid-turning, requires not only an exteroceptive sensory feedback, but also an accurate internal representation of the relationships between the mechanical state of the body and the contextual environment (predictive internal models; [42]). Similarly, during final turning, for successful rotation a proactive neuronal control in relation to the subsequent stand to sit task is necessary. Rebuilding an internal model of the body is one of the greatest difficulties for subjects with an amputation since the missing part and the prosthesis upset the biomechanical and temporal relationships between body segments. This issue was recently addressed by Saimpont et al. [19] using a motor imagery protocol applied to amputees performing 7-m walking and the TUG test. These authors found timing and accuracy discrepancies in amputees comparing the actual execution and the mental imagery of the TUG test or walking. It was suggested that these discrepancies were due to dysfunctions in the internal predictive models that need to be updated. The motor imagery of the total time of the TUG test tended to be slower than the real execution with respect to 7-m walking. However, in the study of Saimpont et al. [19], no parameter measurements were recorded for the TUG subcomponents.

In summary, the basic biomechanics and neuronal processes to accomplish the transition from linear to curved walking during the TUG test suggest that velocity reductions in TF amputees and the negative influence of gait cycle and pelvis asymmetries in both amputee groups, may be attributable, at least in part, to the difficulty of the amputees in updating forward internal models capable of preventing the effects of perturbations acting on the mediolateral axis of the body during turning tasks.

## 5. Conclusions and Practical Implications

The main message from the current study is that the idea of asymmetry occurring in amputee motions to compensate for the differences between the two limbs is not applicable to the various motor actions. Gait cycle and pelvis asymmetries are appropriate for balancing linear walking, but they can be detrimental for turning performance. Considering that predictive control is a specific and critical factor for the successful transition from linear to curved walking, we suggest reinforcing those rehabilitative protocols addressed to stimulating the sensory feedback from the residual limb segments so as to trigger an update of the internal body representation and the forward models. For example, considering that amputees have no neurological injury, using the appropriate accommodations, the use of a tool such as a balance board can be important to stimulate the proprioceptive afferences and produce learning of new postural skills, as was demonstrated for healthy individuals [43,44]. Moreover, recent studies have demonstrated that the use of new tools, such as motor imagery [19,45] and virtual reality [46], can provide an important support in training and monitoring the development and updating of internal models, in subjects with a lower limb amputation.

Finally, as the use of a single wearable sensor contributed significantly in this study to detecting measures from different motor contexts, similarly the use of these devices can improve the quality of rehabilitation protocols, making the measurements of demanding motor tasks easier and more reliable [47,48].

## 6. Limitations

The reduced sample size, made up of male participants, may represent only a portion of the population. This is a limitation to consider for the interpretation of the results reported in the current study. In addition, some characteristics, such as the type of prothesis or the time elapsed since amputation, are not homogenous over the amputees. However, we tried to adopt the most suitable statistical procedures to partially compensate for these limitations. We believe that the level of significance and the magnitude of the effect size should guarantee the reliability of the main results. Further studies including a larger sample of people with a lower limb amputation could reinforce the general character of these results.

**Author Contributions:** M.C. and A.V. contributed to the study design; M.C., S.L. and I.S. were involved in data collection; M.C., M.S.V. and A.C. performed data analysis and interpreted the results; M.C., M.S.V., L.L. and A.C. were involved in the preparation of the manuscript. All authors have read and agreed to the published version of the manuscript.

**Funding:** This research study was funded by the University of Catania, grant "fondi di ateneo 2020–2022" PIA.CE.RI, linea Open Access, n°20722142133.

**Institutional Review Board Statement:** The study was approved by the local ethics committee of Catania University Hospital "Policlinico Vittorio Emanuele-San Marco" (BIOART Project, n° 72/2019/PO), and all participants provided written consent to participate, before starting the research project. All procedures were in accordance with the Helsinki Declaration of 1975, as revised in 2000.

**Informed Consent Statement:** Informed consent was obtained from all participants involved in the study.

**Data Availability Statement:** The data that support the findings of this study are available on request from the corresponding author, M.S.V. The data are not publicly available due to the fact that they contain information that could compromise the privacy of research participants.

**Acknowledgments:** The authors thank the amputees and the volunteers for their participation in this study. Furthermore, we wish to thank the Scientific Bureau of the University of Catania for language support.

**Conflicts of Interest:** A.V. is an engineer belonging to BTS Bioengineering, which commercializes the inertial wearable sensor.

## References


1. Jaegers, S.M.; Arendzen, J.H.; de Jongh, H.J. Prosthetic gait of unilateral transfemoral amputees: A kinematic study. *Arch. Phys. Med. Rehabil.* **1995**, *76*, 736–743. [CrossRef]
2. Sagawa, Y., Jr.; Turcot, K.; Armand, S.; Thevenon, A.; Vuillerme, N.; Watelain, E. Biomechanics and physiological parameters during gait in lower-limb amputees: A systematic review. *Gait Posture* **2011**, *33*, 511–526. [CrossRef]
3. Devan, H.; Hendrick, P.; Ribeiro, D.C.; Hale, L.A.; Carman, A. Asymmetrical movements of the lumbopelvic region: Is this a potential mechanism for low back pain in people with lower limb amputation? *Med. Hypotheses* **2014**, *82*, 77–85. [CrossRef]
4. Winiarski, S.; Rutkowska-Kucharska, A.; Kowal, M. Symmetry function—An effective tool for evaluating the gait symmetry of trans-femoral amputees. *Gait Posture* **2021**, *90*, 9–15. [CrossRef]
5. Rabuffetti, M.; Recalcati, M.; Ferrarin, M. Trans-femoral amputee gait: Socket-pelvis constraints and compensation strategies. *Prosthet. Orthot. Int.* **2005**, *29*, 183–192. [CrossRef]
6. Hak, L.; van Dieën, J.H.; van der Wurff, P.; Houdijk, H. Stepping asymmetry among individuals with unilateral transtibial limb loss might be functional in terms of gait stability. *Phys. Ther.* **2014**, *94*, 1480–1488. [CrossRef]
7. Armannsdottir, A.; Tranberg, R.; Halldorsdottir, G.; Briem, K. Frontal plane pelvis and hip kinematics of transfemoral amputee gait. Effect of a prosthetic foot with active ankle dorsiflexion and individualized training—A case study. *Disabil. Rehabil. Assist. Technol.* **2018**, *13*, 388–393. [CrossRef]
8. Mohamed, A.; Sexton, A.; Simonsen, K.; McGibbon, C.A. Development of a Mechanistic Hypothesis Linking Compensatory Biomechanics and Stepping Asymmetry during Gait of Transfemoral Amputees. *Appl. Bionics Biomech.* **2019**, *2019*, 4769242. [CrossRef]
9. Devan, H.; Carman, A.; Hendrick, P.; Hale, L.; Cury Ribeiro, D. Spinal, pelvic, and hip movement asymmetries in people with lower-limb amputation: Systematic review. *J. Rehabil. Res. Dev.* **2015**, *52*, 1–19. [CrossRef]
10. Podsiadlo, D.; Richardson, S. The timed "Up & Go": A test of basic functional mobility for frail elderly persons. *J. Am. Geriatr. Soc.* **1991**, *39*, 142–148.
11. Pondal, M.; del Ser, T. Normative data and determinants for the timed "up and go" test in a population-based sample of elderly individuals without gait disturbances. *J. Geriatr. Phys. Ther.* **2008**, *31*, 57–63. [CrossRef]
12. Cioni, M.; Amata, O.; Seminara, M.R.; Marano, P.; Palermo, F.; Corallo, V.; Brugliera, L. Responsiveness to sensory cues using the Timed Up and Go test in patients with Parkinson's disease: A prospective cohort study. *J. Rehabil. Med.* **2015**, *47*, 824–829. [CrossRef]
13. Kaya, P.; Alemdaroğlu, İ.; Yılmaz, Ö.; Karaduman, A.; Topaloğlu, H. Effect of muscle weakness distribution on balance in neuromuscular disease. *Pediatr. Int.* **2015**, *57*, 92–97. [CrossRef]
14. Hafsteinsdóttir, T.B.; Rensink, M.; Schuurmans, M. Clinimetric properties of the Timed Up and Go Test for patients with stroke: A systematic review. *Top. Stroke Rehabil.* **2014**, *21*, 197–210. [CrossRef]
15. Schoppen, T.; Boonstra, A.; Groothoff, J.W.; de Vries, J.; Göeken, L.N.; Eisma, W.H. The Timed "up and go" test: Reliability and validity in persons with unilateral lower limb amputation. *Arch. Phys. Med. Rehabil.* **1999**, *80*, 825–828. [CrossRef]
16. Sions, J.M.; Beisheim, E.H.; Seth, M. Selecting, Administering, and Interpreting Outcome Measures among Adults with Lower-Limb Loss: An Update for Clinicians. *Curr. Phys. Med. Rehabil. Rep.* **2020**, *8*, 92–109. [CrossRef]
17. Dite, W.; Connor, H.J.; Curtis, H.C. Clinical identification of multiple fall risk early after unilateral transtibial amputation. *Arch. Phys. Med. Rehabil.* **2007**, *88*, 109–114. [CrossRef]

18. Claret, C.R.; Herget, G.W.; Kouba, L.; Wiest, D.; Adler, J.; von Tscharner, V.; Stieglitz, T.; Pasluosta, C. Neuromuscular adaptations and sensorimotor integration following a unilateral transfemoral amputation. *J. Neuroeng. Rehabil.* **2019**, *16*, 115. [CrossRef]

19. Saimpont, A.; Malouin, F.; Durand, A.; Mercier, C.; di Rienzo, F.; Saruco, E.; Collet, C.; Guillot, A.; Jackson, P.L. The effects of body position and actual execution on motor imagery of locomotor tasks in people with a lower-limb amputation. *Sci. Rep.* **2021**, *11*, 13788. [CrossRef]

20. Mangano, G.R.A.; Valle, M.S.; Casabona, A.; Vagnini, A.; Cioni, M. Age-Related Changes in Mobility Evaluated by the Timed Up and Go Test Instrumented through a Single Sensor. *Sensors* **2020**, *20*, 719. [CrossRef]

21. Clemens, S.M.; Gailey, R.S.; Bennett, C.L.; Pasquina, P.F.; Kirk-Sanchez, N.J.; Gaunaurd, I.A. The Component Timed-Up-and-Go test: The utility and psychometric properties of using a mobile application to determine prosthetic mobility in people with lower limb amputations. *Clin. Rehabil.* **2018**, *32*, 388–397. [CrossRef]

22. Borrenpohl, D.; Kaluf, B.; Major, M.J. Survey of US Practitioners on the Validity of the Medicare Functional Classification Level System and Utility of Clinical Outcome Measures for Aiding K-Level Assignment. *Arch. Phys. Med. Rehabil.* **2016**, *97*, 1053–1063. [CrossRef]

23. Ottonello, M.; Ferriero, G.; Benevolo, E.; Sessarego, P.; Dughi, D. Psychometric evaluation of the Italian version of the Berg Balance Scale in rehabilitation inpatients. *Eur. Med.* **2003**, *39*, 181–189.

24. Wong, C.K.; Chen, C.C.; Welsh, J. Preliminary assessment of balance with the Berg Balance Scale in adults who have a leg amputation and dwell in the community: Rasch rating scale analysis. *Phys. Ther.* **2013**, *93*, 1520–1529. [CrossRef]

25. Negrini, S.; Serpelloni, M.; Amici, C.; Gobbo, M.; Silvestro, C.; Buraschi, R.; Borboni, A.; Crovato, D.; Lopomo, N.F. Use of wearable inertial sensor in the assessment of Timed-Up-and-Go Test: Influence of device placement on temporal variable estimation. In *Lecture Notes of the Institute for Computer Sciences, Social-Informatics and Telecommunications Engineering, Proceedings of the Wireless Mobile Communication and Healthcare, Mobihealth 2016, Milan, Italy, 14–16 November 2016*; Springer International Publishing: Cham, Switzerland, 2017; Volume 192, pp. 310–317.

26. Viteckova, S.; Kutilek, P.; Svoboda, Z.; Krupicka, R.; Kauler, J.; Szabo, Z. Gait symmetry measures: A review of current and prospective methods. *Biomed. Signal Process. Control* **2018**, *42*, 89–100. [CrossRef]

27. Albers, C.; Lakens, D. When power analyses based on pilot data are biased: Inaccurate effect size estimators and follow-up bias. *J. Exp. Soc. Psychol.* **2018**, *74*, 187–195. [CrossRef]

28. Patla, A.E.; Prentice, S.D.; Robinson, C.; Neufeld, J. Visual control of locomotion: Strategies for changing direction and for going over obstacles. *J. Exp. Psychol. Hum. Percept. Perform.* **1991**, *17*, 603–634. [CrossRef]

29. Hase, K.; Stein, R.B. Turning strategies during human walking. *J. Neurophysiol.* **1999**, *81*, 2914–2922. [CrossRef]

30. Taylor, M.J.; Dabnichki, P.; Strike, S.C. A three-dimensional biomechanical comparison between turning strategies during the stance phase of walking. *Hum. Mov. Sci.* **2005**, *24*, 558–573. [CrossRef]

31. Golyski, P.R.; Hendershot, B.D. Trunk and pelvic dynamics during transient turns among individuals with unilateral traumatic lower limb amputation. *Hum. Mov. Sci.* **2018**, *58*, 41–54. [CrossRef]

32. Orendurff, M.S.; Segal, A.D.; Berge, J.S.; Flick, K.C.; Spanier, D.; Klute, G.K. The kinematics and kinetics of turning: Limb asymmetries associated with walking a circular path. *Gait Posture* **2006**, *23*, 106–111. [CrossRef]

33. Segal, A.D.; Orendurff, M.S.; Czerniecki, J.M.; Shofer, J.B.; Klute, G.K. Local dynamic stability in turning and straight-line gait. *J. Biomech.* **2008**, *41*, 1486–1493. [CrossRef]

34. Segal, A.D.; Klute, G.K. Lower-limb amputee recovery response to an imposed error in mediolateral foot placement. *J. Biomech.* **2014**, *47*, 2911–2918. [CrossRef]

35. Beltran, E.J.; Dingwell, J.B.; Wilken, J.M. Margins of stability in young adults with traumatic transtibial amputation walking in destabilizing environments. *J. Biomech.* **2014**, *47*, 1138–1143. [CrossRef]

36. Andrews, J.R.; McLeod, W.D.; Ward, T.; Howard, K. The cutting mechanism. *Am. J. Sports Med.* **1977**, *5*, 111–121. [CrossRef]

37. Bauby, C.E.; Kuo, A.D. Active control of lateral balance in human walking. *J. Biomech.* **2000**, *33*, 1433–1440. [CrossRef]

38. O'Connor, S.M.; Kuo, A.D. Direction-dependent control of balance during walking and standing. *J. Neurophysiol.* **2009**, *102*, 1411–1419. [CrossRef]

39. Aruin, A.S.; Nicholas, J.J.; Latash, M.L. Anticipatory postural adjustments during standing in below-the-knee amputees. *Clin. Biomech.* **1997**, *12*, 52–59. [CrossRef]

40. Major, M.J.; Serba, C.K.; Chen, X.; Reimold, N.; Ndubuisi-Obi, F.; Gordon, K.E. Proactive Locomotor Adjustments Are Specific to Perturbation Uncertainty in Below-Knee Prosthesis Users. *Sci. Rep.* **2018**, *8*, 1863. [CrossRef]

41. Laudani, L.; Rum, L.; Valle, M.S.; Macaluso, A.; Vannozzi, G.; Casabona, A. Age differences in anticipatory and executory mechanisms of gait initiation following unexpected balance perturbations. *Eur. J. Appl. Physiol.* **2021**, *121*, 465–478. [CrossRef]

42. Wolpert, D.M.; Flanagan, J.R. Motor prediction. *Curr. Biol.* **2001**, *11*, R729–R732. [CrossRef]

43. Valle, M.S.; Casabona, A.; Cavallaro, C.; Castorina, G.; Cioni, M. Learning Upright Standing on a Multiaxial Balance Board. *PLoS ONE* **2015**, *10*, e0142423. [CrossRef]

44. Casabona, A.; Valle, M.S.; Cavallaro, C.; Castorina, G.; Cioni, M. Selective improvements in balancing associated with offline periods of spaced training. *Sci. Rep.* **2018**, *8*, 7836. [CrossRef]

45. Saruco, E.; Guillot, A.; Saimpont, A.; Di Rienzo, F.; Durand, A.; Mercier, C.; Malouin, F.; Jackson, P. Motor imagery ability of patients with lower-limb amputation: Exploring the course of rehabilitation effects. *Eur. J. Phys. Rehabil. Med.* **2019**, *55*, 634–645. [CrossRef]

46. Abbas, R.L.; Cooreman, D.; Al Sultan, H.; El Nayal, M.; Saab, I.M.; El Khatib, A. The Effect of Adding Virtual Reality Training on Traditional Exercise Program on Balance and Gait in Unilateral, Traumatic Lower Limb Amputee. *Games Health J.* **2021**, *10*, 50–56. [CrossRef]
47. Belluscio, V.; Bergamini, E.; Tramontano, M.; Formisano, R.; Buzzi, M.G.; Vannozzi, G. Does Curved Walking Sharpen the Assessment of Gait Disorders? An Instrumented Approach Based on Wearable Inertial Sensors. *Sensors* **2020**, *20*, 5244. [CrossRef]
48. Simonetti, E.; Bergamini, E.; Vannozzi, G.; Bascou, J.; Pillet, H. Estimation of 3D Body Center of Mass Acceleration and Instantaneous Velocity from a Wearable Inertial Sensor Network in Transfemoral Amputee Gait: A Case Study. *Sensors* **2021**, *21*, 3129. [CrossRef]

*sensors*

*Article*

# Rapid Dynamic Naturalistic Monitoring of Bradykinesia in Parkinson's Disease Using a Wrist-Worn Accelerometer

Jeroen G. V. Habets [1,*], Christian Herff [1], Pieter L. Kubben [1], Mark L. Kuijf [2], Yasin Temel [1], Luc J. W. Evers [3], Bastiaan R. Bloem [3], Philip A. Starr [4], Ro'ee Gilron [4,†] and Simon Little [4,†]

[1] Department of Neurosurgery, School of Mental Health and Neuroscience, Maastricht University, 6229 ER Maastricht, The Netherlands; c.herff@maastrichtuniversity.nl (C.H.); p.kubben@maastrichtuniversity.nl (P.L.K.); y.temel@maastrichtuniversity.nl (Y.T.)

[2] Department of Neurology, School of Mental Health and Neuroscience, Maastricht University, 6229 ER Maastricht, The Netherlands; mark.kuijf@mumc.nl

[3] Department of Neurology, Center of Expertise for Parkinson & Movement Disorders, Donders Institute for Brain, Cognition and Behaviour, Radboud University Medical Center, 6525 GC Nijmegen, The Netherlands; Luc.Evers@radboudumc.nl (L.J.W.E.); bas.bloem@radboudumc.nl (B.R.B.)

[4] Department of Movement Disorders and Neuromodulation, University of California San Francisco, San Francisco, CA 94143, USA; philip.starr@ucsf.edu (P.A.S.); roeegilron@gmail.com (R.G.); Simon.Little@ucsf.edu (S.L.)

* Correspondence: j.habets@maastrichtuniversity.nl; Tel.: +31-433-876-052

† Authors contributed equally.

**Citation:** Habets, J.G.V.; Herff, C.; Kubben, P.L.; Kuijf, M.L.; Temel, Y.; Evers, L.J.W.; Bloem, B.R.; Starr, P.A.; Gilron, R.; Little, S. Rapid Dynamic Naturalistic Monitoring of Bradykinesia in Parkinson's Disease Using a Wrist-Worn Accelerometer. *Sensors* **2021**, *21*, 7876. https://doi.org/10.3390/s21237876

Academic Editors: Paolo Capodaglio and Veronica Cimolin

Received: 29 October 2021
Accepted: 22 November 2021
Published: 26 November 2021

**Publisher's Note:** MDPI stays neutral with regard to jurisdictional claims in published maps and institutional affiliations.



**Abstract:** Motor fluctuations in Parkinson's disease are characterized by unpredictability in the timing and duration of dopaminergic therapeutic benefits on symptoms, including bradykinesia and rigidity. These fluctuations significantly impair the quality of life of many Parkinson's patients. However, current clinical evaluation tools are not designed for the continuous, naturalistic (real-world) symptom monitoring needed to optimize clinical therapy to treat fluctuations. Although commercially available wearable motor monitoring, used over multiple days, can augment neurological decision making, the feasibility of rapid and dynamic detection of motor fluctuations is unclear. So far, applied wearable monitoring algorithms are trained on group data. In this study, we investigated the influence of individual model training on short timescale classification of naturalistic bradykinesia fluctuations in Parkinson's patients using a single-wrist accelerometer. As part of the Parkinson@Home study protocol, 20 Parkinson patients were recorded with bilateral wrist accelerometers for a one hour OFF medication session and a one hour ON medication session during unconstrained activities in their own homes. Kinematic metrics were extracted from the accelerometer data from the bodyside with the largest unilateral bradykinesia fluctuations across medication states. The kinematic accelerometer features were compared over the 1 h duration of recording, and medication-state classification analyses were performed on 1 min segments of data. Then, we analyzed the influence of individual versus group model training, data window length, and total number of training patients included in group model training, on classification. Statistically significant areas under the curves (AUCs) for medication induced bradykinesia fluctuation classification were seen in 85% of the Parkinson patients at the single minute timescale using the group models. Individually trained models performed at the same level as the group trained models (mean AUC both 0.70, standard deviation respectively 0.18 and 0.10) despite the small individual training dataset. AUCs of the group models improved as the length of the feature windows was increased to 300 s, and with additional training patient datasets. We were able to show that medication-induced fluctuations in bradykinesia can be classified using wrist-worn accelerometry at the time scale of a single minute. Rapid, naturalistic Parkinson motor monitoring has the clinical potential to evaluate dynamic symptomatic and therapeutic fluctuations and help tailor treatments on a fast timescale.

**Keywords:** Parkinson's disease; bradykinesia; real-life; naturalistic monitoring; wearable sensors; accelerometer; motor fluctuation

---

## 1. Introduction

Parkinson's disease (PD) is a disabling neurodegenerative disorder characterized by motor and non-motor symptoms that affect patients' motor performance and quality of life (QoL) [1–3]. Symptomatic PD management often initially focuses on dopamine replacement therapies [4]. However, half of PD patients develop wearing-off motor fluctuations during the first decade after diagnosis [5,6]. Wearing-off motor fluctuations are defined as inconsistent therapeutic benefits on symptoms such as bradykinesia and rigidity, despite regular dopaminergic delivery [6]. These motor fluctuations and other dopaminergic-related side effects can markedly impair patients' QoL [7]. Motor fluctuations are therefore a primary indication for consideration of deep brain stimulation (DBS) [1,8]. Adequate monitoring of motor fluctuations is essential for treatment evaluation, both in the presence and absence of DBS, and wearable motion sensing represents an appealing approach to support this quantification [9,10], although several challenges remain to be addressed [11,12].

Ideally, objective motor fluctuation monitoring should accurately measure and decode movement, during real world (naturalistic) activities, and be simple to implement for patients [10,13]. Currently used Parkinson's evaluation tools, such as the Movement Disorders Society Unified Parkinson Disease Rating Scale (MDS-UPDRS) and the Parkinson Disease QoL questionnaire (PDQ-39), are not designed for chronic dynamic naturalistic symptom monitoring [14,15]. They contain questionnaires which capture subjective estimates of retrospective symptoms over a week (MDS-UPDRS II and IV) or a month (PDQ-39). They are also dependent on patient recall, which is often imperfect, particularly in patients with cognitive dysfunction. Observing and scoring motor fluctuations requires trained health providers to perform single time point evaluations (MDS-UPDRS III). Motor diaries, often used as the gold standard for 24-h naturalistic monitoring, require self-reporting every 30 min [16]. This burden causes recall bias and diary fatigue over long-term use [17].

The strong clinical need for continuous symptom tracking, together with the wide availability and presence of affordable accelerometer-based sensors, has led to several academic and commercially available wearable sensor PD monitoring systems [18–22]. Motor fluctuation monitoring with commercially available devices is currently based primarily on summary metrics derived from multiple days of sensor data. Incorporation of these metrics during neurological consultation has led to promising augmentation of clinical decision making [21,23–26]. However, these sensor monitoring systems have thus far been found to be better correlated with PD clinical metrics on a time scale of days rather than hours [21,27], which is a notably longer time window than used in the original development studies [19,21,28–30]. Successful motor fluctuation classification over shorter time periods (minutes or hours) would enable dynamic therapeutic motor response monitoring. Thus, we have suggested individual model training as a methodological improvement to pursue this. To date, motor monitoring algorithms have typically been trained on group data; individual model training is suggested due to inter-subject heterogeneity of PD symptomatology [11,18,20,31]. This hypothesis is strengthened by a recent successful algorithm innovation combining short- and long-time epochs in a deep learning model correlating wrist and ankle-worn accelerometer metrics with total UPDRS III scores on 5 min epochs [32].

In the present work, we investigated the performance of machine learning classification models identifying rapid (single minute resolution), medication-induced motor fluctuations in PD patients. The classification models were trained on unconstrained naturalistic (at home) motion data derived from a unilateral wrist-worn accelerometer. Classification models based on individual data were compared with models based on group data. Further, we analyzed the influence of the number of individuals included in the group model training data, and the length of analyzed accelerometer data epochs (time window lengths), on classification results. We focused symptom decoding on bradykinesia since this cardinal feature of PD has been found to be more challenging to detect with motion sensors than tremor or dyskinesia [1,33]. This is likely due to higher distributional kinematic overlap of bradykinesia fluctuations with normal movements and normal periods of rest [18,34–36].

We hypothesized that single-minute bradykinesia classification would be achievable using machine learning and that individualized motion classification models in PD would demonstrate more reliable short-term classification of naturalistic bradykinesia fluctuations compared to group models.

## 2. Material and Methods

### 2.1. Study Sample

For our analysis, we used data from 20 participants of the Parkinson@Home validation study [37]. Detailed descriptions of the study's protocol and feasibility have been described previously [38,39]. In brief, the Parkinson@Home study recruited 25 patients diagnosed with PD by a movement disorders neurologist. All patients underwent dopaminergic replacement treatment with oral levodopa therapy, experienced wearing-off periods (MDS-UPDRS part IV item 4.3 $\geq$ 1) and had at least slight Parkinson-related gait impairments (MDS-UPDRS part II item 2.12 $\geq$ 1 and/or item 2.13 $\geq$ 1). Participants who were treated with advanced therapies (DBS or infusion therapies) or who suffered significant psychiatric or cognitive impairments which hindered completion of the study protocol were excluded.

For the current subset of PD patients, we included 20 patients who showed a levodopa-induced improvement in unilateral upper extremity bradykinesia at least at one side (equal or less than zero points). Unilateral upper extremity bradykinesia was defined as the sum of MDS-UPDRS part III items 3c, 4b, 5b, and 6b for the left side, and items 3b, 4a, 5a, and 6a for the right side. Sum scores from medication on-states were compared with sum scores from medication off-states. For each included participant, only data from the side with the largest clinical change in upper extremity bradykinesia sub items were included. Patients' activities were recorded on video and annotated as described in the validation study [37].

For our current analysis, only unilateral wrist tri-axial accelerometer data from the side with the largest fluctuation in unilateral upper extremity bradykinesia were included (Gait Up Physilog 4, Gait Up SA, CH). Recordings consisted of two sessions which took place on the same day. First, the pre-medication recording was performed in the morning after overnight withdrawal of dopaminergic medication. Second, the post-medication recording was performed when the participants experienced the full clinical effect after intake of their regular dopaminergic medication. During both recordings, participants performed an hour of unconstrained activities within and around their houses. At the start of both recordings, a formal MDS-UPDRS III and Abnormal Involuntary Movement Scale (AIMS) was conducted by a trained clinician.

### 2.2. Data Pre-Processing and Feature Extraction

Accelerometer data were sampled at 200 Hz and downsampled to a uniform sampling rate of 120 Hertz (Hz) using piecewise cubic interpolation. The effect of gravity was removed from each of the three time series (x-, y-, and z-axes) separately, by applying an l1-trend filter designed to analyze time series with an underlying piecewise linear trend [40]. Time series were low-pass filtered at 3.5 Hz to attenuate frequencies typically associated with Parkinsonian tremor in accelerometer time series [41]. In addition to the three individual accelerometer time series, we computed a composite time series containing the vector magnitude of the three individual accelerometer axes ($x^2 + y^2 + z^2$).

Multiple features previously shown to correlate with bradykinesia were extracted from the four time series (in total 103 features from four time series: x, y, z, and vector magnitude) (see extensive overview including references in Table S1). The features included characteristics from the temporal domain (such as extreme values, variances, jerkiness, number of peaks, and root mean squares), the spectral domain (such as spectral power in specific frequency ranges), and dominant frequencies. The standard window length of analysis for each extracted feature was set at 60 s, meaning one mean value per feature was extracted per time series over every 60 s of data. To explore the influence of varying window lengths (3, 10, 30, 90, 120, 150, and 300 s), separate feature sets were extracted for each sub analysis. All individual feature sets were balanced for medication-status by discarding the surplus of available data in the longest recording (pre- or post-medication). Features were standardized by calculating individual z-scores per feature. To not average out pre- and post-medication differences, the mean of only the pre-medication recordings was extracted from a value, and the result was divided by the standard deviation of only the pre-medication recordings [42]. To test the influence of an activity filter, data windows without motion activity were identified. For this, different methodologies of activity filtering are described in PD monitoring literature [9,43,44]. We applied an activity filter which classified every 60 s window with a coefficient of variation of the vector magnitude less than 0.3 as no activity and discarded them from the analysis. The choice of selected feature was based on previous work [43], and the threshold was chosen pragmatically by group-level observations of video-annotated sections identified as non-active [37]. The activity-filtered data sets were individually balanced for medication-states. For example, if a participant's data set resulted in 50 active minutes of pre-medication data, and only 45 active minutes of post-medication data, the surplus of features from 5 active minutes of pre-medication data was discarded at the end of the data set.

### 2.3. Descriptive Statistics and Analysis of Variance

The demographic and disease characteristics of the included participants are described in Table 1. Unilateral scores are provided only for the side on which accelerometer data was analyzed.

**Table 1.** Demographic and disease specific characteristics of patient population. AIMS: Abnormal Involuntary Movement Scale. MDS-UPDRS: Movement Disorders Society Unified Parkinson Disease Rating Scale. Sd: standard deviation.

| *Characteristics* | |
| --- | --- |
| Total number (% female) | 20 (60%) |
| Age (years, mean (sd)) | 63.4 (6.4) |
| Accelerometer data per medication state (minutes, mean (sd)) | 59.5 (14.3) |
| Accelerometer data per medication state, after activity filtering (minutes, mean (sd)) | 44.5 (13.9) |
| PD duration (years, mean (sd)) | 8.1 (3.5) |
| Levodopa equivalent daily dosage (milligrams, mean (sd)) | 959 (314) |
| MDS-UPDRS III pre-medication | 43.8 (11.6) |
| MDS-UPDRS III post-medication | 27.1 (9.6) |
| AIMS pre-medication | 0.5 (1.8) |
| AIMS post-medication | 3.7 (4.2) |

To test the statistical distinguishability of the pre- and post-medication recordings at the group level before using the entire dataset as an input, four main accelerometer features were chosen a priori. These four features covered the most often used domains of motion metrics applied for naturalistic bradykinesia monitoring (maximum acceleration, coefficient of variation of acceleration over time, root mean square of acceleration over time, and the total spectral power below 4 Hz) [18,43] and were extracted from the vector magnitude time series. Individual averages of each of the four features over the entire dataset (~60 min per condition) were analyzed for statistically significant differences between the medication states with a multivariate analysis of variance (M-ANOVA). Post-hoc repeated measures ANOVA were performed to explore which feature(s) contributed to the pre- versus post-medication difference. An alpha-level of 0.05 was implemented and multiple comparison correction was performed using the false discovery rate (FDR) method described by Benjamini and Hochberg [45].

*2.4. Classification of Medication States*

2.4.1. Individually Trained and Group Trained Models

Supervised classification analyses were performed to test whether differentiation between short-term pre- and post-medication was feasible, based on 60 s accelerometer features (Figure 1). First, this was tested using the four previously mentioned features extracted from the vector magnitude signal. Afterward, the feature set was expanded to include all described features, as well as the x, y, z time series (Table S1). Analyses were performed using a support vector machine (SV) and a random forest (RF) classifier. Classification models trained on individual data and models trained on group data were then compared (Figure 1B).

For individually trained models, 80% of a participant's total balanced data was used as training data, and 20% as test data (Figure S1). Small blocks (2%) of training data which neighbored the test data were discarded (Figure S1) to decrease the temporal dependence between training and test data. To prevent bias caused by the selected block of test data, a 41-fold cross-validation was performed. Each fold (out of 41) included two continual blocks of 10% of total data, one block from the pre- and one block from the post-medication recording as test data (percentiles 0 to 10 and 50–60, percentiles 1 to 11 and 51 to 61, ... , and percentiles 40 to 50 and 90 to 100, see visualization in Figure S2).

For group-trained models, a leave one out cross-validation was performed. For every participant, a model was trained based on all data (balanced for medication status) from the remaining 19 participants and tested on all data (balanced for medication status) of the specific participant (Figure 1B). To assess all models, the area under the receiver operator curve (AUC) and the classification accuracy were calculated as predictive metrics. For the individual models, individual classification outcomes were averaged over the 41 folds. All parameters and details related to the implemented classifiers are available through the GitHub codebase [46].

To test the statistical significance of each individual and group model performance, 5000 permutation tests were performed in which medication state labels were shuffled. The 95th percentile of permutation scores was taken as significance threshold (alpha = 0.05), and FDR multiple comparison corrections were performed [45].

2.4.2. The Influence of Training Data Size, and Feature Window Lengths

To test the impact of the size of the training set on the group models, the training phases were repeated with varying numbers of participants included in the training data. As in the original group model analysis, the test data consisted of all data from one participant. The number of training data participants varied between 1 and 19. To prevent selection bias in the selection of the training participants, the analyses were repeated five times per number of included training participants, with different random selections of training participants. Individual classification models were excluded from this analysis by definition.

To analyze the influence of feature window lengths, we repeated the group model analysis with features extracted from data windows of 3, 10, 30, 90, 120, 150, and 300 s duration. For every analysis, one participant was selected as a test participant, and the other 19 were training participants. This was repeated for all participants and the averages over 20 test participants were reported. This was performed at the group level modelling only, as individual models were limited by total available data size.

2.4.3. Comparing Two Models' Predictive Performance

Equality plots were drawn to compare the AUC scores and accuracies between two models, for example a model using 4 features versus 103 features, a model using an SV classifier versus an RF classifier, a model with versus without activity filtering (Figure S3). All comparisons were performed separately for the individual and group models. For example, model A led to a higher AUC score than model B in 14 out of 20 participants (14 dots above the equality line). Permutation tests plotted 20 random dots on an equality plot and tested whether the permuted distribution generated 14 or more dots (out of 20) above the equality line. This was repeated 5000 times, and the probability that the distribution 14 out of 20 was the result of chance was determined.

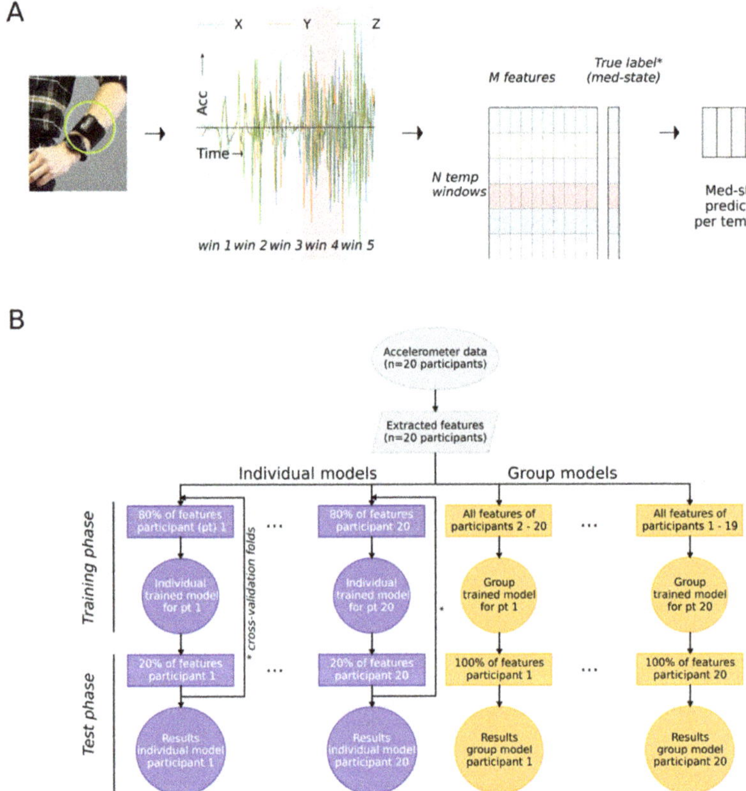



**Figure 1.** Accelerometer-based Parkinsonian motor fluctuation detection workflows. (**A**) Left: A wrist-worn motion sensor (Physilog 4, Gait Up SA, CH; green circled) is used to collect unilateral tri-axial accelerometer data. X, Y, and Z represent acceleration (meters/second per second) in three axes over time (seconds). Temporal windows are determined for data analysis and are indicated in different colors over time (win1, win2, ... ). Center: Signal preprocessing and feature extraction convert the raw tri-axial signal into a dataset containing M features (Table S1), calculated for every temporal window (in total M columns and N rows). For the training phase of the machine learning classification models, the true labels representing medication states (*) are used. Right: In the testing phase, inserting the feature set (M × N) in the trained classification model leads to N medication state predictions. (**B**) Workflow to train and test individual and group models. Identical features are extracted from the raw accelerometer data of the twenty included participants (grey symbols). For the individually trained models (blue), the features from 80% of a participant's epochs are used in the *training phase* (y-axis). The trained individual model is tested with the remaining, unused, 20% of epochs during the *test phase*. The arrows (*) from test phase to training phase represent the multiple cross-validation folds applied to train and test the individual models on different selections of training and test data. For the group models (yellow), each participant is tested in turn, with data from the other 19 participants used in the *training phase*.

### 2.4.4. Predictive Performance and Clinical Assessed Symptom Fluctuations

The influence of clinical bradykinesia, tremor, and abnormal involuntary movement fluctuations on predictive performance was tested at a group level by Spearman R correlations between the fluctuation in individual bradykinesia and tremor sub scores and AIMS scores, and the predictive performance (Table S3). Individual participants were visualized according to descending tremor and AIMS fluctuation ratings to enable visual comparison of predictive performance with and without co-occurring tremor and abnormal involuntary movement fluctuation (respectively Figure S4A,B). The tremor scores consisted of the MDS-UPDRS III items representing unilateral upper extremity tremor (items 15b, 16b, and 17b for the left side, and items 15a, 16a, and 17a for the right side).

### 2.4.5. Software

Raw acceleration time series were down sampled and filtered (for gravity effects) in Matlab. All further preprocessing, feature extraction, and analysis was performed in Jupyter Notebook (Python 3.7). The code used to extract features and analyze data is made publically available [46].

### 2.4.6. Code and Data Availability

The code used to extract features and analyze data is publicly available [46]. The de-identified open source dataset will be made available to the scientific community by the Michael J. Fox Foundation.

## 3. Results

### 3.1. Study Population and Recorded Data

Twenty PD patients from the Parkinson@Home data repository [37] were included in this study. We excluded three participants who did not show a levodopa-induced improvement in unilateral upper extremity bradykinesia and two participants were further excluded because there was less than 40 min of accelerometer data available from their pre- or post-medication recording, resulting in a dataset of 20 patients.

Demographic and disease-specific characteristics are presented in Table 1. In total, 3138 min of accelerometer data were recorded in the 20 included patients. After balancing the data sets for medication status, 2380 min of accelerometer data were included. On average, 59.5 ($\pm$14.3) min of accelerometer data from both pre- and post-medication recordings were included per participant. On average, 44.5 min ($\pm$13.9 min) of features were included after applying the activity filter and balancing the individual data to include equal individual features per medication state.

We extracted multiple features which are described in the current literature to index naturalistic bradykinesia with a wrist accelerometer (see Table S1 for details and references). In total, 103 motion accelerometer features were extracted for every feature window, including both time domain and spectral features from the accelerometer.

### 3.2. Group Level Statistical Analysis of Cardinal Motion Features across Medication States

At the group level, the pre- and post-medication recordings differed significantly regarding the individual means of the four main motion features (maximum acceleration magnitude, coefficient of variation of acceleration magnitude, root mean square of acceleration, and spectral power (below 4 Hz) [18,43]) (MANOVA, Wilk's lamba = 0.389, F-value = 14.2, $p < 0.001$). Post-hoc repeated measures ANOVAs demonstrated that only the individual coefficient of variation averages significantly differed between pre- and post-medication states ($p = 0.0042$) (Figure 2).

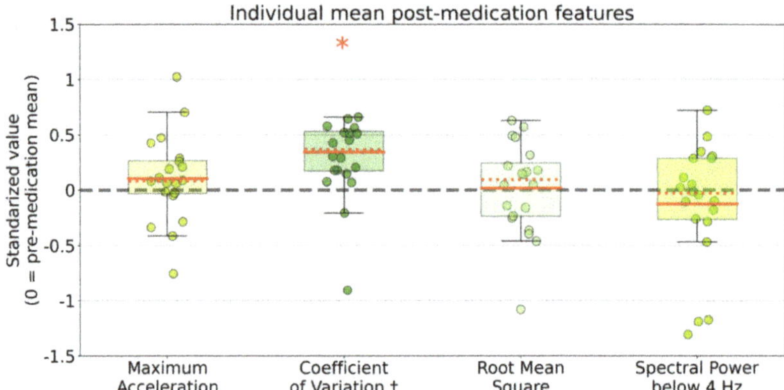


**Figure 2.** Distributions of individual means for four main movement features. Colored dots represent the mean feature values during the whole post-medication recording per participant ($n = 20$). Individual post-medication mean values are standardized as z-scores (individual pre-medication recordings are used as references, and therefore the pre-medication mean values equal 0). The red asterisk indicates a significant difference on group level between mean coefficient of variations of pre- and post-medication means (alpha = 0.05, MANOVA and post-hoc analysis, FDR corrected). † = one positive outlier (1.7) not visualized.

### 3.3. Machine Learning Classification of Short Window Data Epochs

Next, the classification performance over short time windows (one-minute resolution) was tested using support vector (SV) and random forest (RF) machine learning models. All medication state classifications using the four previously selected motion features led to low AUC scores (means per model ranged between 0.49 and 0.64) and low accuracies (means per model ranged between 49% and 60%) (see Table S2 for detailed results).

We therefore repeated the classification analysis, using an expanded kinematic feature set (103 features). With the full feature set, notably higher AUC scores and classification accuracies were seen for all individual and group (SV and RF) models (Table S2 and Figures S3A,B). Mean AUC scores per model ranged between 0.65 and 0.70, and mean accuracies per model ranged between 60% and 65% (Table S2). Most participants yielded AUC scores and accuracies significantly better than chance level (17 out of 20 participants per model), tested through random surrogate dataset generation.

In 90% of participants (18 out of 20), either the best individual or group model classified medication states per 60 s significantly better than chance level based on our surrogate datasets (Figure 3 and Table S2). Group trained models resulted in AUC scores statistically significantly higher than random classification in 17 participants. Individually trained models resulted in AUC scores statistically significantly higher than random classification in 13 participants. Both individual and group models resulted in mean AUC scores of 0.70 (±respectively 0.18 and 0.10), and mean accuracies of respectively 65% (±0.14) and 64% (±0.08) over all 20 participants (Figure 3, Table S2). Notably, the individual models resulted in a larger standard deviation of AUC scores, including several AUC scores higher than 0.9 as well as below chance level (Figure 3).

Overall, these findings confirmed the feasibility of rapid naturalistic bradykinesia classification based on wrist-worn accelerometer metrics. Individual model training resulted in a similar mean AUC with a wider standard deviation compared to group model training.

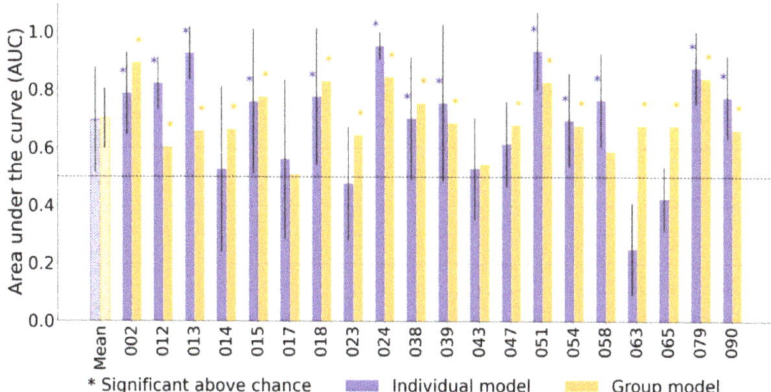



**Figure 3.** Classification of medication induced motor fluctuations on short accelerometer time windows in individual participants. The first pair of bars represents the mean area under the curve (AUC) score over the twenty participants. Each subsequent pair of bars (002 to 090) represents the AUC scores from one participant. The blue bars represent the AUC score for the individual model, and the yellow bars represent the group model. Note that for the individual models, AUC scores are the averages over the multiple cross-validation folds within a participant (Figure 1B). The asterisks indicate whether the corresponding AUC score was significantly better than chance level (5000-repetitions permutation test). Both models have equal mean AUC scores. It is notable that the majority (18 out of 20 of participants) has at least one significant score. Half of the participants yielded a higher AUC score with the individual model than with the group model.

### 3.4. Classification of Bradykinesia-Centred Motor Fluctuations Versus Co-Occurring Symptoms

We found no significant correlations between the individual classification performance and the individual clinically scored fluctuations in bradykinesia, tremor, and AIMS (see Table S3, all $p$-values larger than 0.1). At an individual level, we found significant AUC scores in participants with (13, 24, and 79) and without (39, 51, and 58) tremor fluctuations (Figure S4A). Similarly, we found significant AUC scores in participants with (2, 15, 51, and 79) and without (39, 18, 24, and 90) AIMS fluctuations (Figure S4B).

Individual predictive performance was found not to be proportional to the size of tremor or AIMS fluctuations, suggesting the feasibility of using the applied metrics for PD patients with and without tremor and abnormal involuntary movements. Meanwhile, the severity of bradykinesia did not influence the classification performance, suggesting feasibility of the metrics for patients with even mild-to-moderate bradykinesia fluctuations.

### 3.5. Influence of Training Data Size and Feature Window Length

We found an increase in the predictive performance (AUC) of the group models as the number of patient datasets used during model training was increased (Figure 4A). Above 15 participants, the increase in mean AUC levelled off toward the 19 included participants.

Next, we wanted to investigate the impact of the accelerometer data feature window length on the predictive performance of the group models. Increasing the length of the feature windows up to 300 s improved the mean AUC (Figure 4B). Due to data size limitations, the feature windows were not expanded further than 300 s. These analyses could not be reproduced for the individual models due to data size limitations.

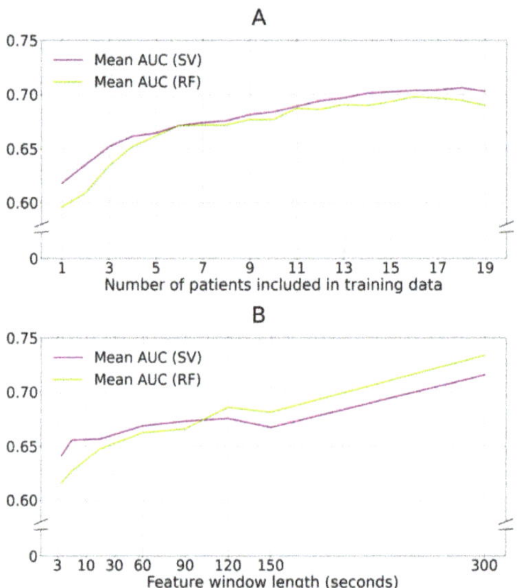

**Figure 4.** Increasing number of training patients and length of data window duration improves classification performance. (**A**): Group models are trained for every patient with a varying number of included training data, (*x*-axis). On the *y*-axis, the AUC is shown for both SV and RF models (both included activity filtering). An increase in AUC is seen for SV and RF models parallel to an increase in included training patients. (**B**): Group models are trained for every patient with various feature window lengths (*x*-axis). On the *y*-axis, the AUC of the SV and RF models are visualized. Due to the longer feature window lengths, the activity filter in this sub-analysis is not applied to data size limitations. Larger window lengths up to 300 s increase classification performance, while smaller window lengths decrease classification performance. AUC: area under the receiver operator characteristic; SV: support vector classifier; RF: random forest classifier.

## 4. Discussion

Our results demonstrated successful classification of naturalistic bradykinesia fluctuations using wrist accelerometer data on different timescales using conventional statistical approaches (over one-hour epochs) and machine learning classification (over one-minute epochs). We found that the coefficient of variation of the accelerometer amplitude was significantly increased following dopaminergic medication when a full 60 min of data was analyzed per medication condition. At shorter timescales (60 s), this feature (complemented with 3 other accelerometer metrics) was not strongly predictive of medication state using machine learning. However, using a larger number of motion metrics (103), statistically significant classification of medication states could be achieved in 90% of participants (18 out of 20) using either group- or individually-trained models (Figure 3). Individual and group models both resulted in a mean AUC of 0.70 on the 60 s epochs, where the individual models' AUC scores had a larger standard deviation (Figure 3, Table S2). Expansion of the data epoch length (from 60 to 300 s), as well as inclusion of more training participants, improved AUC scores in the group models. Limited individual data sizes withheld us from testing individual models with expanded data epochs and may explain the larger standard deviation for individual model AUCs (Figures 1B and S1).

These results represent the first demonstration of classification of Parkinsonian bradykinesia fluctuations using individually-trained models for single-wrist accelerometer data on a rapid timescale. Although we showed statistically significant classification over short time windows, we did not find added value yielded by individual model training based on our current results. Reproduction with longer accelerometer recordings for individuals is, however, likely to improve classification results further.

The significant difference of only the mean individual coefficient of variance between pre- and post-medication recordings (Figure 2) may be explained by a suggested larger discriminative potential of naturalistic sensor features describing a distribution rather than describing extreme values or sum scores [11].

In general, the presented classification models were notable due to the unconstrained naturalistic (real-world) data collection environment and short time scale of classification. Operating at this shorter timescale, the models showed good classification performance compared with benchmark naturalistic medication-state detection models (Figure 3) [34–36]. Although better classification performances have previously been described with models using data over longer timescales or from more constrained recording scenarios, these methodologies improved classification results at the cost of less naturalistic generalizability [18,22,27,28,43]. Adding a second motion sensor would likely increase performance at the cost of user-friendliness and feasibility [32]. Systems using wrist, ankle and/or axial motion sensors have the theoretical advantage of being more sensitive for arm versus leg- and gait-centered symptomatology.

### 4.1. Clinical Relevance and Methodological Challenges of Naturalistic and Rapid PD Motor Monitoring

Wearable accelerometer-based PD monitoring systems have been developed to augment therapeutic decision making, [20,21,23] and to augment clinical assessments in pharmacological trials [11,47]. Previous systems have been validated over the course of days. However, other suggested clinical state tracking applications would require short time scale feedback [22], including fine-grained cycle-by-cycle medication adjustments and conventional [48,49] or adaptive [50–53] deep brain stimulation programming. Moreover, the significant predictive performance in patients with and without both tremor and dyskinesia (AIMS) fluctuations underlines the potential of dynamic naturalistic monitoring for a wide spectrum of PD patients (Figure S4A,B). The latter was important to investigate despite the filtering out of typical tremor bandwidths, since solely band pass filtering cannot rule out any influence of tremor dynamics.

Notably, we observed marked differences in classification accuracy using either 4 accelerometer features or 103 features (Figures 3 and S3). This suggests that bradykinesia classification on shorter timescales, requires rich feature sets. The significance testing with surrogate datasets aimed to rule out any resulting overfitting. However, a thorough comparison of feature sets is often complicated by proprietary algorithms or the lack of open-source code [12]. This underlines the importance of transparent, open source, and reproducible movement metric feature sets for naturalistic PD monitoring [54,55].

Another methodological challenge for rapid, objective, naturalistic short-term PD monitoring is the lack of a high-quality labelling of data on the same time scale. PD clinical assessment tools, currently applied as gold standards, are limited in their applicability for rapid time scales. Multiple longitudinal time windows of the dynamic accelerometer time series are labelled with a single clinical score, which weakens model training and evaluation. In effect, sensor-based outcomes are often aggregated to match clinical evaluation metrics and time scales, which might account for the current upper limit in wearable classification performance [21,24,29]. PD-specific eDiaries [56–59] labelled video-recordings on fine time scales [60], and other virtual telemedicine concepts [61] may contribute to this challenge.

### 4.2. Future Scientific Opportunities to Improve Naturalistic PD Monitoring Development

We predict that the coming expansion of real-world motion data sets, containing long-term data over weeks to years in patients with PD, will support optimization of individually-trained models [62]. These larger datasets will also allow the exploration of alternative, more data-dependent, computational analyses, such as deep neural network classification and learning [35,63]. Moreover, unsupervised machine learning models could also be explored to overcome the issues of lacking temporally matching gold standard for model training and evaluation by surpassing the need for long-term, repetitive, true labels [11,64]. The observed discriminative potential of the coefficient of variation (Figure 2) might be of value in post hoc differentiation of clusters in unsupervised machine learning models.

Additionally, open-source research initiatives should catalyze the development of naturalistic PD monitor models which are not dependent on proprietary software [10,43]. The Mobilize-D consortium, for example, introduced a roadmap to standardize and structure naturalistic PD monitoring by creating specific "unified digital mobility outcomes" [54,55]. During the development of these outcomes, features describing distribution ranges and extreme values—rather than means or medians—should be considered [11]. Parallel to open-source initiatives, other creative collaborations between industry and academia, such as data challenges, might offer valuable (interdisciplinary) cross-fertilization [65]. Furthermore, adding more limb sensors to improve naturalistic PD monitoring is controversial. Although there is evidence supporting the combined use of wrist, ankle [32,66], or insoles [67] sensory tracking, other reports have not shown improved performance and instead described additional burden to the patient [35,68,69].

### 4.3. Limitations

Our study was limited by the individual data set sizes, which restricted inferences that could be made regarding models trained with individual versus group data. Additionally, the unconstrained character of the pre- and post-medication recordings led to an imbalance in terms of captured activities during the two medication states. The applied activity-filter addressed this limitation partly but does not rule out imbalance in exact activities. This imbalance compromises pattern recognition based data analysis [35], but is also inherent to naturalistic PD monitoring [18]. Exploring the boundaries of this limitation is essential for future PD monitor applications. Replication of our methodologies in larger data sets, and inclusion of validated activity classifiers may contribute to overcoming this limitation. Future studies should also aim to detect symptom states beyond a binary differentiation between on- versus off-medication.

## 5. Conclusions

We demonstrated that classification of naturalistic bradykinesia fluctuations at the single-minute time scale was feasible with machine learning models trained on both individual and group data in PD patients using a single wrist-worn accelerometer. At longer timescales (i.e., an hour), a single accelerometer feature, the coefficient of variation, was predictive of bradykinesia at the group level. Extension of short accelerometer time epochs and an increased number of training patients improved classification of group-trained models. Rapid, dynamic monitoring has the potential to support personalized and precise therapeutic optimization with medication and stimulation therapies in Parkinson's patients.

**Supplementary Materials:** The following are available online at https://www.mdpi.com/article/10.3390/s21237876/s1, Table S1: Overview of extracted features. Figure S1: Schematic visualization of data splitting method for individual models. Figure S2: Visualization of activity filter performance versus the parallel raw signal vector magnitude. Table S2: Predictive metrics for all models and approaches. Figure S3: Comparison of different model approaches for short window medication states classification. Figure S4: Classification performance in patients with and without tremor and abnormal involuntary movements. Table S3: Spearman R correlations between symptom fluctuation and predictive performance at an individual level.

**Author Contributions:** Study conception and design: J.G.V.H., S.L., R.G. and P.A.S.; Data acquisition: B.R.B. and L.J.W.E.; Data analysis: J.G.V.H., R.G. and S.L.; Writing manuscript: J.G.V.H., S.L. and R.G.; Critical revision of analysis and manuscript: J.G.V.H., C.H., P.L.K., L.J.W.E., M.L.K., Y.T., B.R.B., P.A.S., R.G. and S.L.; All authors have read and agreed to the published version of the manuscript.

**Funding:** J.G.V.H. received funding from the Dutch Health Research Funding Body ZonMW (PTO2 grant nr. 446001063) and the Dutch Parkinson Foundation for this work. J.G.V.H., P.L.K. and

Y.T. received funding from the Stichting Weijerhorst. S.L. is supported by the Weill Investigators Award Program.

**Institutional Review Board Statement:** The study protocol was approved by the local medical ethics committee (Commissie Mensgebonden Onderzoek, region Arnhem-Nijmegen, file number 2016-1776). All participants received verbal and written information about the study protocol and signed a consent form prior to participation, in line with the Declaration of Helsinki.

**Informed Consent Statement:** All patients gave written Informed Consent according to the above described medical ethical protocol (Commissie Mensgebonden Onderzoek, region Arnhem-Nijmegen, file number 2016-1776).

**Data Availability Statement:** The de-identified open source dataset will be made available to the scientific community by the Michael J. Fox Foundation.

**Acknowledgments:** The authors want to thank Robert Wilt for his organizational support during this project. This research was supported by personal travel grants awarded to J.G.V.H. by the Dutch Science Funding Body ZonMW (Translational Research 446001063) and the Dutch Parkinson FOUndation, by the Weijerhorst Foundation grant awarded to Y.T. and P.L.K., and by the Weill Investigators Award Program supporting S.J.L.

**Conflicts of Interest:** None of the authors have a conflict of interest. J.G.V.H. received funding from the Dutch health care research organization ZonMW (Translational Research 2017–2024 grant nr. 446001063). J.G.V.H., P.T. and Y.T. received funding from the Stichting Weijerhorst. C.H. received a VENI-funding from the Dutch Research Council (WMO). S.L. and research reported in this publication was supported by the National Institute Of Neurological Disorders And Stroke of the National Institutes of Health under Award Number K23NS120037. The content is solely the responsibility of the authors and does not necessarily represent the official views of the National Institutes of Health.

# References

1. Bloem, B.R.; Okun, M.S.; Klein, C. Parkinson's Disease. *Lancet Lond. Engl.* **2021**, *397*, 2284–2303. [CrossRef]
2. Jankovic, J.; Tan, E.K. Parkinson's Disease: Etiopathogenesis and Treatment. *J. Neurol. Neurosurg. Psychiatry* **2020**, *91*, 795–808. [CrossRef] [PubMed]
3. Kuhlman, G.D.; Flanigan, J.L.; Sperling, S.A.; Barrett, M.J. Predictors of Health-Related Quality of Life in Parkinson's Disease. *Park. Relat. Disord.* **2019**, *65*, 86–90. [CrossRef] [PubMed]
4. De Bie, R.M.A.; Clarke, C.E.; Espay, A.J.; Fox, S.H.; Lang, A.E. Initiation of Pharmacological Therapy in Parkinson's Disease: When, Why, and How. *Lancet Neurol.* **2020**, *19*, 452–461. [CrossRef]
5. Fasano, A.; Fung, V.S.C.; Lopiano, L.; Elibol, B.; Smolentseva, I.G.; Seppi, K.; Takáts, A.; Onuk, K.; Parra, J.C.; Bergmann, L.; et al. Characterizing Advanced Parkinson's Disease: OBSERVE-PD Observational Study Results of 2615 Patients. *BMC Neurol.* **2019**, *19*, 50. [CrossRef] [PubMed]
6. Kim, H.-J.; Mason, S.; Foltynie, T.; Winder-Rhodes, S.; Barker, R.A.; Williams-Gray, C.H. Motor Complications in Parkinson's Disease: 13-Year Follow-up of the CamPaIGN Cohort. *Mov. Disord.* **2020**, *35*, 185–190. [CrossRef] [PubMed]
7. Hechtner, M.C.; Vogt, T.; Zöllner, Y.; Schröder, S.; Sauer, J.B.; Binder, H.; Singer, S.; Mikolajczyk, R. Quality of Life in Parkinson's Disease Patients with Motor Fluctuations and Dyskinesias in Five European Countries. *Park. Relat. Disord.* **2014**, *20*, 969–974. [CrossRef]
8. Fox, S.H.; Katzenschlager, R.; Lim, S.Y.; Barton, B.; de Bie, R.M.A.; Seppi, K.; Coelho, M.; Sampaio, C. International Parkinson and Movement Disorder Society Evidence-Based Medicine Review: Update on Treatments for the Motor Symptoms of Parkinson's Disease. *Mov. Disord.* **2018**, *33*, 1248–1266. [CrossRef]
9. Van Hilten, J.J.; Middelkoop, H.A.; Kerkhof, G.A.; Roos, R.A. A New Approach in the Assessment of Motor Activity in Parkinson's Disease. *J. Neurol. Neurosurg. Psychiatry* **1991**, *54*, 976–979. [CrossRef]
10. Espay, A.J.; Hausdorff, J.M.; Sanchez-Ferro, A.; Klucken, J.; Merola, A.; Bonato, P.; Paul, S.S.; Horak, F.B.; Vizcarra, J.A.; Mestre, T.A.; et al. A Roadmap for Implementation of Patient-Centered Digital Outcome Measures in Parkinson's Disease Obtained Using Mobile Health Technologies. *Mov. Disord.* **2019**, *34*, 657–663. [CrossRef]
11. Warmerdam, E.; Hausdorff, J.M.; Atrsaei, A.; Zhou, Y.; Mirelman, A.; Aminian, K.; Espay, A.J.; Hansen, C.; Evers, L.J.; Keller, A. Long-Term Unsupervised Mobility Assessment in Movement Disorders. *Lancet Neurol.* **2020**, *19*, 462–470. [CrossRef]
12. Fasano, A.; Mancini, M. Wearable-Based Mobility Monitoring: The Long Road Ahead. *Lancet Neurol.* **2020**, *19*, 378–379. [CrossRef]
13. Odin, P.; Chaudhuri, K.R.; Volkmann, J.; Antonini, A.; Storch, A.; Dietrichs, E.; Pirtosek, Z.; Henriksen, T.; Horne, M.; Devos, D.; et al. Viewpoint and Practical Recommendations from a Movement Disorder Specialist Panel on Objective Measurement in the Clinical Management of Parkinson's Disease. *NPJ Park. Dis.* **2018**, *4*, 14. [CrossRef]

14. Goetz, C.G.; Tilley, B.C.; Shaftman, S.R.; Stebbins, G.T.; Fahn, S.; Martinez-Martin, P.; Poewe, W.; Sampaio, C.; Stern, M.B.; Dodel, R.; et al. Movement Disorder Society-Sponsored Revision of the Unified Parkinson's Disease Rating Scale (MDS-UPDRS): Scale Presentation and Clinimetric Testing Results. *Mov. Disord.* **2008**, *23*, 2129–2170. [CrossRef]

15. Hagell, P.; Nygren, C. The 39 Item Parkinson's Disease Questionnaire (PDQ-39) Revisited: Implications for Evidence Based Medicine. *J. Neurol. Neurosurg. Psychiatry* **2007**, *78*, 1191–1198. [CrossRef]

16. Hauser, R.A.; McDermott, M.P.; Messing, S. Factors Associated with the Development of Motor Fluctuations and Dyskinesias in Parkinson Disease. *Arch. Neurol.* **2006**, *63*, 1756–1760. [CrossRef]

17. Papapetropoulos, S.S. Patient Diaries as a Clinical Endpoint in Parkinson's Disease Clinical Trials. *CNS Neurosci. Ther.* **2012**, *18*, 380–387. [CrossRef] [PubMed]

18. Thorp, J.E.; Adamczyk, P.G.; Ploeg, H.L.; Pickett, K.A. Monitoring Motor Symptoms During Activities of Daily Living in Individuals With Parkinson's Disease. *Front Neurol.* **2018**, *9*, 1036. [CrossRef] [PubMed]

19. Horne, M.K.; McGregor, S.; Bergquist, F. An Objective Fluctuation Score for Parkinson's Disease. *PLoS ONE* **2015**, *10*, e0124522. [CrossRef] [PubMed]

20. Sama, A.; Perez-Lopez, C.; Rodriguez-Martin, D.; Catala, A.; Moreno-Arostegui, J.M.; Cabestany, J.; de Mingo, E.; Rodriguez-Molinero, A. Estimating Bradykinesia Severity in Parkinson's Disease by Analysing Gait through a Waist-Worn Sensor. *Comput. Biol. Med.* **2017**, *84*, 114–123. [CrossRef]

21. Powers, R.; Etezadi-Amoli, M.; Arnold, E.M.; Kianian, S.; Mance, I.; Gibiansky, M.; Trietsch, D.; Alvarado, A.S.; Kretlow, J.D.; Herrington, T.M.; et al. Smartwatch Inertial Sensors Continuously Monitor Real-World Motor Fluctuations in Parkinson's Disease. *Sci. Transl. Med.* **2021**, *13*, eabd7865. [CrossRef] [PubMed]

22. Sica, M.; Tedesco, S.; Crowe, C.; Kenny, L.; Moore, K.; Timmons, S.; Barton, J.; O'Flynn, B.; Komaris, D.-S. Continuous Home Monitoring of Parkinson's Disease Using Inertial Sensors: A Systematic Review. *PLoS ONE* **2021**, *16*, e0246528. [CrossRef] [PubMed]

23. Pahwa, R.; Isaacson, S.H.; Torres-Russotto, D.; Nahab, F.B.; Lynch, P.M.; Kotschet, K.E. Role of the Personal KinetiGraph in the Routine Clinical Assessment of Parkinson's Disease: Recommendations from an Expert Panel. *Expert Rev. Neurother.* **2018**, *18*, 669–680. [CrossRef]

24. Santiago, A.; Langston, J.W.; Gandhy, R.; Dhall, R.; Brillman, S.; Rees, L.; Barlow, C. Qualitative Evaluation of the Personal KinetiGraph TM Movement Recording System in a Parkinson's Clinic. *J. Park. Dis.* **2019**, *9*, 207–219. [CrossRef]

25. Joshi, R.; Bronstein, J.M.; Keener, A.; Alcazar, J.; Yang, D.D.; Joshi, M.; Hermanowicz, N. PKG Movement Recording System Use Shows Promise in Routine Clinical Care of Patients with Parkinson's Disease. *Front. Neurol.* **2019**, *10*, 1027. [CrossRef]

26. Farzanehfar, P.; Woodrow, H.; Horne, M. Assessment of Wearing Off in Parkinson's Disease Using Objective Measurement. *J. Neurol.* **2020**, *268*, 914–922. [CrossRef]

27. Ossig, C.; Gandor, F.; Fauser, M.; Bosredon, C.; Churilov, L.; Reichmann, H.; Horne, M.K.; Ebersbach, G.; Storch, A. Correlation of Quantitative Motor State Assessment Using a Kinetograph and Patient Diaries in Advanced PD: Data from an Observational Study. *PLoS ONE* **2016**, *11*, e0161559. [CrossRef] [PubMed]

28. Rodriguez-Molinero, A.; Sama, A.; Perez-Martinez, D.A.; Perez Lopez, C.; Romagosa, J.; Bayes, A.; Sanz, P.; Calopa, M.; Galvez-Barron, C.; de Mingo, E.; et al. Validation of a Portable Device for Mapping Motor and Gait Disturbances in Parkinson's Disease. *JMIR Mhealth Uhealth* **2015**, *3*, e9. [CrossRef]

29. Rodriguez-Molinero, A. Monitoring of Mobility of Parkinson's Patients for Therapeutic Purposes—Clinical Trial (MoMoPa-EC). Available online: https://clinicialtrials.gov/ct2/show/NCT04176302 (accessed on 3 May 2021).

30. Great Lake Technologies Kinesia 360 Parkinson's Monitoring Study 2018. Available online: https://clinicialtrials.gov/ct2/show/NCT02657655 (accessed on 3 May 2021).

31. Van Halteren, A.D.; Munneke, M.; Smit, E.; Thomas, S.; Bloem, B.R.; Darweesh, S.K.L. Personalized Care Management for Persons with Parkinson's Disease. *J. Park. Dis.* **2020**, *10*, S11–S20. [CrossRef]

32. Hssayeni, M.D.; Jimenez-Shahed, J.; Burack, M.A.; Ghoraani, B. Ensemble Deep Model for Continuous Estimation of Unified Parkinson's Disease Rating Scale III. *Biomed. Eng. Online* **2021**, *20*, 32. [CrossRef]

33. Clarke, C.E.; Patel, S.; Ives, N.; Rick, C.E.; Woolley, R.; Wheatley, K.; Walker, M.F.; Zhu, S.; Kandiyali, R.; Yao, G.; et al. *UK Parkinson's Disease Society Brain Bank Diagnostic Criteria*; NIHR Journals Library; 2016. Available online: www.ncbi.nlm.nih.gov/books/NBK37 (accessed on 3 May 2021).

34. Shawen, N.; O'Brien, M.K.; Venkatesan, S.; Lonini, L.; Simuni, T.; Hamilton, J.L.; Ghaffari, R.; Rogers, J.A.; Jayaraman, A. Role of Data Measurement Characteristics in the Accurate Detection of Parkinson's Disease Symptoms Using Wearable Sensors. *J. NeuroEng. Rehabil.* **2020**, *17*, 52. [CrossRef]

35. Lonini, L.; Dai, A.; Shawen, N.; Simuni, T.; Poon, C.; Shimanovich, L.; Daeschler, M.; Ghaffari, R.; Rogers, J.A.; Jayaraman, A. Wearable Sensors for Parkinson's Disease: Which Data Are Worth Collecting for Training Symptom Detection Models. *Npj Digit. Med.* **2018**, *1*, 64. [CrossRef]

36. Fisher, J.M.; Hammerla, N.Y.; Ploetz, T.; Andras, P.; Rochester, L.; Walker, R.W. Unsupervised Home Monitoring of Parkinson's Disease Motor Symptoms Using Body-Worn Accelerometers. *Park. Relat. Disord.* **2016**, *33*, 44–50. [CrossRef]

37. Evers, L.J.; Raykov, Y.P.; Krijthe, J.H.; Silva de Lima, A.L.; Badawy, R.; Claes, K.; Heskes, T.M.; Little, M.A.; Meinders, M.J.; Bloem, B.R. Real-Life Gait Performance as a Digital Biomarker for Motor Fluctuations: The Parkinson@Home Validation Study. *J. Med. Internet Res.* **2020**, *22*, e19068. [CrossRef] [PubMed]

38. De Lima, A.L.S.; Hahn, T.; de Vries, N.M.; Cohen, E.; Bataille, L.; Little, M.A.; Baldus, H.; Bloem, B.R.; Faber, M.J. Large-Scale Wearable Sensor Deployment in Parkinson's Patients: The Parkinson@Home Study Protocol. *JMIR Res. Protoc.* **2016**, *5*, e5990. [CrossRef]

39. De Lima, A.L.S.; Hahn, T.; Evers, L.J.W.; de Vries, N.M.; Cohen, E.; Afek, M.; Bataille, L.; Daeschler, M.; Claes, K.; Boroojerdi, B.; et al. Feasibility of Large-Scale Deployment of Multiple Wearable Sensors in Parkinson's Disease. *PLoS ONE* **2017**, *12*, e0189161. [CrossRef]

40. Kim, S.J.; Koh, K.; Boyd, S.; Gorinevsky, D.L. Trend Filtering. *SIAM Rev.* **2009**, *51*, 339–360. [CrossRef]

41. Van Brummelen, E.M.J.; Ziagkos, D.; de Boon, W.M.I.; Hart, E.P.; Doll, R.J.; Huttunen, T.; Kolehmainen, P.; Groeneveld, G.J. Quantification of Tremor Using Consumer Product Accelerometry Is Feasible in Patients with Essential Tremor and Parkinson's Disease: A Comparative Study. *J. Clin. Mov. Disord.* **2020**, *7*, 4. [CrossRef]

42. Pedregosa, F.; Varoquaux, G.; Gramfort, A.; Michel, V.; Thirion, B.; Grisel, O.; Blondel, M.; Prettenhofer, P.; Weiss, R.; Dubourg, V. Scikit-Learn: Machine Learning in Python. *J. Mach. Learn. Res.* **2011**, *12*, 2825–2830.

43. Mahadevan, N.; Demanuele, C.; Zhang, H.; Volfson, D.; Ho, B.; Erb, M.K.; Patel, S. Development of Digital Biomarkers for Resting Tremor and Bradykinesia Using a Wrist-Worn Wearable Device. *NPJ Digit. Med.* **2020**, *3*, 5. [CrossRef] [PubMed]

44. Keijsers, N.L.; Horstink, M.W.; Gielen, S.C. Ambulatory Motor Assessment in Parkinson's Disease. *Mov. Disord.* **2006**, *21*, 34–44. [CrossRef]

45. Korthauer, K.; Kimes, P.K.; Duvallet, C.; Reyes, A.; Subramanian, A.; Teng, M.; Shukla, C.; Alm, E.J.; Hicks, S.C. A Practical Guide to Methods Controlling False Discoveries in Computational Biology. *Genome Biol.* **2019**, *20*, 118. [CrossRef]

46. Jgvhabets/Brady_reallife: First Release for Short-Term, Individual and Group Modelling Analyses | Zenodo. Available online: https://zenodo.org/record/4734199#.YJAOZRQza3J (accessed on 3 May 2021).

47. Khodakarami, H.; Ricciardi, L.; Contarino, M.F.; Pahwa, R.; Lyons, K.E.; Geraedts, V.J.; Morgante, F.; Leake, A.; Paviour, D.; De Angelis, A.; et al. Prediction of the Levodopa Challenge Test in Parkinson's Disease Using Data from a Wrist-Worn Sensor. *Sensors* **2019**, *19*, 5153. [CrossRef]

48. Vibhash, D.S.; Safarpour, D.; Mehta, S.H.; Vanegas-Arroyave, N.; Weiss, D.; Cooney, J.W.; Mari, Z.; Fasano, A. Telemedicine and Deep Brain Stimulation—Current Practices and Recommendations. *Park. Relat. Disord.* **2021**. [CrossRef]

49. Van den Bergh, R.; Bloem, B.R.; Meinders, M.J.; Evers, L.J.W. The State of Telemedicine for Persons with Parkinson's Disease. *Curr. Opin. Neurol.* **2021**, *34*, 589–597. [CrossRef] [PubMed]

50. Velisar, A.; Syrkin-Nikolau, J.; Blumenfeld, Z.; Trager, M.H.; Afzal, M.F.; Prabhakar, V.; Bronte-Stewart, H. Dual Threshold Neural Closed Loop Deep Brain Stimulation in Parkinson Disease Patients. *Brain Stimulat.* **2019**, *12*, 868–876. [CrossRef]

51. Castaño-Candamil, S.; Ferleger, B.I.; Haddock, A.; Cooper, S.S.; Herron, J.; Ko, A.; Chizeck, H.J.; Tangermann, M. A Pilot Study on Data-Driven Adaptive Deep Brain Stimulation in Chronically Implanted Essential Tremor Patients. *Front. Hum. Neurosci.* **2020**, *14*, 421. [CrossRef] [PubMed]

52. Gilron, R.; Little, S.; Perrone, R.; Wilt, R.; de Hemptinne, C.; Yaroshinsky, M.S.; Racine, C.A.; Wang, S.S.; Ostrem, J.L.; Larson, P.S.; et al. Long-Term Wireless Streaming of Neural Recordings for Circuit Discovery and Adaptive Stimulation in Individuals with Parkinson's Disease. *Nat. Biotechnol.* **2021**, *39*, 1078–1085. [CrossRef] [PubMed]

53. Habets, J.G.V.; Heijmans, M.; Kuijf, M.L.; Janssen, M.L.F.; Temel, Y.; Kubben, P.L. An Update on Adaptive Deep Brain Stimulation in Parkinson's Disease. *Mov. Disord.* **2018**, *33*, 1834–1843. [CrossRef] [PubMed]

54. Rochester, L.; Mazzà, C.; Mueller, A.; Caulfield, B.; McCarthy, M.; Becker, C.; Miller, R.; Piraino, P.; Viceconti, M.; Dartee, W.P.; et al. A Roadmap to Inform Development, Validation and Approval of Digital Mobility Outcomes: The Mobilise-D Approach. *Digit. Biomark.* **2020**, *4*, 13–27. [CrossRef]

55. Kluge, F.; Del Din, S.; Cereatti, A.; Gaßner, H.; Hansen, C.; Helbostadt, J.L.; Klucken, J.; Küderle, A.; Müller, A.; Rochester, L.; et al. Consensus Based Framework for Digital Mobility Monitoring. *medRxiv* **2020**. [CrossRef] [PubMed]

56. Heijmans, M.; Habets, J.G.V.; Herff, C.; Aarts, J.; Stevens, A.; Kuijf, M.L.; Kubben, P.L. Monitoring Parkinson's Disease Symptoms during Daily Life: A Feasibility Study. *NPJ Park. Dis.* **2019**, *5*, 21. [CrossRef]

57. Vizcarra, J.A.; Sanchez-Ferro, A.; Maetzler, W.; Marsili, L.; Zavala, L.; Lang, A.E.; Martinez-Martin, P.; Mestre, T.A.; Reilmann, R.; Hausdorff, J.M.; et al. The Parkinson's Disease e-Diary: Developing a Clinical and Research Tool for the Digital Age. *Mov. Disord.* **2019**. [CrossRef] [PubMed]

58. Habets, J.; Heijmans, M.; Herff, C.; Simons, C.; Leentjens, A.F.; Temel, Y.; Kuijf, M.; Kubben, P. Mobile Health Daily Life Monitoring for Parkinson Disease: Development and Validation of Ecological Momentary Assessments. *JMIR Mhealth Uhealth* **2020**, *8*, e15628. [CrossRef]

59. Habets, J.G.V.; Heijmans, M.; Leentjens, A.F.G.; Simons, C.J.P.; Temel, Y.; Kuijf, M.L.; Kubben, P.L.; Herff, C. A Long-Term, Real-Life Parkinson Monitoring Database Combining Unscripted Objective and Subjective Recordings. *Data* **2021**, *6*, 22. [CrossRef]

60. Bourke, A.K.; Ihlen, E.A.F.; Bergquist, R.; Wik, P.B.; Vereijken, B.; Helbostad, J.L. A Physical Activity Reference Data-Set Recorded from Older Adults Using Body-Worn Inertial Sensors and Video Technology-The ADAPT Study Data-Set. *Sensors* **2017**, *17*, 559. [CrossRef]

61. Alberts, J.L.; Koop, M.M.; McGinley, M.P.; Penko, A.L.; Fernandez, H.H.; Shook, S.; Bermel, R.A.; Machado, A.; Rosenfeldt, A.B. Use of a Smartphone to Gather Parkinson's Disease Neurological Vital Signs during the COVID-19 Pandemic. *Park. Dis.* **2021**, *2021*, 5534282. [CrossRef]

62. Bloem, B.R.; Marks, W.J., Jr.; Silva de Lima, A.L.; Kuijf, M.L.; van Laar, T.; Jacobs, B.P.F.; Verbeek, M.M.; Helmich, R.C.; van de Warrenburg, B.P.; Evers, L.J.W.; et al. The Personalized Parkinson Project: Examining Disease Progression through Broad Biomarkers in Early Parkinson's Disease. *BMC Neurol.* **2019**, *19*, 160. [CrossRef]
63. Watts, J.; Khojandi, A.; Vasudevan, R.; Ramdhani, R. Optimizing Individualized Treatment Planning for Parkinson's Disease Using Deep Reinforcement Learning. In Proceedings of the 2020 42nd Annual International Conference of the IEEE Engineering in Medicine & Biology Society (EMBC), Montréal, QC, Canada, 20–24 July 2020; pp. 5406–5409.
64. Matias, R.; Paixão, V.; Bouça, R.; Ferreira, J.J. A Perspective on Wearable Sensor Measurements and Data Science for Parkinson's Disease. *Front. Neurol.* **2017**, *8*, 677. [CrossRef]
65. MJFF, S. BEAT-PD DREAM Challenge (by Sage Bionetworks; Michael J. Fox Foundation). Available online: https://synapse.org/#!synapse:syn20825169/wiki/600904 (accessed on 3 May 2021).
66. Pulliam, C.L.; Heldman, D.A.; Brokaw, E.B.; Mera, T.O.; Mari, Z.K.; Burack, M.A. Continuous Assessment of Levodopa Response in Parkinson's Disease Using Wearable Motion Sensors. *IEEE Trans. Biomed. Eng.* **2018**, *65*, 159–164. [CrossRef]
67. Hua, R.; Wang, Y. Monitoring Insole (MONI): A Low Power Solution Toward Daily Gait Monitoring and Analysis. *IEEE Sens. J.* **2019**, *19*, 6410–6420. [CrossRef]
68. Daneault, J.; Lee, S.I.; Golabchi, F.N.; Patel, S.; Shih, L.C.; Paganoni, S.; Bonato, P. Estimating Bradykinesia in Parkinson's Disease with a Minimum Number of Wearable Sensors. In Proceedings of the 2017 IEEE/ACM International Conference on Connected Health: Applications, Systems and Engineering Technologies (CHASE), Philadelphia, PA, USA, 17–19 July 2017; pp. 264–265.
69. Daneault, J.-F.; Vergara-Diaz, G.; Parisi, F.; Admati, C.; Alfonso, C.; Bertoli, M.; Bonizzoni, E.; Carvalho, G.F.; Costante, G.; Fabara, E.E.; et al. Accelerometer Data Collected with a Minimum Set of Wearable Sensors from Subjects with Parkinson's Disease. *Sci. Data* **2021**, *8*, 48. [CrossRef]

*Article*

# Kinematics Adaptation and Inter-Limb Symmetry during Gait in Obese Adults

Massimiliano Pau [1], Paolo Capodaglio [2,3], Bruno Leban [1], Micaela Porta [1], Manuela Galli [4] and Veronica Cimolin [4,*]

[1] Department of Mechanical, Chemical and Materials Engineering, University of Cagliari, 09123 Cagliari, Italy; massimiliano.pau@dimcm.unica.it (M.P.); bruno.leban@dimcm.unica.it (B.L.); porta.micaela.ib@gmail.com (M.P.)

[2] Orthopaedic Rehabilitation Unit and Clinical Lab for Gait Analysis and Posture, Istituto Auxologico Italiano IRCCS, San Giuseppe Hospital, 28824 Verbania, Italy; p.capodaglio@auxologico.it

[3] Department Surgical Sciences, Physical and Rehabilitation Medicine, University of Turin, 10124 Turin, Italy

[4] Department of Electronics, Information and Bioengineering, Politecnico di Milano, Piazza Leonardo da Vinci 32, 20133 Milano, Italy; manuela.galli@polimi.it

* Correspondence: veronica.cimolin@polimi.it; Tel.: +39-02-2399-3359; Fax: +39-02-2399-3360

**Abstract:** The main purpose of this study is to characterize lower limb joint kinematics during gait in obese individuals by analyzing inter-limb symmetry and angular trends of lower limb joints during walking. To this purpose, 26 obese individuals (mean age 28.5 years) and 26 normal-weight age- and sex-matched were tested using 3D gait analysis. Raw kinematic data were processed to derive joint-specific angle trends and angle-angle diagrams (synchronized cyclograms) which were characterized in terms of area, orientation and trend symmetry parameters. The results show that obese individuals exhibit a kinematic pattern which significantly differs from those of normal weight especially in the stance phase. In terms of inter-limb symmetry, higher values were found in obese individuals for all the considered parameters, even though the statistical significance was detected only in the case of trend symmetry index at ankle joint. The described alterations of gait kinematics in the obese individuals and especially the results on gait asymmetry are important, because the cyclic uneven movement repeated for hours daily can involve asymmetrical spine loading and cause lumbar pain and could be dangerous for overweight individuals.

**Keywords:** angle-angle diagrams; cyclograms; gait; kinematics; obesity; symmetry

**Citation:** Pau, M.; Capodaglio, P.; Leban, B.; Porta, M.; Galli, M.; Cimolin, V. Kinematics Adaptation and Inter-Limb Symmetry during Gait in Obese Adults. *Sensors* **2021**, *21*, 5980. https://doi.org/10.3390/s21175980

Academic Editor: Marco Iosa

Received: 27 July 2021
Accepted: 1 September 2021
Published: 6 September 2021

**Publisher's Note:** MDPI stays neutral with regard to jurisdictional claims in published maps and institutional affiliations.

## 1. Introduction

Obesity is a pathological condition that has a profound effect on disability and quality of life [1]. The abnormal amount of fat, which modifies the body geometry by adding passive mass to different regions, causes relevant alterations in skeletal statics and dynamics. In particular, the mass excess has been recognized to influence the biomechanics of several movements and activities of daily living, such as walking, standing up, and bending [2–5], causing functional limitations, and possibly predisposing individuals to injuries [6]. Investigating these capacities quantitatively appears necessary to define the functional profile in the obese population and then plan appropriate rehabilitation interventions.

As locomotion is one of the most important and frequent tasks in daily life, it is not surprising that the features of gait in obese individuals have been extensively investigated. Indeed, the quantification of the way obesity affects the biomechanics of gait provides important insights about the relationship between metabolic and mechanical energetics, mechanical loading (in particular at lower limb joints), and the associated risk of musculoskeletal injuries and/or pathologies. Our understanding of how obesity affects gait biomechanics is increasing, and currently a rich body of literature and several reviews [7–10] are available. However, it is noteworthy that the findings related to the effects of obesity on the kinematics and kinetics of walking are mixed. While some studies reported that obesity induces slower velocity [5,8,11,12], lower cadence [5,11,12], reduced stride length [4,5,8,12] and swing time [12,13], increased stance time [8,12,13], decreased single support time [4,8], and increased double support time [8,14], other studies failed in detecting significant changes in velocity [4,13–16], cadence [13,14,17], step length [11,14,16], stride length [17], stance time [13], single and double leg support time and swing phase duration [11]. However, the literature is consistent as regards the increased step width [4,5,14,16,17]. In addition, increased peak hip joint flexion [18], extension [12], sagittal plane range of motion (ROM) [19], and increased ankle eversion from mid-stance to pre-swing [8] have been described. Conversely, no changes in hip joint sagittal plane ROM, ankle joint peak, and ROM of eversion [18] have been found. Finally, in some cases increased hip adduction during terminal stance and pre-swing and increased knee adduction in stance and swing [8,18] have been observed as well as increased ankle plantar flexion and reduced knee flexion [12].

In summary, although the main alterations of gait pattern in obese individuals have been extensively investigated, there are some aspects which remain mostly unexplored. First, previous researches were typically conducted using discrete parameters obtained by gait analysis (angle values at specific instants of the gait cycle, range of motion, ...) while no comparisons between the angle variations of hip, knee, and ankle joints in obese and normal weight individuals have been ever performed on a point-by-point basis. Comprehensive analysis of the whole angular trends during the gait cycle may provide a broader view of the gait alterations, thus representing a sound basis to plan suitable training and rehabilitative programs. Second, no data are available as regards inter-limb symmetry of obese individuals at hip, knee, and ankle joints during walking. The concept of symmetry in movement is quite controversial, as some researchers consider the human nature intrinsically asymmetrical and, as such, perfect symmetry does not exist in humans [20,21]. Nonetheless, it is commonly assumed that when a certain threshold of asymmetry is exceeded, its existence is indicative of gait alterations, which can originate from impaired motor control or from structural damage in the musculoskeletal system.

So far, different approaches have been proposed to quantify lower limb asymmetry during gait. Among discrete methods, that is those which consider single values of selected gait cycle parameters (i.e., spatio-temporal parameters or ground reaction force data [22]), the symmetry index (SI) [23–26] is one of the most commonly used. To the best of our knowledge, only one study exists about the application of SI in overweight individuals [20]; in this case, the SI was used to describe the difference between the left and right loadings considering the vertical components of the ground reaction force. It was demonstrated that a significant and high correlation is present between the SI and BMI of overweight subjects, thus suggesting that higher asymmetry of lower limb loading is associated with overweight, which implies greater risk to health of those people.

More recently, the techniques which make use of the entire angular waveforms have become widespread. In this case, inter-limb asymmetry is computed starting from continuous joint angle using bilateral cyclograms, [27–30], representing mutual dependencies between contralateral joint during the entire gait cycle [31]. Since asymmetry is usually associated with several pathologies, some studies have been conducted on this topic, in musculoskeletal, orthopaedic, and neurological diseases [27,28,32–34]. However, to the author's knowledge, this approach has never been employed to obese individuals. In literature, Stodolka and Sobera [20] demonstrated that the higher postural asymmetry of the lower limb loading is associated with overweight, leading to greater risk to health of those people. Repeated asymmetry of loading both legs for hours every day can involve asymmetrical spine loading and lumbar pain [35]; this effect could be more dangerous for health in the case of overweight or obese people. A better understanding of abnormalities in gait functionality of obese individuals may result in a more detailed understanding of biomechanical factors that influence their kinematics and could give suggestions for a more appropriate and effective rehabilitation and exercise prescription. Thus, the primary goal of the present study was to investigate the existence of possible alterations in lower limb joint kinematics in obese individuals during gait using two approaches: (1) assessment of inter-limb symmetry on the basis of the angular trend of each joint calculated for the whole gait cycle and (2) assessment of the existence of possible differences, in terms of lower limb joint kinematics, with respect to normal weight individuals by means of a point-by-point comparison of the angular trends acquired during the gait cycle.

## 2. Methods

### 2.1. Participants

A convenience sample of 26 obese individuals (OW, 11 male, 15 female, mean age 28.5 years, BMI > 30 kg/m$^2$, median 39.0 kg/m$^2$, range 34.9–51.6 kg/m$^2$) admitted for an integrated bodyweight reduction and rehabilitation program at the Istituto Auxologico Italiano, Piancavallo (VB, Italy), were recruited for the study on a voluntary basis. At the time of the experimental tests, all of them were free from any acute musculoskeletal, neuromuscular, psychological, and/or cardiopulmonary conditions able to significantly affect their walking abilities and postural control. Gait analysis data were taken from retrospective studies performed at Istituto Auxologico Italiano, Piancavallo (VB, Italy). An equal size number of normal weight individuals (BMI median 21.4 kg/m$^2$, range 17.0–26.5 kg/m$^2$) recruited among the hospital and University of Cagliari staff matched for age, sex, and height served as control group (NW). All participants (whose anthropometric and clinical features are reported in Table 1) were required to sign a written informed consent form, in which the details of the experimental tests were reported. The study was carried out in compliance with the World Medical Association Declaration of Helsinki and its later amendments.

**Table 1.** Anthropometric and clinical features of participants. Values are expressed as mean (SD).

|  | Normal Weight (NW) | Obese (OW) |
|---|---|---|
| Participants (M, F) | 26 (11M, 15F) | 26 (11M, 15F) |
| Age (years) | 28.5 (7.8) | 28.7 (7.6) |
| Body mass (kg) | 60.2 (11.9) | 109.8 (15.8) |
| Height (cm) | 165.6 (8.3) | 165.5 (9.0) |
| Body Mass Index (kg m$^{-2}$) | 21.8 (2.8) | 40.4 (0.8) |

### 2.2. Spatio-Temporal and Kinematic Data Collection and Processing

Spatio-temporal and kinematic parameters of gait were acquired by means of a 6-camera motion-capture system (VICON, Oxford Metrics Ltd., Oxford, UK) with a sampling rate of 100 Hz, and two force platforms (Kistler, CH). Prior to the experimental tests, the following anthropometric data were collected: height, body mass, anterior superior iliac spine distance, pelvis thickness, knee and ankle width, and leg length. Then, 22 spherical retro-reflective passive markers were placed on the individual' skin at specific landmarks according to the protocol proposed by Davis et al. [36]. Once this preparation phase was completed, participants were requested to walk along a 10 m long walkway at their self-selected speed in the most natural manner. A trial was considered valid, and subsequently processed, if the marker trajectories were not lost during the subject's gait and included at least one cycle per limb. At least six trials were acquired for each participant in order to guarantee reproducibility of the results. The raw 3D markers' trajectories were thus processed using the dedicated software (Polygon Application, version 2.4, VICON, Oxford Metrics Ltd., Oxford, UK), to obtain the following variables:

- spatiotemporal parameters of gait (i.e., speed, stride length, cadence, stance, swing, and double support phase duration);
- dynamic range of motion (ROM) of hip, knee and ankle joints, calculated as difference between the minimum and the maximum angle of flexion-extension (hip and knee) and dorsi-plantarflexion (ankle) observed during the gait cycle;
- angular trends of hip, knee, and ankle joints on the sagittal plane across the gait cycle. Such curves will be subsequently used to calculate the indexes of interlimb symmetry as described later in detail.

### 2.3. Gait Symmetry Quantification by Means of Cyclograms

Synchronized bilateral cyclograms were obtained from the sagittal kinematics properly processed with a custom routine developed under Matlab environment following the procedure proposed by Goswami et al. [37]. Briefly, using right and left limb angle values acquired during the gait cycle, angle-angle diagrams for hip, knee, and ankle joints were built and, on their basis, the following symmetry parameters were extracted:

- Cyclogram area (degrees$^2$): the area enclosed by the curve obtained from each angle-angle diagram [38]. In the ideal case of perfectly symmetrical gait, the cyclogram area is null, as left and right joint angles assume the same value for each time period of the gait cycle and thus all cyclogram points lie on a 45° line. Increasing values of area indicate larger interlimb asymmetry.
- Cyclogram orientation (degrees): the absolute value of the angle $\phi$ formed by the 45° line (i.e., perfect interlimb symmetry) and the principal axis of inertia of the cyclogram [37]. A zero value indicates perfect symmetry, while increasing values of $\phi$ denote higher interlimb asymmetry.
- Trend symmetry: this dimensionless parameter, calculated according to the procedure described by Crenshaw and Richards [39] is obtained by eigenvector analysis on time-normalized right and left limb gait cycles. Even in this case a null value indicates perfect symmetry, while increasing interlimb asymmetry corresponds to higher trend symmetry values.

### 2.4. Statistical Analysis

We preliminarily checked all data separately acquired for left and right limb to verify the presence of statistically significant differences between them. As they were not found, in the subsequent analysis for each participant we considered the mean value of left and right joint/limb.

The existence of differences between OW and NW groups was assessed using three different statistical approaches. In particular, one-way multivariate analysis of variance (MANOVA) was carried out to investigate the possible differences introduced by obesity in spatio-temporal parameters of gait and dynamic ROM, while one-way multivariate analysis of covariance (MANCOVA) was used in case of interlimb symmetry parameters, including gait speed as a covariate. This design allows observers to take into account in the analysis any possible differences of speed between groups which might affect, to some extent, the lower limb kinematics. In both analyses group (OW/NW) was set as the independent variable, while the dependent variables were respectively the six previously listed spatio-temporal parameters, the 3 dynamic ROM at hip, knee and ankle joints and the 3 symmetry parameters calculated for each joint. The level of significance was set at $p = 0.05$ and the effect sizes were assessed using the eta-squared ($\eta^2$) coefficient. Univariate analysis of variance (ANOVA) was carried out as a post-hoc test by reducing the level of significance according to the Bonferroni correction.

Instead, the differences associated with the presence of obesity in joint kinematic data were assessed by comparing the "angle vs. time" curves of both groups, for each of the 3 joints of interest, on a point-by-point basis using a one-way ANOVA, using an approach previously proposed in the literature to characterize sex-related differences in kinematic patterns among population of unaffected adults (i.e., men vs. women, Bruening et al. [40]) or between healthy individuals and those affected by neurologic and orthopedic conditions [27,41,42]. In this way it was possible to define the time intervals of the gait cycle characterized by significant differences associated with obesity.

We also explored the existence of possible relationships between BMI and interlimb symmetry parameters, while controlling for gait speed, by calculating partial correlation coefficients. All statistical analyses were performed using the SPSS version 20$_b$ software (IBM SPSS Statistics, Armonk, NY, USA).

## 3. Results

The results of the comparison between obese and normal weight individuals with regard to spatio-temporal gait parameters, dynamic ROM and interlimb symmetry are summarized in Tables 2–4; an example of cyclograms calculated for obese and normal weight participants are reported in Figure 1, and the angle variations in the sagittal plane during the gait cycle for hip, knee, and ankle joints are shown in Figure 2.

The statistical analysis revealed a significant effect of obesity on spatio-temporal parameters. [$F(6,45)$ = 17.87, $p < 0.001$, Wilks $\lambda$ = 0.30, $\eta^2$ = 0.704 95% CI [0.525–0.751]]. In particular, the follow-up ANOVA showed that obese individuals were characterized by significant lower gait speed, stride length and swing phase duration and increased stance and double support phases duration.

Similarly, in case of dynamic ROM a main effect of group was detected [$F(3,48)$ = 5.49, $p$ = 0.003, Wilks $\lambda$ = 0.74, $\eta^2$ = 0.255 95% CI [0.063–0.377]]. The subsequent follow-up ANOVA revealed that obese individuals exhibited significantly reduced dynamic ankle and knee (but not hip) ROM with respect to normal weight individuals.

Finally, even in the case of symmetry parameters, after controlling for gait speed, a significant effect associated with obesity was found at knee and ankle joints (knee $F(3,47)$ = 3.23, $p$ = 0.031, Wilks $\lambda$ = 0.83, $\eta^2$ = 0.171 95% CI [0.009–0.290], ankle $F(3,47)$ = 2.96, $p$ = 0.042, Wilks $\lambda$ = 0.84, $\eta^2$ = 0.159 95% CI [0.003–0.276]). The post-hoc analysis revealed that both trend symmetry parameter for the ankle joint and cyclogram orientation at knee joint were significantly higher in the OW groups thus indicating the presence of relevant interlimb asymmetry.

The results of the correlation analysis between BMI and symmetry parameters, which are reported in Table 5, showed the existence of significant moderate positive relationship between BMI and trend symmetry at ankle joint and cyclogram orientation at knee joint.

*Point-by-Point Analysis of Kinematic Curves*

The results of the analysis of hip, knee, and ankle kinematics in the sagittal plane, performed (Figure 2) on a point-by-point basis, show that significant differences between obese and normal weight individuals exist at all the three considered joints. In particular, obese individuals exhibited

- significantly increased hip flexion through all the stance phase (0 to 59% of the gait cycle) and at the end of the swing phase (90 to 100%);
- significantly reduced knee flexion through all the gait cycle;
- significantly reduced ankle dorsi-flexion for initial and mid-stance (3 to 41% of the gait cycle) and increased dorsi-flexion at the terminal stance and initial swing (48 to 64% of the gait cycle).

**Table 2.** Comparison between the spatio-temporal gait parameters of normal weight and obese individuals. Values are expressed as mean (SD).

|  | Normal Weight | Obese |
| --- | --- | --- |
| Gait speed (m s$^{-1}$) | 1.30 (0.20) | 1.16 (0.13) * |
| Stride length (m) | 1.38 (0.13) | 1.23 (0.10) * |
| Cadence (steps min$^{-1}$) | 114.54 (10.78) | 112.65 (7.51) |
| Stance phase (% of the gait cycle) | 59.15 (1.56) | 61.87 (1.32) * |
| Swing phase (% of the gait cycle) | 40.85 (1.56) | 38.13 (1.35) * |
| Double support (% of the gait cycle) | 18.44 (2.86) | 47.40 (5.24) * |

The symbol * denotes statistically significant difference vs. normal weight after Bonferroni correction ($p < 0.008$).

**Table 3.** Comparison between dynamic ROM of normal weight and obese individuals. Values are expressed as mean (SD).

|  | Normal Weight | Obese |
| --- | --- | --- |
| Ankle ROM (°) | 32.4 (5.7) | 28.5 (5.4) * |
| Knee ROM (°) | 61.9 (4.5) | 56.0 (6.0) * |
| Hip ROM (°) | 46.7 (5.2) | 43.7 (5.3) |

The symbol * denotes statistically significant difference vs. normal weight after Bonferroni correction ($p < 0.016$).

**Table 4.** Comparison between interlimb symmetry parameters of gait of normal weight and obese individuals. Values are expressed as mean (SD).

| Parameter | Joint | Normal Weight | Obese |
| --- | --- | --- | --- |
| Cyclogram area (°$^2$) |  | 77.68 (59.59) | 103.84 (63.05) |
| Cyclogram orientation $\phi$ (°) | Ankle | 2.63 (2.04) | 3.94 (4.83) |
| Trend Symmetry |  | 1.34 (1.08) | 2.51 (1.63) * |
| Cyclogram area (°$^2$) |  | 228.50 (178.97) | 312.71 (221.37) |
| Cyclogram orientation $\phi$ (°) | Knee | 1.09 (0.81) | 2.02 (1.98) * |
| Trend Symmetry |  | 0.39 (0.30) | 0.70 (0.62) |
| Cyclogram area (°$^2$) |  | 97.22 (87.41) | 124.43 (72.95) |
| Cyclogram orientation $\phi$ (°) | Hip | 1.74 (1.22) | 1.94 (20.6) |
| Trend Symmetry |  | 0.20 (0.17) | 0.68 (1.16) |

The symbol * denotes a significant difference with respect to the normal weight group.

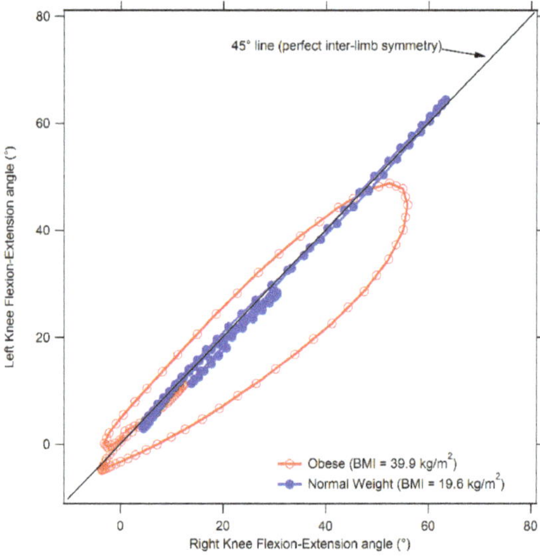


**Figure 1.** Comparison between cyclograms of obese and normal weight individuals. The diagram refers to the knee joint.

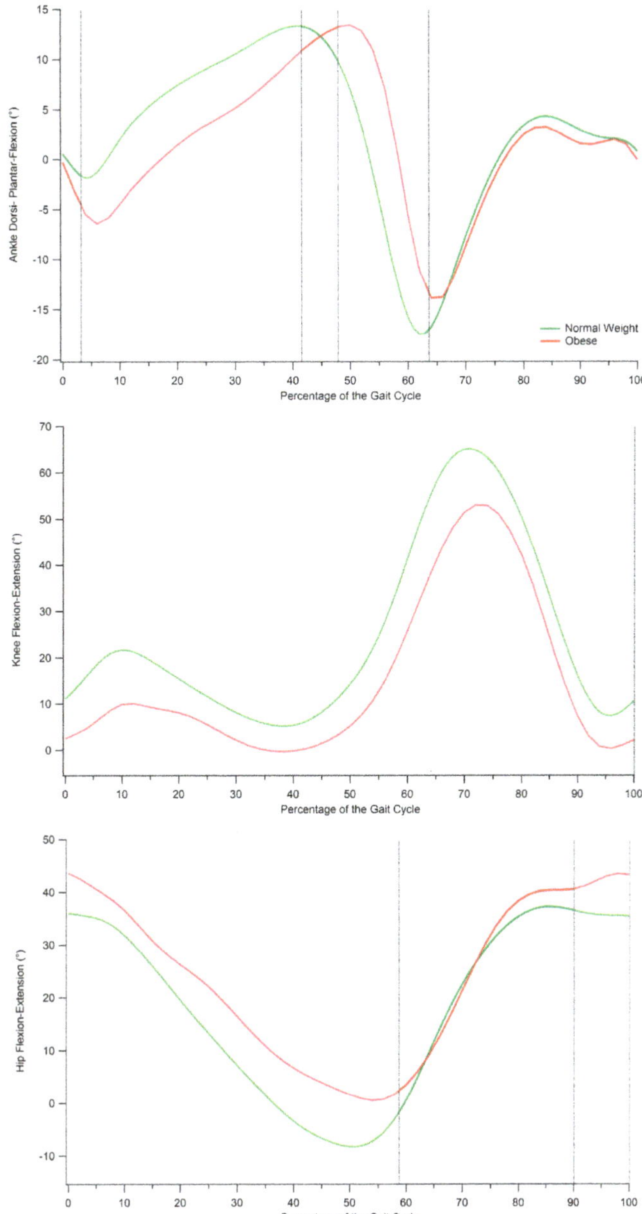



**Figure 2.** Gait kinematics in the sagittal plane for normal weight and obese individuals. From top to bottom: ankle dorsi-plantar-flexion, knee flexion-extension, and hip flexion-extension angles during gait cycle. Grey-shaded areas denote the periods of the gait cycle in which a significant difference between groups was detected ($p < 0.05$).

Table 5. Partial correlation coefficients between BMI and interlimb symmetry parameters.

| Joint | Symmetry Parameter | BMI | $p$ |
|---|---|---|---|
| Ankle | Cyclogram area ($°^2$) | 0.242 | N.S. |
| | Cyclogram orientation $\phi$ ($°$) | 0.252 | N.S. |
| | Trend Symmetry | 0.441 | 0.001 |
| Knee | Cyclogram area ($°^2$) | 0.163 | N.S. |
| | Cyclogram orientation $\phi$ ($°$) | 0.320 | 0.022 |
| | Trend Symmetry | 0.245 | N.S. |
| Hip | Cyclogram area ($°^2$) | 0.194 | N.S. |
| | Cyclogram orientation $\phi$ ($°$) | −0.053 | N.S. |
| | Trend Symmetry | 0.093 | N.S. |

N.S. = Not Significant.

## 4. Discussion

The aim of the present study was to characterize the main alterations in gait kinematics in obese individuals, with special focus on interlimb asymmetry and detailed point-by-point comparison of the angular trends of hip, knee, and ankle joints between obese and normal weight individuals.

Our data indicated that most of the spatio-temporal parameters differ significantly among obese participants and controls, particularly in terms of gait speed, stride length, stance and swing phase and double support duration. Overall, such results are consistent with previous studies focused on characterizing gait patterns in obesity [4,5,8,11–14]. Taken together, the observed changes suggest the existence of a strategy specifically aimed to improve stabilization and balance control, which are threatened by the biomechanical alterations associated with mass excess. The longer duration of double support and stance phase are probably the result of a strategy aimed to allow a safer locomotion through a better optimized balanced distribution of the weight on both limbs and thus reduce the risk of instability and falls [4,43].

The kinematic pattern on the sagittal plane, as defined by the point-by-point analysis, indicates that obesity mainly affects the stance phase of gait, as most significant differences with respect to normal weight individuals were observed during that part of the gait cycle. In particular the results show that obese individuals exhibit reduced hip extension, knee flexion and ankle plantarflexion [17,44], which overall lead to a significant reduction of ankle and knee dynamic ROMs. Such alterations, which were reported (individually or in combination) in previous studies on adults and adolescents [12,45–47], might represent, together with reduced walking speed and longer stance phase duration, a strategy to reduce the articular stress and to compensate for the reduced muscular strength and the altered joint proprioception. Walking speed certainly plays a crucial role in defining the sagittal kinematics of gait in obese individuals and can be considered the main cause for the combination of increased knee flexion and increased plantarflexion at toe off [12]. At the same time, some researchers pointed out that walking at slower speeds may also represent the expression of a compensatory strategy aimed to limit the magnitude of the forces acting on lower extremity joints [45] and thus to reduce the risk of musculoskeletal diseases [12,17].

The significantly reduced dynamic ROMs that we detected, in particular at knee and ankle joints, may also be connected to the continuous search for stability typical of obese individuals. As they need to keep both limbs in contact with the ground, this condition increases the amount of time spent in a closed lower-limb kinematic chain condition. In this situation, the degrees of freedom of the rigid lower body system are reduced and the constraint, especially on the knee joint, increases. At last, we observe that another factor involved in the ROM reduction at knee joint might be due to the excess fat on the thigh and shank, which mechanically encumbers intersegmental rotation and counteracts the antigravity action exerted by the knee flexors [48]. As for the ankle joint, the deficits in plantar- and dorsiflexion might be due to a reduced strength of the ankle muscles, which were already reported by previous studies [49]. In this context, it is noteworthy that such effect could be reduced through suitable physical and rehabilitative intervention [50,51].

The results of the inter-limb symmetry analysis show a well-defined trend characterized by higher values of all considered parameters in obese individuals even though the statistical significance was achieved only in case of trend symmetry at ankle joint. To tour knowledge, no previous studies investigated inter-limb symmetry of lower limb joint kinematics in obese individuals, and thus there are no available data for direct comparison. However, it is noticeable that few previous studies reported the existence of significant asymmetries in spatio-temporal parameters of gait, such as step length [52] and stance phase duration [53]. Moreover, Stodolka et al. [20] calculated the symmetry index (SI) in a group of overweight individuals to quantitatively characterize the existence of possible differences between the left and right limb loading during quiet stance by analyzing the vertical components of the ground reaction force. They found that asymmetry was strongly correlated with body mass index and suggested that increased body weight may be a disadvantageous determinant of dynamic stability.

The presence of asymmetry during gait is a well-known phenomenon in neurologic conditions (i.e., multiple sclerosis, stroke [28,54] as well as in musculoskeletal disorders with a marked unilateral presentation [27]. However, in all these cases there are clear factors (e.g., either damage in a specific location of the central nervous system or injury of one limb) that justify the lack of symmetry. In case of obesity, neither of these two factors is present, and thus the reasons of the observed asymmetry should be found elsewhere. Several recent studies reported that obese individuals are characterized by uneven fat distribution in the left and right side of the body [55] and, among a group of adolescents, a larger proportion of individuals characterized by asymmetric lower limb force/power was found among obese with respect to normal weight peers [56]. At last, Bell et al. [57] suggested that lean mass asymmetries represent a co-factor in force/power asymmetry during jumping. Although we don't have any direct evidence regarding the existence of fat/lean mass or muscular strength among left and right limb in our sample, it appears reasonable to hypothesize that asymmetry of lower limb kinematics is due to "mechanical" factors associated with differences in body composition and muscular performance of the two legs.

The described alterations of gait in our sample of obese individuals could be informative from a clinical and rehabilitative point of view. It is known that walking at a constant intensity for a prolonged time is a useful and frequently employed strategy to achieve body mass reduction in obese individuals because it is a convenient type of physical activity which involves a significant amount of metabolic energy expenditure [14]. Therefore, walking abnormalities should be carefully assessed and considered to avoid overload and possible musculoskeletal problems which would prolong the rehabilitation phase and possibly introduce negative effects. In particular, the investigation on gait asymmetry seems to be important, because the cyclic uneven movement daily repeated for hours can involve asymmetrical spine loading and cause lumbar pain [35] and this effect could be certainly more dangerous in case of individuals overweight. Even a relatively low degree of asymmetry of weight-bearing repeated every day for years could represent a co-factor for the onset of either low back pain or hip joints issues [20]. Thus, an appropriate and effective rehabilitation and exercise prescription parallel to weight loss could be tailored to recover gait pattern and reduce its asymmetry.

This study has several limitations. Firstly, the tested sample was composed of young adults. Previous research carried out on older adults obese reported that they often exhibit articular problems (such as osteoarthritis) and severe gait alterations [58,59] that could be due to the progressive/cumulative effect of excessive joint loads over the years. Our results, which refer to young adults, could have been influenced by age-factor both in terms of gait pattern and of asymmetry which revealed a moderate severity of gait modifications with respect to controls, confirming that obesity does not determine major and immediate changes in the learned motor strategy in young adult patients. In other words, the effect of obesity on joint biomechanics seems to be not immediate, but progressive [4]. Another limitation is due to the heterogeneity of the participants in terms of severity of obesity, which makes more difficult the comparison with existing data on the literature. Finally, it is important to highlight that in this study the gait pattern was quantified using a standard marker set [36] which, in case of obese individuals, might suffer from reduced accuracy in terms of marker placement and soft tissue artefact [60,61]. Of particular concern is the marker positioning on anterior superior iliac spine (ASIS) and/or greater trochanter markers to establish the pelvis and hip joint centers. This might lead to inaccurate estimates of joint centers and, consequently, errors of the resultant kinematic/kinetic parameters, particularly as regards hip and knee [62]. However, it is also known that parameters like dynamic ROMs during gait are only slightly affected by these issues [63] and thus, even in obese, their values can be considered reliable. Future studies are needed to clearly identify the optimal marker placement as well as suitable skeletal model development procedures, to properly remove (or at very least greatly reduce) the errors possibly associated with marker placement. In this context, it is noteworthy that a combination of dual-energy X-ray absorptiometry images (to exactly assess the inter-ASIS distance and estimating segment inertial parameters [64]) with a sacral marker cluster and digitized pelvic anatomical landmarks [65] have been suggested to improve the accuracy of marker-based motion capture. Future studies should also be directed towards the investigation of specific classes of obesity to better understand its effects on gait, as the previously mentioned negative issues might be either dependent or independent by the magnitude of the mass excess.

**Author Contributions:** M.P. (Massimiliano Pau): conceptualization, methodology, writing—original draft; P.C.: clinical assessment of the participants, review and editing; B.L., M.P. (Micaela Porta): formal analysis, software; M.G.: review and editing; V.C.: conceptualization, methodology, data curation, writing, review. All authors have read and agreed to the published version of the manuscript.

**Funding:** This research received no external funding.

**Institutional Review Board Statement:** The study was conducted according to the guidelines of the Declaration of Helsinki, and approved by the Institutional Ethics Committee of Istituto Auxologico Italiano IRCCS (GAITNEW 31C302, protocol code CE 2013_04_23_03).

**Informed Consent Statement:** Informed consent was obtained from all subjects involved in the study.

**Data Availability Statement:** Data available on request due to restrictions, e.g., privacy or ethical.

**Acknowledgments:** The authors would like to acknowledge Eng. Martina Bona, Carlo Maria Canossi, Emma Confortola, Giovanni Distante, for their valuable contribution in data processing.

**Conflicts of Interest:** The authors declare no conflict of interest.

# References

1. Bray, G.A. Medical Consequences of Obesity. *J. Clin. Endocrinol. Metab.* **2004**, *89*, 2583–2589. [CrossRef] [PubMed]
2. Sibella, F.; Galli, M.; Romei, M.; Montesano, A.; Crivellini, M. Biomechanical analysis of sit-to-stand movement in normal and obese subjects. *Clin. Biomech.* **2003**, *18*, 745–750. [CrossRef]
3. Saibene, F.; Minetti, A.E. Biomechanical and physiological aspects of legged locomotion in humans. *Eur. J. Appl. Physiol.* **2003**, *88*, 297–316. [CrossRef]
4. Vismara, L.; Romei, M.; Galli, M.; Montesano, A.; Baccalaro, G.; Crivellini, M.; Grugni, G. Clinical implications of gait analysis in the rehabilitation of adult patients with "Prader-Willi" Syndrome: A cross-sectional comparative study ("Prader-Willi" Syndrome vs matched obese patients and healthy subjects). *J. Neuroeng. Rehabil.* **2007**, *4*, 14. [CrossRef] [PubMed]
5. De Souza, S.A.F.; Faintuch, J.; Valezi, A.C.; Anna, A.F.S.; Gama-Rodrigues, J.J.; Fonseca, I.C.D.B.; Souza, R.B.; Senhorini, J. Gait Cinematic Analysis in Morbidly Obese Patients. *Obes. Surg.* **2005**, *15*, 1238–1242. [CrossRef]
6. Wearing, S.C.; Hennig, E.; Byrne, N.; Steele, J.; Hills, A.P. The biomechanics of restricted movement in adult obesity. *Obes. Rev.* **2006**, *7*, 13–24. [CrossRef] [PubMed]
7. Gilleard, W.L. Functional Task Limitations in Obese Adults. *Curr. Obes. Rep.* **2012**, *1*, 174–180. [CrossRef]
8. Lai, P.P.; Leung, K.L.; Li, A.N.; Zhang, M. Three-dimensional gait analysis of obese adults. *Clin. Biomech.* **2008**, *23*, S2–S6. [CrossRef] [PubMed]
9. Runhaar, J.; Koes, B.; Clockaerts, S.; Bierma-Zeinstra, S.M.A. A systematic review on changed biomechanics of lower extremities in obese individuals: A possible role in development of osteoarthritis. *Obes. Rev.* **2011**, *12*, 1071–1082. [CrossRef]

10. Gushue, D.L.; Houck, J.; Lerner, A. Effects of Childhood Obesity on Three-Dimensional Knee Joint Biomechanics During Walking. *J. Pediatr. Orthop.* **2005**, *25*, 763–768. [CrossRef] [PubMed]

11. Benedetti, M.G.; Di Gioia, A.; Conti, L.; Berti, L.; Degli Esposti, L.; Tarrini, G.; Melchionda, N.; Giannini, S. Physical activity monitoring in obese people in the real life environment. *J. Neuroeng. Rehabil.* **2009**, *6*, 47. [CrossRef] [PubMed]

12. DeVita, P.; Hortobágyi, T. Obesity is not associated with increased knee joint torque and power during level walking. *J. Biomech.* **2003**, *36*, 1355–1362. [CrossRef]

13. Błaszczyk, J.W.; Cieślińska-Świder, J.; Plewa, M.; Zahorska-Markiewicz, B.; Markiewicz, A. Effects of excessive body weight on postural control. *J. Biomech.* **2009**, *42*, 1295–1300. [CrossRef]

14. Browning, R.C.; McGowan, C.P.; Kram, R. Obesity does not increase external mechanical work per kilogram body mass during walking. *J. Biomech.* **2009**, *42*, 2273–2278. [CrossRef]

15. Browning, R.C.; Kram, R. Energetic Cost and Preferred Speed of Walking in Obese vs. Normal Weight Women. *Obes. Res.* **2005**, *13*, 891–899. [CrossRef]

16. Wu, X.; Lockhart, T.E.; Yeoh, H.T. Effects of obesity on slip-induced fall risks among young male adults. *J. Biomech.* **2012**, *45*, 1042–1047. [CrossRef] [PubMed]

17. Browning, R.C.; Kram, R. Effects of Obesity on the Biomechanics of Walking at Different Speeds. *Med. Sci. Sports Exerc.* **2007**, *39*, 1632–1641. [CrossRef] [PubMed]

18. Russell, E.M.; Hamill, J. Lateral wedges decrease biomechanical risk factors for knee osteoarthritis in obese women. *J. Biomech.* **2011**, *44*, 2286–2291. [CrossRef]

19. Gilleard, W.; Smith, T. Effect of obesity on posture and hip joint moments during a standing task, and trunk forward flexion motion. *Int. J. Obes.* **2006**, *31*, 267–271. [CrossRef]

20. Stodółka, J.; Sobera, M. Symmetry of lower limb loading in healthy adults during normal and abnormal stance. *Acta Bioeng. Biomech.* **2017**, *19*. [CrossRef]

21. Sadeghi, H.; Allard, P.; Prince, F.; Labelle, H. Symmetry and limb dominance in able-bodied gait: A review. *Gait Posture* **2000**, *12*, 34–45. [CrossRef]

22. Herzog, W.; Nigg, B.M.; Read, L.J.; Olsson, E. Asymmetries in ground reaction force patterns in normal human gait. *Med. Sci. Sports Exerc.* **1989**, *21*, 110–114. [CrossRef] [PubMed]

23. Robinson, R.O.; Herzog, W.; Nigg, B.M. Use of force platform variables to quantify the effects of chiropractic manipulation on gait symmetry. *J. Manip. Physiol. Ther.* **1987**, *10*, 172–176.

24. Patterson, K.K.; Gage, W.H.; Brooks, D.; Black, S.; McIlroy, W.E. Evaluation of gait symmetry after stroke: A comparison of current methods and recommendations for standardization. *Gait Posture* **2010**, *31*, 241–246. [CrossRef]

25. Noble, J.W.; Prentice, S.D. Adaptation to unilateral change in lower limb mechanical properties during human walking. *Exp. Brain Res.* **2005**, *169*, 482–495. [CrossRef]

26. Haddad, J.M.; van Emmerik, R.E.; Whittlesey, S.N.; Hamill, J. Adaptations in interlimb and intralimb coordination to asymmetrical loading in human walking. *Gait Posture* **2006**, *23*, 429–434. [CrossRef]

27. Porta, M.; Pau, M.; Leban, B.; Deidda, M.; Sorrentino, M.; Arippa, F.; Marongiu, G. Lower Limb Kinematics in Individuals with Hip Osteoarthritis during Gait: A Focus on Adaptative Strategies and Interlimb Symmetry. *Bioengineering* **2021**, *8*, 47. [CrossRef] [PubMed]

28. Pau, M.; Leban, B.; Deidda, M.; Putzolu, F.; Porta, M.; Coghe, G.; Cocco, E. Kinematic Analysis of Lower Limb Joint Asymmetry During Gait in People with Multiple Sclerosis. *Symmetry* **2021**, *13*, 598. [CrossRef]

29. Kwek, J.L.; Williams, G.K. Age-based comparison of gait asymmetry using unilateral ankle weights. *Gait Posture* **2021**, *87*, 11–18. [CrossRef]

30. Kutilek, P.; Viteckova, S.; Svoboda, Z.; Smrcka, P. Kinematic quantification of gait asymmetry in patients with peroneal nerve palsy based on bilateral cyclograms. *J. Musculoskelet. Neuronal Interact.* **2013**, *13*, 244–250.

31. Viteckova, S.; Kutilek, P.; Svoboda, Z.; Krupicka, R.; Kauler, J.; Szabo, Z. Gait symmetry measures: A review of current and prospective methods. *Biomed. Signal Process. Control* **2018**, *42*, 89–100. [CrossRef]

32. Patterson, K.; Parafianowicz, I.; Danells, C.J.; Closson, V.; Verrier, M.C.; Staines, W.; Black, S.; McIlroy, W.E. Gait Asymmetry in Community-Ambulating Stroke Survivors. *Arch. Phys. Med. Rehabil.* **2008**, *89*, 304–310. [CrossRef]

33. Laroche, D.P.; Cook, S.B.; Mackala, K. Strength Asymmetry Increases Gait Asymmetry and Variability in Older Women. *Med. Sci. Sports Exerc.* **2012**, *44*, 2172–2181. [CrossRef] [PubMed]

34. Yogev, G.; Plotnik, M.; Peretz, C.; Giladi, N.; Hausdorff, J.M. Gait asymmetry in patients with Parkinson's disease and elderly fallers: When does the bilateral coordination of gait require attention? *Exp. Brain Res.* **2006**, *177*, 336–346. [CrossRef] [PubMed]

35. Fortin, C.; Grunstein, E.; Labelle, H.; Parent, S.; Feldman, D.E. Trunk imbalance in adolescent idiopathic scoliosis. *Spine J.* **2016**, *16*, 687–693. [CrossRef] [PubMed]

36. Davis, R.B.; Õunpuu, S.; Tyburski, D.; Gage, J.R. A gait analysis data collection and reduction technique. *Hum. Mov. Sci.* **1991**, *10*, 575–587. [CrossRef]

37. Goswami, A. A new gait parameterization technique by means of cyclogram moments: Application to human slope walking. *Gait Posture* **1998**, *8*, 15–36. [CrossRef]

38. Hershler, C.; Milner, M. Angle—Angle diagrams in the assessment of locomotion. *Am. J. Phys. Med.* **1980**, *59*, 109–125. [PubMed]

39. Crenshaw, S.J.; Richards, J.G. A method for analyzing joint symmetry and normalcy, with an application to analyzing gait. *Gait Posture* **2006**, *24*, 515–521. [CrossRef]

40. Bruening, D.A.; Frimenko, R.E.; Goodyear, C.D.; Bowden, D.R.; Fullenkamp, A.M. Sex differences in whole body gait kinematics at preferred speeds. *Gait Posture* **2015**, *41*, 540–545. [CrossRef]

41. Pau, M.; Corona, F.; Pilloni, G.; Porta, M.; Coghe, G.; Cocco, E. Do gait patterns differ in men and women with multiple sclerosis? *Mult. Scler. Relat. Disord.* **2017**, *18*, 202–208. [CrossRef] [PubMed]

42. Porta, M.; Pilloni, G.; Arippa, F.; Casula, C.; Cossu, G.; Pau, M. Similarities and Differences of Gait Patterns in Women and Men With Parkinson Disease With Mild Disability. *Arch. Phys. Med. Rehabil.* **2019**, *100*, 2039–2045. [CrossRef]

43. Fjeldstad, C.; Fjeldstad, A.S.; Acree, L.S.; Nickel, K.J.; Gardner, A.W. The influence of obesity on falls and quality of life. *Dyn. Med.* **2008**, *7*, 4. [CrossRef]

44. Lerner, Z.F.; Shultz, S.P.; Board, W.J.; Kung, S.; Browning, R.C. Does adiposity affect muscle function during walking in children? *J. Biomech.* **2014**, *47*, 2975–2982. [CrossRef]

45. Pamukoff, D.N.; Lewek, M.D.; Blackburn, J.T. Greater vertical loading rate in obese compared to normal weight young adults. *Clin. Biomech.* **2016**, *33*, 61–65. [CrossRef]

46. Martz, P.; Bourredjem, A.; Maillefert, J.F.; Binquet, C.; Baulot, E.; Ornetti, P.; Laroche, D. Influence of body mass index on sagittal hip range of motion and gait speed recovery six months after total hip arthroplasty. *Int. Orthop.* **2019**, *43*, 2447–2455. [CrossRef] [PubMed]

47. Summa, S.; De Peppo, F.; Petrarca, M.; Caccamo, R.; Carbonetti, R.; Castelli, E.; Adorisio, D.O. Gait changes after weight loss on adolescent with severe obesity after sleeve gastrectomy. *Surg. Obes. Relat. Dis.* **2019**, *15*, 374–381. [CrossRef] [PubMed]

48. Park, K.; Hur, P.; Rosengren, K.S.; Horn, G.P.; Hsiao-Wecksler, E.T. Effect of load carriage on gait due to firefighting air bottle configuration. *Ergonomics* **2010**, *53*, 882–891. [CrossRef]

49. Koushyar, H.; Nussbaum, M.A.; Davy, K.P.; Madigan, M.L. Relative Strength at the Hip, Knee, and Ankle Is Lower Among Younger and Older Females Who Are Obese. *J. Geriatr. Phys. Ther.* **2017**, *40*, 143–149. [CrossRef]

50. Zhao, X.; Tsujimoto, T.; Kim, B.; Katayama, Y.; Wakaba, K.; Wang, Z.; Tanaka, K. Effects of increasing physical activity on foot structure and ankle muscle strength in adults with obesity. *J. Phys. Ther. Sci.* **2016**, *28*, 2332–2336. [CrossRef]

51. Zhao, X.; Tsujimoto, T.; Kim, B.; Katayama, Y.; Ogiso, K.; Takenaka, M.; Tanaka, K. Does Weight Reduction Affect Foot Structure and the Strength of the Muscles That Move the Ankle in Obese Japanese Adults? *J. Foot Ankle Surg.* **2018**, *57*, 281–284. [CrossRef]

52. Hills, A.P.; Parker, A.W. Gait characteristics of obese children. *J. Pediatr. Orthop.* **1991**, *11*, 811. [CrossRef]

53. Wiik, A.V.; Aqil, A.; Brevadt, M.; Jones, G.; Cobb, J. Abnormal ground reaction forces lead to a general decline in gait speed in knee osteoarthritis patients. *World J. Orthop.* **2017**, *8*, 322–328. [CrossRef] [PubMed]

54. Séléna, L.; Betschart, M.; Aissaoui, R.; Nadeau, S. Understanding spatial ans temporal gait asymmetries in individuals post stroke. *Int. J. Phys. Med. Rehabil.* **2014**, *2*. [CrossRef]

55. Koidou, I.; Karelani, A.; Grouios, G. Distribution of subcutaneous body fat on the left and right side of the body in adult overweight right handed women. *Clin. Nutr. ESPEN* **2016**, *13*, e59. [CrossRef]

56. Tishukaj, F.; Shalaj, I.; Gjaka, M.; Wessner, B.; Tschan, H. Lower Limb Force and Power Production and Its Relation to Body Composition in 14-to 15-Year-Old Kosovan Adolescents. *Adv. Phys. Educ.* **2021**, *11*, 61–81. [CrossRef]

57. Bell, D.R.; Sanfilippo, J.L.; Binkley, N.; Heiderscheit, B.C. Lean Mass Asymmetry Influences Force and Power Asymmetry During Jumping in Collegiate Athletes. *J. Strength Cond. Res.* **2014**, *28*, 884–891. [CrossRef]

58. Syed, I.; Davis, B. Obesity and osteoarthritis of the knee: Hypotheses concerning the relationship between ground reaction forces and quadriceps fatigue in long-duration walking. *Med. Hypotheses* **2000**, *54*, 182–185. [CrossRef]

59. Messier, S.P.; Gutekunst, D.; Davis, C.; DeVita, P. Weight loss reduces knee-joint loads in overweight and obese older adults with knee osteoarthritis. *Arthritis Rheum.* **2005**, *52*, 2026–2032. [CrossRef]

60. Baker, R. Gait analysis methods in rehabilitation. *J. Neuroeng. Rehabil.* **2006**, *3*, 4. [CrossRef] [PubMed]

61. Della Croce, U.; Leardini, A.; Chiari, L.; Cappozzo, A. Human movement analysis using stereophotogrammetry: Part 4: Assessment of anatomical landmark misplacement and its effects on joint kinematics. *Gait Posture* **2005**, *21*, 226–237. [CrossRef] [PubMed]

62. Stagni, R.; Leardini, A.; Cappozzo, A.; Benedetti, M.G.; Cappello, A. Effects of hip joint centre mislocation on gait analysis results. *J. Biomech.* **2000**, *33*, 1479–1487. [CrossRef]

63. Kirtley, C. Sensitivity of the Modified Helen Hayes Model to Marker Placement Errors. In Proceedings of the Seventh International Symposium on the 3-D Analysis of Human Movement, Newcastle, UK, 10–12 July 2002.

64. Chambers, A.J.; Sukits, A.L.; McCrory, J.L.; Cham, R. The effect of obesity and gender on body segment parameters in older adults. *Clin. Biomech.* **2010**, *25*, 131–136. [CrossRef] [PubMed]

65. Segal, N.; Yack, H.J.; Khole, P. Weight, Rather Than Obesity Distribution, Explains Peak External Knee Adduction Moment During Level Gait. *Am. J. Phys. Med. Rehabil.* **2009**, *88*, 180–191. [CrossRef]

*Article*

# Measurement of Ankle Joint Movements Using IMUs during Running

Byong Hun Kim [1,2], Sung Hyun Hong [3], In Wook Oh [4], Yang Woo Lee [4], In Ho Kee [4] and Sae Yong Lee [1,2,5,*]

[1] Department of Physical Education, Yonsei University, Seoul 03722, Korea; bh_kim@yonsei.ac.kr
[2] International Olympic Committee Research Centre Korea, Yonsei University, Seoul 03722, Korea
[3] Department of Sports Industry Studies, Yonsei University, Seoul 03722, Korea; hyun-ny-love@hanmail.net
[4] Department of Mechanical Engineering, Yonsei University, Seoul 03722, Korea;
inwookoh@yonsei.ac.kr (I.W.O.); ywlee0305@naver.com (Y.W.L.); kee.inho0@gmail.com (I.H.K.)
[5] Institute of Convergence Science, Yonsei University, Seoul 03722, Korea
* Correspondence: sylee1@yonsei.ac.kr; Tel.: +82-2-2123-6189; Fax: +82-2-2123-8375

**Abstract:** Gait analysis has historically been implemented in laboratory settings only with expensive instruments; yet, recently, efforts to develop and integrate wearable sensors into clinical applications have been made. A limited number of previous studies have been conducted to validate inertial measurement units (IMUs) for measuring ankle joint kinematics, especially with small movement ranges. Therefore, the purpose of this study was to validate the ability of available IMUs to accurately measure the ankle joint angles by comparing the ankle joint angles measured using a wearable device with those obtained using a motion capture system during running. Ten healthy subjects participated in the study. The intraclass correlation coefficient (ICC) and standard error of measurement were calculated for reliability, whereas the Pearson coefficient correlation was performed for validity. The results showed that the day-to-day reliability was excellent (0.974 and 0.900 for sagittal and frontal plane, respectively), and the validity was good in both sagittal ($r = 0.821$, $p < 0.001$) and frontal ($r = 0.835$, $p < 0.001$) planes for ankle joints. In conclusion, we suggest that the developed device could be used as an alternative tool for the 3D motion capture system for assessing ankle joint kinematics.

**Keywords:** validation; kinematic; inertial measurement units; motion analysis; gait

**Citation:** Kim, B.H.; Hong, S.H.; Oh, I.W.; Lee, Y.W.; Kee, I.H.; Lee, S.Y. Measurement of Ankle Joint Movements Using IMUs during Running. *Sensors* **2021**, *21*, 4240. https://doi.org/10.3390/s21124240

Academic Editors: Paolo Capodaglio and Veronica Cimolin

Received: 28 May 2021
Accepted: 17 June 2021
Published: 21 June 2021

**Publisher's Note:** MDPI stays neutral with regard to jurisdictional claims in published maps and institutional affiliations.

## 1. Introduction

The ankle joint is the most frequently involved in human lower body movements, and it plays a vital role in supporting body weight by distributing gravitational and inertial loads. Once injuries, such as strain or sprain, by an external force occur in the ankle joint, they cause deformities in its structure. Impairments of the ankle joint can result in chronic ankle instability; therefore, irregular loading on one side could provoke pain on the ankle. Ankle sprains are common injuries in the general population, as well as among professional athletes [1–3]. The characteristics range from structural deficits such as joint laxity to functional impairments in gait [4]. In terms of rehabilitation, measuring the ankle joint movement pattern during ambulating or running can help clinicians determine the optimal care level a patient should receive.

Many clinical settings for gait training and rehabilitation in patients with motor impairments use a three-dimensional (3D) motion capture system considered the gold standard measurement of joint kinematics [5,6]. The 3D motion capture system is one of the measurement tools with high accuracy, e.g., mean absolute marker-tracking errors of 0.15 mm during static trials [6] and 0.2 mm (with corresponding angle errors of 0.3) during dynamic trials [7]. A VICON system, showing high validity and reliability in measuring joint kinematics, has been used as a suitable comparison tool to examine whether alternative systems, e.g., inertial measurement unit (IMU)-based systems, provide a sufficiently accurate method for motion analysis [8,9]. Although this sophisticated system allows the assessment of kinetic and kinematic data from complicated human movements, it has several limitations. The fact that the 3D motion capture system is a marker-based system requiring many cameras is considered the primary limitation. The high cost of the instrument makes it impractical to use in various settings, such as a clinic, the field, or patients' homes. Furthermore, the system cannot be used to measure and track movements simultaneously.

To overcome the limitations of the 3D motion capture system, many efforts to develop a device that can be simply conducted with a concise process have been made by researchers. Recently, IMUs—a markerless motion capture technology—have been developed as an alternative measurement tool to 3D motion capture devices. An IMU is a wearable-designed device that allows motion measurement data to be sent to a computer in real time and immediate feedback (5) to be received. It collects 3D data (x, y, and z) using a combination of accelerometers, gyroscopes, and magnetometers; it is lighter, smaller, and easier to use than the 3D motion capture system. Collecting and combing raw data from multiple individual sensors are enabled by sensor fusion algorithms, and thus, the estimation of 3D spherical coordinates and Euler angles in a global reference domain can be made [10].

IMUs have been evaluated and shown to be promising in estimating the angular kinematics of lower limb joints, including the hip, knee, and ankle [11–13], as well as upper body posture [14]. However, as IMUs are not easily available to all professionals, due to movement complexity, sensor placement, biomechanical model, and calibration procedure that could increase the risk of error of the measurement, most researchers have tried analyzing the movements of joints conducted in the sagittal plane, such as flexion, extension, and hyperextension movements.

Especially, the errors of measurement values for the ankle in the transverse and frontal planes for gait analysis were large, which might be due to the small range of motion in these planes or the differences in the anatomical or biomechanical definitions between the two systems [15]. However, to the best of our knowledge, the number of previous studies that have investigated the angular kinematics of the ankle joint are limited, and it is necessary to establish validity as a clinical tool to aid in the diagnosis of gait impairment and treatment. Therefore, the purpose of this study was to verify whether the newly developed device can be simply operated with a high accuracy and concise calibration process.

## 2. Materials and Methods

### 2.1. Participants and Data Collection

Ten healthy male participants of $30.2 \pm 5.3$ years, $171 \pm 15.3$ cm, and $73.6 \pm 12.4$ kg body mass were recruited in this study. Exclusion criteria for the study were as follows: individuals who had ankle surgery or nervous system damage or disorder and those with any injuries to the lower limbs within the past three months that could affect the neuromuscular function. The study protocol was approved by the Office of Research Ethics at Yonsei University (IRB No. 7001988-202101-HR-1076-03) and all subjects provided an informed consent, which were compliant with the Declaration of Helsinki.

### 2.2. IMU System and Sensor Placements

A Raspberry Pi 3 Model B+ computer and two Adafruit BNO055 IMU sensors were used (Adafruit, New York, NY, USA) for data collection. Sensor data were collected at a constant frequency of 100 Hz. One IMU sensor was placed on top of the instep of the right foot (Sensor 1), and the second IMU sensor was tightly fixed on the right shin (Sensor 2) using a specially designed holder. The part of holder meeting the right shin has a round shape so it does not move horizontally. The sensors were required to be perfectly parallel to each other, as well as the ground, for accurate calculations. Thus, we designed a shoe mount for Sensor 1 and a sensor holder with a strap for Sensor 2. Both parts comprised a set of an acrylic sensor slide plate and acrylic holder for easy detachment during the calibration process. In addition, we mounted four-corner leveling systems on both holders for leveling. Sensor 2 was fixed as reference coordinates by built-in coordinates (ref-coordinates) of the

BNO055 IMU sensor. Each sensor was directly wired to the single board computer using 3,4 buses for I2C communication. This whole setup was powered by a portable lithium battery with a capacity of 5000 mAh, lasting more than 3 h, which is sufficient for most IMU-based trainings. Eulerian displacements were calculated by subtracting Sensor 1 coordinate data (test-coordinate) from ref-coordinate values. Displacement values of each axis were referred to the yaw, pitch, and roll status.

### 2.3. Vicon System and Marker Placement

The kinematic data were collected at 100 Hz and their positions targeted the capture volume. The calibration of the Vicon system was conducted before each data collection. The Plug-in-Gait (PiG) lower body model was used to analyze movement at the ankle joints. A total of 16 reflective markers were placed on the participants before testing, and a static calibration trial was initially collected to form a musculoskeletal model based on (Figure 1) an 8-camera motion analysis system (VICON, Oxford, UK). The place of markers was attached to the following landmarks: ASIS, PSIS, mid-lateral thigh, lateral knee joint line, lateral mid-shank, lateral malleoli, calcaneal tuberosity, and head of the second metatarsal.

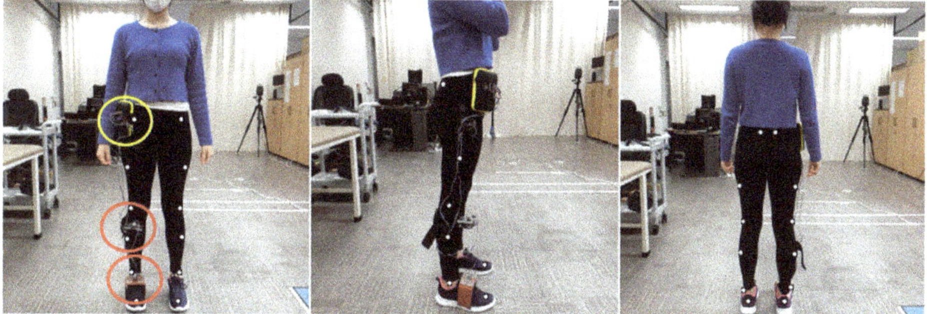

**Figure 1.** Participants' setup.

The participants' specific information of weight, height, ankle width, knee width, and leg length were measured in the lower body model. Figure 1 shows the participants' setup of the anterior, lateral, and posterior views with the markers in place. The PiG model of Vicon was used to evaluate all parameters. The lower body was modeled as seven segments (one pelvis, two thighs, two shanks, and two feet). A normal gait cycle was defined from the initial heel-to-heel contact with the same limb.

### 2.4. IMU Joint Angle Calculations

The proposed IMU sensor includes internal algorithms to calibrate the gyroscope, accelerometer, and magnetometer inside the device. The calibrations of gyroscope, accelerometer, and magnetometer were conducted at the same time the investigator held the device with their hand and shook it in the shape of 8. However, the IMU sensor did not contain any internal electrically erasable programmable read-only memory, so we had to perform the formal calibration process every time the device started up.

After the calibration, raw sensor orientation data were received as types of quaternions. These quaternions needed to be converted to a Euler angle, commonly used units, for easy comprehension. Euler angles were obtained from the quaternions via the following equations [16]:

$$\begin{bmatrix} \varphi \\ \theta \\ \psi \end{bmatrix} = \begin{bmatrix} arctan\frac{2(q_0q_1+q_2q_3)}{1-2(q_1{}^2+q_2{}^2)} \\ arcsin(2(q_0q_2 - q_3q_1)) \\ arctan\frac{2(q_0q_3+q_1q_2)}{1-2(q_2{}^2+q_3{}^2)} \end{bmatrix}, \quad \begin{aligned} q_0 &= q_w = cos(\alpha/2) \\ q_1 &= q_x = sin(\alpha/2)\cos(\beta_x) \\ q_2 &= q_y = sin(\alpha/2)\cos(\beta_y) \\ q_3 &= q_z = sin(\alpha/2)\cos(\beta_z) \end{aligned}$$

where $\varphi, \theta$, and $\psi$ are Euler angles and $q_0, q_1$, $q_2$, and $q_3$ are quaternions. $\alpha$ is a simple rotation angle and $\cos(\beta_x)$, $\cos(\beta_y)$, and $\cos(\beta_z)$ are the direction cosines (Euler's rotation theorem).

*Sensor 1 Euler angle*
$$S_1 = [\varphi_1, \, \theta_1, \, \psi_1]$$
*Sensor 2 Euler angle*
$$S_2 = [\varphi_2, \theta_2, \, \psi_2]$$
*Static Euler angle*
$$S_{st} = [\varphi_{st}, \theta_{st}, \, \psi_{st}]$$
*Ankle joint angle*
$$= [\varphi_1 - \varphi_2 + \varphi_{st}, \quad \theta_1 - \theta_2 + \theta_{st}, \quad \psi_1 - \psi_3 + \psi_{st}]$$

The coordinate system of the IMU sensor was aligned parallel to the floor, and the angle started at $0°$ based on that state. As the sensors and ground were started parallelly (Figure 2), ankle motion was generated by simply subtracting Sensor 1 (ref-coordinate) and Sensor 2 (test-coordinate) angles. The static angle value was added to the subtracted value of the IMU sensor.

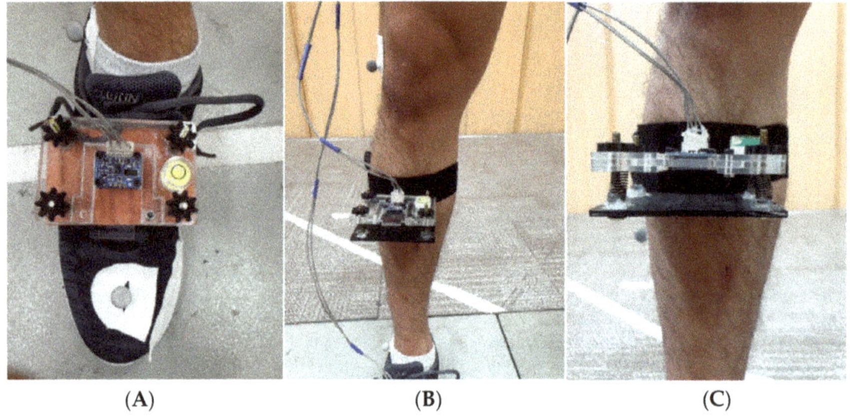

(A)       (B)       (C)

**Figure 2.** Sensor placement: (**A**) Sensor 1, (**B**) Sensor 2, and (**C**) Sensor 2 holder with strap and leveling.

Description of the location of each IMU sensor (red), Raspberry Pi (yellow), and PiG body model marker location for the: (left) anterior view; (middle) lateral view; and (right) posterior view.

### 2.5. Vicon Joint Angle Calculations

Kinematics of the ankle joint were measured using the Vicon PiG model. Sagittal plane motion of the ankle was taken between the shank anterior to posterior axis and the projection of the axis formed by the heel and toe markers into the sagittal plane of the foot.

Furthermore, frontal plane motion of the ankle was taken between the ankle medial to lateral axis, and the projection of the axis formed both malleoli.

Additional information of the PiG angle calculations can be found on Vicon's website.

### 2.6. Experiment Protocol

To evaluate the validity between VICON and IMUs for ankle movements, a functional movement protocol was generated. Along with the reflective markers, two wearable IMU sensors were attached to participants. Participants were asked to perform a running task. Initially, they were instructed to naturally walk to synchronize the position of the markers and sensors as the zero spots and then to try running. The peak point (maximum dorsiflexion (Max DF) to Max DF) of this movement was detected to synchronize the two systems. Participants performed the running task. The data recording protocol consisted of five trials of running (2.68 m/s). Prior to the test, all participants were given time for a 10 min warm up and familiarization session, and they were asked to have a rest of 2 min between each trial.

### 2.7. Data Processing and Statistical Analysis

The motion capture data were considered the gold standard reference for kinematic data for this study. Data from the IMUs and VICON were synchronized by matching them based on the positive peak of the measure by each system [17]. The marker trajectories were imported to Matlab, and joint angles were computed and filtered with Matlab. The five cycles from Vicon and IMUs were synchronized using the positive peak value in the sagittal and frontal planes. The raw data were filtered by a fourth-order Butterworth low-pass filter with a cut-off frequency of 6 Hz, following the recommendation of previous studies [18], to attenuate unwanted noise. Data analysis was performed in Matlab software for running for both sagittal and frontal planes of movement. All data were calculated as averages of all repetitions before being averaged across all participants. All statistical analyses were conducted using SPSS ver. 25.0 (IBM, Armonk, NY, USA). For the test–retest, the intraclass correlation coefficient (ICC) was calculated for each plane of the ankle joint during running for each of the two systems [19]. Pearson's coefficient correlation was performed to verify the relationship of the ankle angle between IMUs and VICON in the sagittal and frontal planes.

## 3. Results

### 3.1. Demographics and Description

Ten male participants (mean $\pm$ standard deviation age: 30.2 $\pm$ 5.3 years; height: 171 $\pm$ 15.3 cm; body mass: 73.6 $\pm$ 12.4 kg) were enrolled in the study. Confirmed consent forms were given from all the participants. A total of 50 trials (running task; five trials per subject) were conducted and analyzed.

### 3.2. Reliability (Test–Retest)

The test–retest reliability of the IMUs in measuring the sagittal and frontal planes with ICC, and its standard error of measurement (SEM), is described in Table 1. A high correlation with ICC (2, 1) values of 0.974 and 0.9 for the sagittal and frontal planes were observed, respectively.

**Table 1.** Intraclass correlation coefficient and SEM of VICON and IMUs for each plane.

| Static Measurement | Sagittal Plane (ICC) | Frontal Plane (ICC) | SEM |
|:---:|:---:|:---:|:---:|
| VICON | 0.978 | 0.969 | 0.39 |
| IMUs | 0.974 | 0.9 | 4.89 |

ICC: intraclass correlation coefficient; SEM: standard error of measurement.

### 3.3. Validity (Pearson's Coefficient Correlation)

The validity test for ankle dorsiflexion/plantarflexion and eversion/inversion is shown in Table 2. Figures 3 and 4 present the sagittal and frontal angles obtained from VICON and IMU systems during the running task, respectively. All planes showed high validity between the pattern of sagittal ($r = 0.821$, $p < 0.001$) and frontal ($r = 0.835$, $p < 0.001$) angles provided by the two systems.

**Table 2.** Pearson's coefficient correlation of sagittal and frontal planes (IMU-based system).

| Measurement | Sagittal Plane | Frontal Plane |
|:---:|:---:|:---:|
| VICON vs. IMUs | 0.821 ** | 0.835 ** |

** $p < 0.001$.

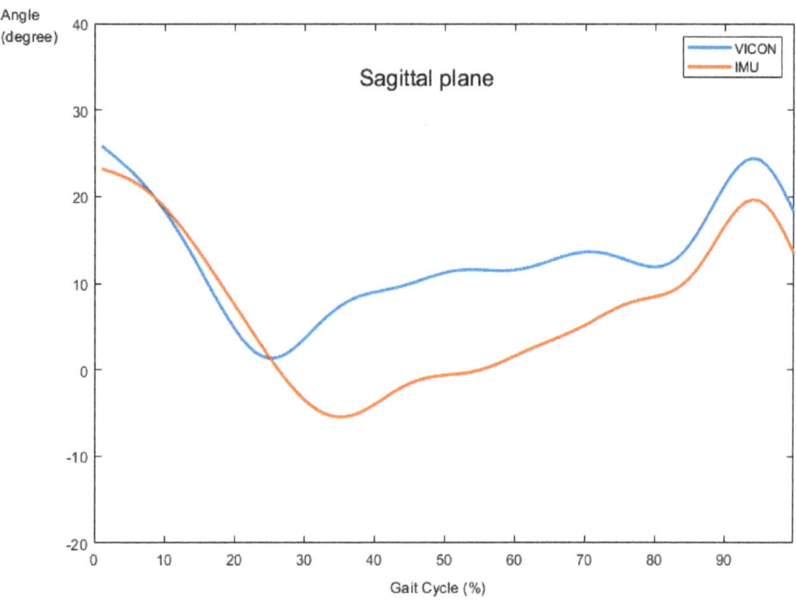

**Figure 3.** Comparison of ankle angle between VICON and IMUs in the sagittal plane.

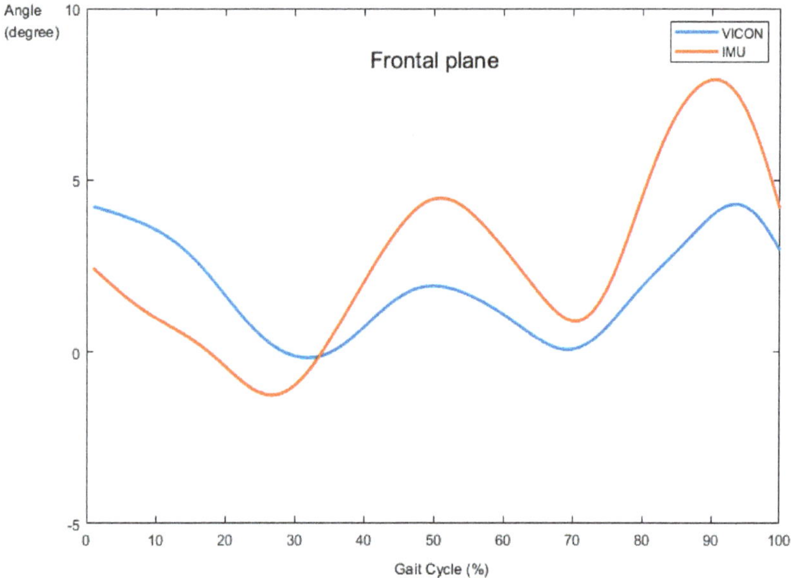

**Figure 4.** Comparison of ankle angle between VICON and IMUs in the frontal plane.

## 4. Discussion

The primary aim of this study was to validate IMU measurement in the sagittal and frontal plane joint kinematics with the VICON system during running. The newly developed IMUs showed excellent reliability between the test and re-test measurements (ICC = X; 0.974, Y; 0.9). The ICC values for kinematic parameters were generally higher or equal to those in other studies, which only assessed the reliability during simple planar movements, such as the sagittal plane [12,15,20–22]. In addition, the validity described as a correlation of the joint angles measured by the two systems was significantly high in the sagittal plane ($r = 0.821$, $p < 0.01$) and frontal plane ($r = 0.835$, $p < 0.01$) during running.

Previous studies have reported a moderate to high validity of IMUs for measuring simple movement but showed low and varied to moderate values for complex movements, such as jumping and running, which may be because the complexity of the movements causes a problem for devices in transmitting and receiving data. Previous studies have suggested that utilizing Blooth, having the frequency-hopping spread spectrum function [23], and low speed transmitter of Wi-Fi [24] might cause technical problems which finally result in lower validity of IMUs in measuring the ankle joint kinematics during fast and complex movements. In this matter, we adopted a direct sequence spread spectrum with Wi-Fi and increased transmitting and receiving speed in our device. Although many efforts to improve the accuracy of data transmission functions of sensors have been made, the limitation of the place where it could be applied persists. However, with Raspberry Pi, the function of acquiring and saving a rapidly varying time-signal with high frequency [25,26], the system-on-chip, which enables the device to save data to its memory room, made it possible to be used in various outdoor activities.

In terms of the validity related to the specific movement—running, in this study—we chose the conventional gait model (PiG) to investigate the relationship between the newly developed device and VICON. One possible limitation of the proposed model is the different location of marker placement on a calculated joint angle, which was used to define the internal and external rotation of the tibia against the line of the ankle joint center, which could cause appreciable errors in ankle joint kinematics, especially the frontal and transverse planes [27]. In this regard, we devised a similar environment as a marker-based system to reduce errors between devices. The sensors were positioned perfectly parallel to each other, as well as the ground, by using a mini-inclinometer for accurate calculations. The results showed that IMUs seemed to be a suitable alternative to motion capture systems in both dorsiflexion/plantarflexion and eversion/inversion movements at the ankle joint during the running task [X: 0.821, Y: 0.835].

High accuracy for assessing the ankle joint movements in the sagittal and frontal planes was a different result from other previous studies [12,22]. According to the previous studies, the poor correlation between VICON and IMUs in measuring inversion and eversion was higher than dorsiflexion and plantarflexion due to its smaller range of motion. Specifically, they reported that if the complexity of movement increases, validity would decrease. In addition, they used only simple planar movement protocols, such as isolated flexion-extension, which may limit the generalizability of their conclusions. Our results extend these previous findings by considering a more challenging task: running.

As complex movements occurred at more than a single plane and with irregular movement velocities affecting system performance [28,29], an accurate method for proper calibration (proper alignment of the IMU axes with the anatomical segment axes) is considered as an essential factor contributing to reliability due to different calibration protocols may potentially result in substantially different consequences [30]. IMUs have been suggested as an alternative tool to the 3D motion capture system, because it provides real-time data in functional tasks within the same error range compared to classical measurement devices. Providing convenience to clinicians in kinematic measurements, it may be useful in clinical settings. Pathological patients such as cerebral palsy, Parkinson's disease, and stroke have been shown to have ankle joint problems during gait, meaning that the analysis of ankle kinematics may be important in prescriptions. Our study showed the high validity of IMUs in measuring the ankle joint (against to VICON), showing the tendency of changes in the degree of the ankle. Therefore, our newly developed device may be useful for clinicians to detect the dysfunctions of their patients.

We acknowledge that our study has two limitations. First, as our sensors could not be rigidly fixed to the shank, a difference in the ankle joint degree between IMU and Vicon occurred due to the fluctuation in sensor motion. Second, although the possibility of application to the measurement in clinical settings has been suggested, we conducted the validation study with healthy individuals.

The development of device will enable us to provide valid data to assess the range of motion and joint orientation, and therefore, rehabilitation research and healthcare services will benefit from IMUs. Although more time and technical resources may be required from users to assess the patients until the system becomes more user-friendly, it will offer convenience with higher accuracy of kinematic measurements in clinical settings. We conducted our validation study with healthy subjects to reduce the error of validity; yet, IMUs need to ultimately benefit pathological populations and clinicians by guiding the clinical decision-making [31]. Therefore, in a future study, special considerations will be needed in pathological populations, as most calibration procedures require specific posture or movement [32].

## 5. Conclusions

We developed a system to measure the ankle joint angle using IMU sensors that are concurrent, convenient, inexpensive (approximately USD 300), light, and portable. Furthermore, it has a function of communicating with a computer via Bluetooth, and the computer is able to immediately calculate the data with Python. In order to validate the device, we compared the ankle X and Y angles data obtained from the IMUs with those acquired from the VICON system. The result of the comparison indicates that the IMUs and motion capture systems deviated the level precision to those well below normal measurements performed in a clinical setting. In the future, we will extend this approach to the pathological populations and, thus, apply it to the IMU-based training that provides multiple body joint angle kinematics in real time.

**Author Contributions:** Conceptualization, B.H.K. and S.Y.L.; Formal analysis, B.H.K.; Investigation, B.H.K. and S.H.H.; Methodology, B.H.K. and S.Y.L.; Project administration, B.H.K.; Resources, B.H.K., S.H.H. and S.Y.L.; Software, I.W.O., Y.W.L. and I.H.K.; Supervision, B.H.K.; Validation, B.H.K.; Writing—original draft, B.H.K.; Writing—review & editing, S.H.H. and S.Y.L. All authors have read and agreed to the published version of the manuscript.

**Funding:** This research received no external funding.

**Institutional Review Board Statement:** The study was conducted according to the guidelines of the Declaration of Helsinki, and approved by the Institutional Review Board of Yonsei University.

**Informed Consent Statement:** Informed consent was obtained from all subjects involved in the study.

**Data Availability Statement:** Not applicable.

**Conflicts of Interest:** The authors declare no conflict of interest.

## References

1. Swenson, D.M.; Collins, C.L.; Fields, S.K.; Comstock, R.D. Epidemiology of US high school sports-related ligamentous ankle injuries, 2005/06–2010/11. *Clin. J. Sport Med. Off. J. Can. Acad. Sport Med.* **2013**, *23*, 190. [CrossRef]
2. Roos, K.G.; Kerr, Z.Y.; Mauntel, T.C.; Djoko, A.; Dompier, T.P.; Wikstrom, E. The epidemiology of lateral ligament complex ankle sprains in National Collegiate Athletic Association sports. *Am. J. Sports Med.* **2017**, *45*, 201–209. [CrossRef]
3. Waterman, B.R.; Owens, B.D.; Davey, S.; Zacchilli, M.A.; Belmont, J.P.J. The epidemiology of ankle sprains in the United States. *JBJS* **2010**, *92*, 2279–2284. [CrossRef] [PubMed]
4. Hertel, J. Sensorimotor deficits with ankle sprains and chronic ankle instability. *Clin. Sports Med.* **2008**, *27*, 353–370. [CrossRef]
5. Wong, W.Y.; Wong, M.S.; Lo, K.H. Clinical applications of sensors for human posture and movement analysis: A review. *Prosthet. Orthot. Int.* **2007**, *31*, 62–75. [CrossRef] [PubMed]
6. Merriaux, P.; Dupuis, Y.; Boutteau, R.; Vasseur, P.; Savatier, X. A study of vicon system positioning performance. *Sensors* **2017**, *17*, 1591. [CrossRef]
7. Smith, A.C. Coach informed biomechanical analysis of the golf swing. Ph.D. Thesis, Loughborough University, Loughborough, UK, 2013; p. 329.
8. Eichelberger, P.; Ferraro, M.; Minder, U.; Denton, T.; Blasimann, A.; Krause, F.; Baur, H. Analysis of accuracy in optical motion capture–A protocol for laboratory setup evaluation. *J. Biomech.* **2016**, *49*, 2085–2088. [CrossRef]

9.  Sessa, S.; Zecca, M.; Lin, Z.; Bartolomeo, L.; Ishii, H.; Takanishi, A. A methodology for the performance evaluation of inertial measurement units. *J. Intell. Robot. Syst.* **2012**, *71*, 143–157. [CrossRef]
10. Filippeschi, A.; Schmitz, N.; Miezal, M.; Bleser, G.; Ruffaldi, E.; Stricker, D. Survey of motion tracking methods based on inertial sensors: A focus on upper limb human motion. *Sensors* **2017**, *17*, 1257. [CrossRef]
11. Bolink, S.; Naisas, H.; Senden, R.; Essers, H.; Heyligers, I.; Meijer, K.; Grimm, B. Validity of an inertial measurement unit to assess pelvic orientation angles during gait, sit–stand transfers and step-up transfers: Comparison with an optoelectronic motion capture system. *Med. Eng. Phys.* **2016**, *38*, 225–231. [CrossRef] [PubMed]
12. Zhang, J.-T.; Novak, A.; Brouwer, B.; Li, Q. Concurrent validation of Xsens MVN measurement of lower limb joint angular kinematics. *Physiol. Meas.* **2013**, *34*, N63–N69. [CrossRef]
13. Cooper, G.; Sheret, I.; McMillian, L.; Siliverdis, K.; Sha, N.; Hodgins, D.; Kenney, L.; Howard, D. Inertial sensor-based knee flexion/extension angle estimation. *J. Biomech.* **2009**, *42*, 2678–2685. [CrossRef]
14. Kang, G.E.; Gross, M.M. Concurrent validation of magnetic and inertial measurement units in estimating upper body posture during gait. *Measurement* **2016**, *82*, 240–245. [CrossRef]
15. Al-Amri, M.; Nicholas, K.; Button, K.; Sparkes, V.; Sheeran, L.; Davies, J.L. Inertial measurement units for clinical movement analysis: Reliability and concurrent validity. *Sensors* **2018**, *18*, 719. [CrossRef] [PubMed]
16. Blanco, J.-L. A tutorial on se (3) transformation parameterizations and on-manifold optimization. *Tech. Rep.* **2010**, *3*, 6.
17. Najafi, B.; Lee-Eng, J.; Wrobel, J.S.; Goebel, R. Estimation of Center of Mass Trajectory using Wearable Sensors during Golf Swing. *J. Sports Sci. Med.* **2015**, *14*, 354–363. [PubMed]
18. Żuk, M.; Pezowicz, C. Kinematic analysis of a six-degrees-of-freedom model based on ISB recommendation: A repeatability analysis and comparison with conventional gait model. *Appl. Bionics Biomech.* **2015**, *2015*, 1–9. [CrossRef] [PubMed]
19. McGraw, K.O.; Wong, S. "Forming inferences about some intraclass correlations coefficients": Correction. *Psychol. Methods* **1996**, *1*, 390. [CrossRef]
20. Kumar, Y.; Yen, S.; Tay, A.; Lee, W.; Gao, F.; Zhao, Z.; Li, J.; Hon, B.; Xu, T.T.; Cheong, A.; et al. Wireless wearable range-of-motion sensor system for upper and lower extremity joints: A validation study. *Healthc. Technol. Lett.* **2015**, *2*, 12–17. [CrossRef] [PubMed]
21. Bergmann, J.H.M.; E Mayagoitia, R.; Smith, C. A portable system for collecting anatomical joint angles during stair ascent: A comparison with an optical tracking device. *Dyn. Med.* **2009**, *8*, 1–7. [CrossRef] [PubMed]
22. Akins, J.S.; Heebner, N.R.; Lovalekar, M.; Sell, T.C. Reliability and validity of instrumented soccer equipment. *J. Appl. Biomech.* **2015**, *31*, 195–201. [CrossRef]
23. Cho, Y.-S.; Jang, S.-H.; Cho, J.-S.; Kim, M.-J.; Lee, H.D.; Lee, S.Y.; Moon, S.-B. Evaluation of Validity and Reliability of Inertial Measurement Unit-Based Gait Analysis Systems. *Ann. Rehabil. Med.* **2018**, *42*, 872–883. [CrossRef]
24. Chhabra, N. Comparative analysis of different wireless technologies. *Int. J. Sci. Res. Netw. Secur. Commun.* **2013**, *1*, 3–4.
25. Tivnan, M.; Gurjar, R.; Wolf, D.E.; Vishwanath, K. High frequency sampling of TTL pulses on a Raspberry Pi for diffuse correlation spectroscopy applications. *Sensors* **2015**, *15*, 19709–19722. [CrossRef]
26. Ambrož, M. Raspberry Pi as a low-cost data acquisition system for human powered vehicles. *Measurement* **2017**, *100*, 7–18. [CrossRef]
27. Ferrari, A.; Benedetti, M.G.; Pavan, E.; Frigo, C.; Bettinelli, D.; Rabuffetti, M.; Crenna, P.; Leardini, A. Quantitative comparison of five current protocols in gait analysis. *Gait Posture* **2008**, *28*, 207–216. [CrossRef] [PubMed]
28. Mifsud, N.L.; Kristensen, N.H.; Villumsen, M.; Hansen, J.; Kersting, U.G. Portable inertial motion unit for continuous assessment of in-shoe foot movement. *Procedia Eng.* **2014**, *72*, 208–213. [CrossRef]
29. Rouhani, H.; Favre, J.; Crevoisier, X.; Aminian, K. Measurement of multi-segment foot joint angles during gait using a wearable system. *J. Biomech. Eng.* **2012**, *134*, 061006. [CrossRef]
30. Bouvier, B.; Duprey, S.; Claudon, L.; Dumas, R.; Savescu, A. Upper limb kinematics using inertial and magnetic sensors: Comparison of sensor-to-segment calibrations. *Sensors* **2015**, *15*, 18813–18833. [CrossRef]
31. Porciuncula, F.; Roto, A.V.; Kumar, D.; Davis, I.; Roy, S.; Walsh, C.J.; Awad, L.N. Wearable movement sensors for rehabilitation: A focused review of technological and clinical advances. *PM&R* **2018**, *10*, S220–S232.
32. Kok, M.; Schön, T. Magnetometer calibration using inertial sensors. *IEEE Sens. J.* **2016**, *16*, 5679–5689. [CrossRef]

Article

# Sensor Network for Analyzing Upper Body Strategies in Parkinson's Disease versus Normative Kinematic Patterns

Paola Romano [1], Sanaz Pournajaf [1], Marco Ottaviani [1,*], Annalisa Gison [1], Francesco Infarinato [1], Claudia Mantoni [1], Maria Francesca De Pandis [2], Marco Franceschini [1,3] and Michela Goffredo [1]

[1] IRCCS San Raffaele Roma, 00166 Rome, Italy; paola.romano@sanraffaele.it (P.R.); sanaz.pournajaf@sanraffaele.it (S.P.); annalisa.gison@sanraffaele.it (A.G.); francesco.infarinato@sanraffaele.it (F.I.); claudia.mantoni@sanraffaele.it (C.M.); marco.franceschini@sanraffaele.it (M.F.); michela.goffredo@sanraffaele.it (M.G.)

[2] San Raffaele Cassino, 03043 Cassino, Italy; maria.depandis@sanraffaele.it

[3] Department of Human Sciences and Promotion of the Quality of Life, San Raffaele University, 00166 Rome, Italy

* Correspondence: marco.ottaviani@sanraffaele.it; Tel.: +39-06-5225-3788

**Abstract:** In rehabilitation, the upper limb function is generally assessed using clinical scales and functional motor tests. Although the Box and Block Test (BBT) is commonly used for its simplicity and ease of execution, it does not provide a quantitative measure of movement quality. This study proposes the integration of an ecological Inertial Measurement Units (IMUs) system for analysis of the upper body kinematics during the execution of a targeted version of BBT, by able-bodied persons with subjects with Parkinson's disease (PD). Joint angle parameters (mean angle and range of execution) and hand trajectory kinematic indices (mean velocity, mean acceleration, and dimensionless jerk) were calculated from the data acquired by a network of seven IMUs. The sensors were applied on the trunk, head, and upper limb in order to characterize the motor strategy used during the execution of BBT. Statistics revealed significant differences ($p < 0.05$) between the two groups, showing compensatory strategies in subjects with PD. The proposed IMU-based targeted BBT protocol allows to assess the upper limb function during manual dexterity tasks and could be used in the future for assessing the efficacy of rehabilitative treatments.

**Keywords:** upper limb; Parkinson's disease; Box and Block test; inertial sensors network; biomechanics analysis; kinematic data; hand trajectories

**Citation:** Romano, P.; Pournajaf, S.; Ottaviani, M.; Gison, A.; Infarinato, F.; Mantoni, C.; De Pandis, M.F.; Franceschini, M.; Goffredo, M. Sensor Network for Analyzing Upper Body Strategies in Parkinson's Disease versus Normative Kinematic Patterns. *Sensors* **2021**, *21*, 3823. https://doi.org/10.3390/s21113823

Academic Editors: Paolo Capodaglio and Veronica Cimolin

Received: 5 May 2021
Accepted: 28 May 2021
Published: 31 May 2021

**Publisher's Note:** MDPI stays neutral with regard to jurisdictional claims in published maps and institutional affiliations.

## 1. Introduction

Upper limb impairment can result from a number of different conditions or pathologies, including stroke, Parkinson's disease, musculoskeletal disorders, infantile cerebral palsy, etc. People who undergo rehabilitation treatments of the upper limb are generally assessed using functional and motor scales [1–4] in order to characterize the efficacy of a specific therapy or the evolution of the disease over time. The performance related to dexterity, strength, upper limb function, and Activities of Daily Living (ADLs) is typically evaluated via a set of validated clinical tests [5,6].

The recovery of manual dexterity is particularly important because the ability to use the hands in a skillful, coordinated way to grasp and manipulate objects is correlated to a good level of quality of life [7]. One of the most used tests to assess manual dexterity is the Box and Blocks Test (BBT) [8], which has been applied in different pathologies such as stroke [9], multiple sclerosis [2], traumatic brain injuries [2], Parkinson's disease [10], and upper limb amputation [11]. The test provides an essential measure for upper limb dexterity and motor coordination and consists of moving, one by one, the maximum number of blocks from one compartment of a box to another of equal size within 60 s. The BBT is commonly used in clinical practice because it is a quick, simple, and inexpensive test [8]. Moreover, it is a well-validated timed performance measure of upper-limb function with good reliability [2]. Specifically, in subjects with Parkinson's disease (PD), the BBT is a good predictor of physical performance in daily living [10]. However, the BBT returns a global score representing the motor task and does not include any assessments of upper limb movement quality. In some cases, in addition to counting the number of cubes moved, clinicians observe a video recorded during the execution of the BBT and qualitatively describe the patient's motor performance. However, in these circumstances, the clinical analysis is subjective, with low inter-rater reliability, and it is time-consuming. To this extent, the instrumented motion analysis during the BBT would be interesting to integrate the assessment of manual dexterity with the study of upper limb movement quality. Specifically, the kinematic analysis during the BBT could allow obtaining an accurate and objective assessment of the movements of the upper limb and trunk, and thus to find potential compensatory strategies used by the subject to perform the task [12,13]. The literature on the instrumented motion analysis during upper limb clinical tests is wide but heterogeneous in terms of the technology employed for the analysis and of the typology of tasks analyzed [14–28]. The most common technologies used for analyzing upper limb kinematics in the clinical setting are stereophotogrammetry and Inertial Measurement Units systems.

The stereophotogrammetry based on reflective markers and optoelectronic sensors has been used in different protocols for the upper limb analysis [15], including modified versions of BBT [16,17]. Specifically, Hebert et al. [16] collected data in 16 able-bodied participants to establish normative kinematics during the BBT. The subjects performed the motor tasks with both arms in standing and seated positions and the results highlighted significant differences between the two conditions in axial trunk rotation, medial-lateral sternum displacement, and anterior-posterior hand displacement. Kontson et al. [17], on the other hand, assessed both upper body kinematics and postural control with an integrated movement analysis framework based on stereophotogrammetry and ground force data. The analysis of 19 able-bodied subjects conducting a modified version of the BBT demonstrated the feasibility of the experimental protocol measure and the average trends of the analyzed population.

Inertial Measurement Units (IMUs) systems have been used as an alternative to stereophotogrammetry because of a series of features that make them easier to use, especially in the clinical setting: ecological environment (outside the movement analysis laboratory), simple application of the sensors (through Velcro strips over the patient's clothing, unlike the stereophotogrammetry where the reflective markers are applied to the skin), and low costs. For these reasons, the literature includes several studies on the use of IMUs for the kinematic analysis of upper limb movement [19–22]. However, the IMU-based quantitative evaluation of clinical arm tests is limited to the Action Research Arm Test [23], the Fugl–Meyer, and the Wolf Motor Function Test in post-stroke subjects [24,25] and three items of the Unified Parkinson's Disease Rating Scale (UPDRS) in subjects with PD [26]. To our best knowledge, only Zhang et al. attempted to automatically assess dexterity with a multimodal wearable sensors-based BBT system [18] based on both electromyography and IMUs. Results from both healthy subjects and people with mild cognitive impairment showed that the multimodal instrumented BBT was feasible and accurate. In this context, although the analysis of upper limb kinematics during a motor task of manual dexterity, such as the BBT, could be particularly relevant in subjects having typical impairment in grasping and manipulating objects, such as PD [13,29], the literature lacks studies on this topic.

This study aims to assess the upper body kinematics during the BBT with an ecological IMU-based system. Specifically, the protocol aims to characterize the movement of the upper body in subjects with PD, comparing them with the data obtained from able-bodied subjects. We hypothesize that the IMU-based BBT would allow us to characterize the quality of the movements and thus quantify the compensatory strategies typical of subjects with PD [13,30,31]. By meeting these objectives, the IMU-based BBT could be a potential system for the standardized assessment of the upper limb.

## 2. Materials and Methods

### 2.1. Study Design

This was an observational single-session-assessment pilot study assessing upper body kinematics during the execution of a BBT motor task, comparing able-bodied persons with subjects with PD. The study was carried out in the neurorehabilitation research laboratory and rehabilitation bioengineering laboratory of IRCCS San Raffaele Roma (Rome, Italy).

### 2.2. Participants

#### 2.2.1. Control Group

Adult able-bodied subjects between 60 and 80 years old without upper limb pathologies (peripheral neurological damage, serious inflammatory degenerative joint diseases, fracture, or trauma results), cognitive and/or severe visual deficit were recruited as the Control Group (CG).

#### 2.2.2. Parkinson's Disease Group

Individuals with idiopathic PD consecutively referred for counseling and outpatient rehabilitation management were included as PD Group (PDG) if they meet the following inclusion criteria: diagnosis of idiopathic PD by UK Brain Bank criteria; Hoehn and Yahr-H&Y stage 2–3; aged between 50 and 80 years old; able to maintain a sitting position on a chair without support for at least 30 min (Trunk Control Test, TCT $\geq$ 48) [32]; moderate disease-related upper limb motor performance deficit (i.e., Unified Parkinson's Disease Rating Scale, UPDRS Part II, items 8, 9, 10, 11, 12, 16 = 2–3; Part III, items 20, 21, 22, 23, 24, 25 = 2–3); stable symptomatic medications during the month before enrollment; and provided written informed consent. We excluded individuals with left-side motor symptom predominance; inability to understand study instructions (Informed Consent Test of Comprehension); cognitive impairment (Montreal Cognitive Impairment Assessment, MoCA < 26 [33]); severe visual deficit; alcohol or drug abuse (including dopamine dysregulation syndrome); active depression; anxiety or psychosis interfering with the use of the equipment or testing; coexisting disabling neurological or orthopedic disorders at upper limb; and previous brain surgery (including pallidotomy, thalamotomy, or deep brain stimulation).

### 2.3. Clinical Assessments

Overall disease-related disability was assessed by the total UPDRS and subtotal UPDRS part II and III scores [34] the trunk stability by TCT, and gross manual dexterity by standard BBT of both dominant and non-dominant sides [5]. All clinical measures were collected in the "ON medication" phase (i.e., 1 h after oral consumption of the usual Levodopa dose and always in the morning to minimize variability). The assessments were by trained professionals. The UPDRS was scored by clinicians specialized in movement disorders and trained for its administration and interpretation.

### 2.4. Experimental Setup

The study took place at the laboratories of IRCCS San Raffaele Roma equipped with the IMU sensors network MOVIT (Captiks srl, Rome, Italy). The experimental setup is shown in Figure 1.

The subject was seated on a stable chair (without backrest and armrests) adjustable in height so that hip and knee angles equal to 90° are formed. A height-adjustable table was placed in front of the subject; the heights of the table and seat were adjusted so that the subject formed 90-degree elbow angles by resting the forearms on the table. A standardized BBT box (53.7 × 25.4 × 8.5 cm³) was placed on the table so that the 15.2 cm high division was in correspondence with the median-lateral axis of the subject and at a distance such that the subject reached the vertex of the box with the distal point of the metacarpal bone of the middle finger.

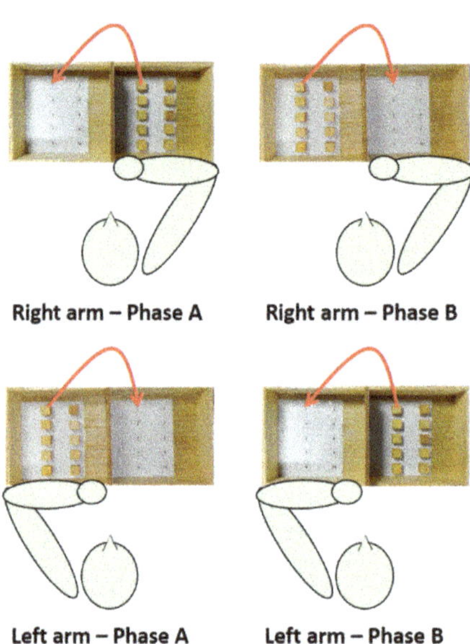


**Figure 1.** Experimental setup of the IMU-based targeted BBT.

After the IMU sensors network calibration phase, seven IMUs were applied using elastic bands fastened with Velcro strip on the following anatomical points: front head; C5; T10; L5; mid arm; mid forearm; and hand (III° metacarpus). Data were collected at a rate of 60 Hz. A digital video camera was also incorporated into the system to capture frontal recordings of the subjects performing each task. After the measurement of anthropometric data (i.e., distances between the following: spinous processes of C5–T10; T10–L5; acromion processes; acromion process–olecranon process; olecranon process–styloid process), the participants were asked to execute the motor task with both the dominant and the non-dominant arm at self-paced velocity.

The motor task was a modified version of the BBT (namely targeted BBT) and consisted of transporting each block over the partition starting with the innermost left block (n° 1), and moving across the rows following the numbering, and placing it in the corresponding position as accurately as possible. Each IMU-based targeted BBT task was composed of two phases, phase A (ipsilateral subtask) and phase B (contralateral subtask). Each task was executed twice with each arm; since the first execution allowed the subject to become familiarized with the experiments, data analysis was conducted on the second execution only.

### 2.5. Data Processing and Statistical Analysis

The Captiks Motion Analyzer software returned the joint angles curves in the sagittal, frontal, transverse planes, and the calibrated quaternions. The following angles were analyzed in the study: wrist Flexion-Extension (F-E); Ulnar Radial Deviation (URD); forearm Prone-Supination (P-S); elbow F-E; shoulder F-E; shoulder Abduction-Adduction (A-A); shoulder Rotation (R); trunk F-E; and trunk R. The data were segmented into ten trials, where the trial start was defined as the initiation of the approach to pick up a block, and the trial end was defined as the release of the block. The angles were analyzed by with an in-house software developed in MATLAB R2020a (The MathWorks, Natick MA, USA). The following joint angle parameters were calculated for each trial and for each subject: the mean angle and the Range of Execution (ROE). The ROE was defined as the difference between the maximum and the minimum values of each joint angle during the motor task. Moreover, the mean temporal trends of each joint angle were plotted with respect to the trial completion percentage.

The 3D hand trajectory was estimated by using calibrated quaternions and anthropometric data. The first objective was to obtain the spatial orientation of each body segment with respect to the absolute reference system acquired during the calibration phase; following software specifications, the calibrated quaternion coefficients allowed to derive the elements of rotation matrix ($R$) of the reference system of each device integral with a body segment with respect to the absolute reference system. Secondly, vector coordinates $\vec{v}$ of distance between consecutive sensors were derived from the anthropometric data. The transformation matrices between two consecutive coordinated systems of each proximal and distal segment couples were obtained by placing column $\vec{v}$ into rotation matrices multiplication, as detailed in the following formula:

$$^{proximal}T_{distal} = \begin{pmatrix} ^{proximal}R * R_{distal} & \vec{v} \\ 0 \quad 0 \quad 0 & 1 \end{pmatrix}$$

Considering the pelvis as a motionless segment during the task, the pose of the hand relative to the pelvis was obtained by concatenating the transformation matrices connecting distal and proximal segments, in accordance with the following formula [35]:

$$^{pelvis}T_{hand} = {}^{pelvis}T_{T10} * {}^{T10}T_{C7} * {}^{C7}T_{arm} * {}^{arm}T_{forearm} * {}^{forearm}T_{hand}$$

The hand trajectory was achieved by selecting the x, y and z axis coordinates from the resulting transformation matrix and the following parameters were calculated: mean velocity ($V_m$); mean acceleration ($A_m$); and DimensionLess Jerk index (DLJ). The DLJ is a measure of the movement smoothness, i.e., as an assessment of the quality of the gesture related to its continuity and interruptions absence [36,37]. In this study, we calculated the DLJ index to estimate the shape of trajectory, considered as the most effective and common smoothness measure [38]. It is defined as follows:

$$\text{DLJ} = -\frac{(t_2 - t_1)^5}{v_{peak}^2} \int_{t_1}^{t_2} \left| \frac{d^2v(t)}{dt^2} \right|^2 dt,$$

where $t_1$ and $t_2$ are the instants of gesture start and end respectively, $v(t)$ is the movement speed and $v_{peak}$ is its maximum in the interval $[t_1, t_2]$. Values of DLJ closer to 0 correspond to a smoother movement shape.

All estimated parameters were averaged within-subject among blocks and then statistical analysis was conducted. Since data were non-normally distributed (Shapiro-Wilk test), the Mann–Whitney test between CG and PDG for each parameter was applied with a significance level set to $p < 0.05$ (IBM SPSS Statistics for Windows, Version 26.0. Armonk, NY, USA: IBM Corp).

### 2.6. Ethical Aspects

This study was conducted in accordance with the Declaration of Helsinki and was approved by the local ethics committee (no. PR 19/34 of December 2019). Participants were included in the study after signing informed consent.

## 3. Results

Thirteen subjects with PD (in the PDG) and eleven able-bodied subjects (in the CG) were enrolled in the study. Two patients in the PDG were excluded from the analysis because of the presence of artifacts in the IMU data. Table 1 describes the clinical and demographic characteristics of the participants included in the study.

**Table 1.** Clinical and demographic characteristics of the enrolled subjects.

|  | CG (N = 11) | PDG (N = 11) |
|---|---|---|
| Age (years) | 66.90 ± 5.80 | 72.00 ± 8.20 |
| Gender Male, n (%) | 6 (54.5%) | 6 (54.5%) |
| BBT—dominant side (n° cubes) | 66.90 ± 10.25 | 56.73 ± 12.97 |
| BBT—non dominant side (n° cubes) | 64.18 ± 8.46 | 52.36 ± 11.27 |
| Affected side dx, n (%) | - | 4 (36.3%) |
| Hoehn&Yahr | - | 2.5 (2–3) |
| UPDRS I | - | 5 (0–8) |
| UPDRS II | - | 19 (13–22) |
| UPDRS III | - | 20 (18–32) |
| UPDRS VI | - | 5 (0–10) |
| UPDRS TOT | - | 51 (38–67) |
| TCT | - | 61 (42–87) |

*Abbreviations: BBT, Box and Blocks Test; UPDRS, Unified Parkinson's Disease Rating Scale; TCT, Trunk Control Test; CG, Control Group; PDG, Parkinson's Disease Group. Notes: Data are reported as mean ± standard deviation or frequency with percentage (%) or median (min–max).*

All participants conducted the IMU-based targeted BBT tasks without any difficulties. The data analysis calculated the joint angle parameters shown in Table 2.

**Table 2.** Joint angles parameters calculated for each phase (A and B) of the targeted BBT tasks.

| | | | Dominant Arm | | | | Non-Dominant Arm | | | |
|---|---|---|---|---|---|---|---|---|---|---|
| | | | Phase A | | Phase B | | Phase A | | Phase B | |
| Joint | | Group | Mean Angle | ROE | Mean Angle | ROE | Mean Angle | ROE | Mean Angle | ROE |
| Wrist | F-E | CG | 7.7 ± 17.5 | 19.4 ± 8.3 | 8.7 ± 18.2 | 16.1 ± 6.4 | 14.2 ± 13.7 | 18.8 ± 8.9 | 11.9 ± 13.2 | 15.5 ± 7.2 |
| | | PDG | **22.4** ± **23.6** | **26.9** ± **12.5** | **24.5** ± **23.5** | 22.3 ± **13.8** | 18.8 ± 14.4 | 21.5 ± 7.6 | **22.7** ± 14.9 | 17.6 ± 7.8 |
| | URD | CG | 6.0 ± 9.6 | 20.3 ± 8.4 | −1.9 ± 9.8 | 11.6 ± 6.4 | 2.2 ± 13.5 | 23.9 ± 10.6 | −4.7 ± 13.9 | 17.8 ± 11.1 |
| | | PDG | 9.4 ± 18.7 | 20.7 ± 11.8 | 7.3 ± 20.4 | 18.8 ± 13.0 | 10.7 ± 19.7 | 16.5 ± 6.7 | 6.8 ± 20.7 | 14.3 ± 10.3 |
| Forearm | P-S | CG | **105.1** ± 22.6 | 13.9 ± 9.0 | **108.7** ± 23.2 | 11.5 ± 6.7 | **104.8** ± 12.9 | 11.7 ± 3.8 | **106.8** ± 14.3 | 12.0 ± 3.7 |
| | | PDG | **92.8** ± 15.2 | 19.9 ± 16.1 | **95.7** ± 14.8 | 18.6 ± 12.5 | **97.8** ± 16.5 | 13.6 ± 9.2 | **100.9** ± 16.3 | 13.5 ± 7.8 |
| Elbow | F-E | CG | 85.6 ± 24.2 | 17.7 ± 7.0 | 87.7 ± 26.1 | 19.4 ± 7.8 | 76.5 ± 14.4 | 19.0 ± 8.2 | 77.4 ± 15.4 | 19.8 ± 9.1 |
| | | PDG | 89.5 ± 41.5 | 17.6 ± 7.7 | 90.9 ± 41.6 | 19.5 ± 8.6 | 79.3 ± 27.0 | 25.3 ± 46.6 | 74.8 ± 44.0 | 38.4 ± 76.3 |
| Shoulder | F-E | CG | 29.5 ± 12.4 | **35.7** ± 11.5 | 26.5 ± 13.4 | **38.5** ± 12.0 | 29.0 ± 10.3 | **38.2** ± 9.7 | 27.3 ± 10.1 | **39.7** ± 9.9 |
| | | PDG | 27.2 ± 27.8 | **26.2** ± 11.4 | 24.3 ± 27.3 | **28.3** ± 13.2 | **42.5** ± 28.5 | 34.8 ± 12.0 | **38.8** ± 33.3 | 34.7 ± 12.8 |
| | A-A | CG | 40.3 ± 14.3 | **25.6** ± 7.0 | 44.4 ± 14.2 | **29.3** ± 8.0 | 37.1 ± 10.2 | **26.4** ± 7.9 | 39.5 ± 10.2 | **29.8** ± 7.6 |
| | | PDG | 39.1 ± 20.3 | **21.2** ± 8.5 | 41.2 ± 18.5 | **25.2** ± 9.5 | 36.1 ± 14.6 | **21.7** ± 12.9 | 37.4 ± 14.2 | **26.7** ± 14.1 |
| | R | CG | −39.8 ± 13.1 | 38.6 ± 12.2 | −44.0 ± 15.3 | 41.1 ± 12.3 | −39.9 ± 10.0 | 40.9 ± 10.1 | −41.8 ± 10.9 | 41.9 ± 9.2 |
| | | PDG | −52.8 ± 32.0 | 28.5 ± 12.3 | −56.5 ± 30.9 | 31.2 ± 11.8 | −34.9 ± 20.8 | 26.8 ± 9.3 | −41.0 ± 26.0 | 28.7 ± 11.2 |
| Trunk | F-E | CG | 10.8 ± 7.9 | **2.8** ± 1.6 | 12.0 ± 7.8 | 2.5 ± 1.7 | 8.6 ± 4.3 | 3.0 ± 1.4 | 9.6 ± 4.3 | 3.0 ± 1.6 |
| | | PDG | 11.7 ± 6.1 | **4.3** ± 2.3 | 11.7 ± 6.6 | 4.6 ± 3.7 | **15.2** ± 8.1 | 4.0 ± 2.3 | **15.6** ± 8.1 | 3.4 ± 2.2 |
| | R | CG | **3.0** ± 6.6 | 6.9 ± 3.8 | **3.5** ± 7.3 | 7.4 ± 3.5 | −10.5 ± 9.0 | 8.0 ± 3.1 | −11.9 ± 10.8 | 8.6 ± 2.9 |
| | | PDG | **6.0** ± 19.0 | 8.1 ± 4.3 | **5.7** ± 19.8 | 9.8 ± 4.8 | −5.9 ± 10.2 | 7.4 ± 3.9 | −6.2 ± 13.1 | 8.4 ± 3.6 |

*Abbreviations: F-E, Flexion-Extension; URD, Ulnar Radial Deviation; P-S, Prone-Supination; A-A, Abduction-Adduction; R, Rotation. CG, Control Group; PDG, Parkinson's Disease Group. Notes: Data are reported as mean ± standard deviation. The data marked in bold denotes significant inter-group difference (p < 0.05).*

The mean joint angles registered significant differences between the PDG and the CG in the following angles: wrist F-E (both arms, both phases) and URD (both arms, both phases); forearm P-S (both arms, both phases); shoulder F-E (non-dominant arm, both phases) and R (both arms, both phases); trunk F-E (non-dominant arm, both phases) and R (both arms, both phases).

The ROE index exhibits statistically significant inter-group differences in the following angles: wrist F-E (both arms, both phases) and URD (dominant arm phase B; non-dominant arm both phases); forearm P-S (dominant arm phase A); shoulder F-E (both arms, both phases), A-A (both arms, both phase) and R (both arms, both phases); trunk F-E (dominant arm both phases; non-dominant arm phase A) and R (non-dominant arm phase A).

Figure 2 depicts the mean joint angle trajectories (dominant arm) over the trial complexion % for both phase A and B (the highlighted line and the shaded color represent the averaged trajectory among blocks and subjects and its standard error, respectively). The analysis of the angle trends revealed the proposed IMU-based targeted BBT protocol is able to detect different motor strategies employed during the movement execution. Specifically, the kinematics of the PDG is characterized by a limited range of movement of the shoulder and a compensatory strategy of the trunk.

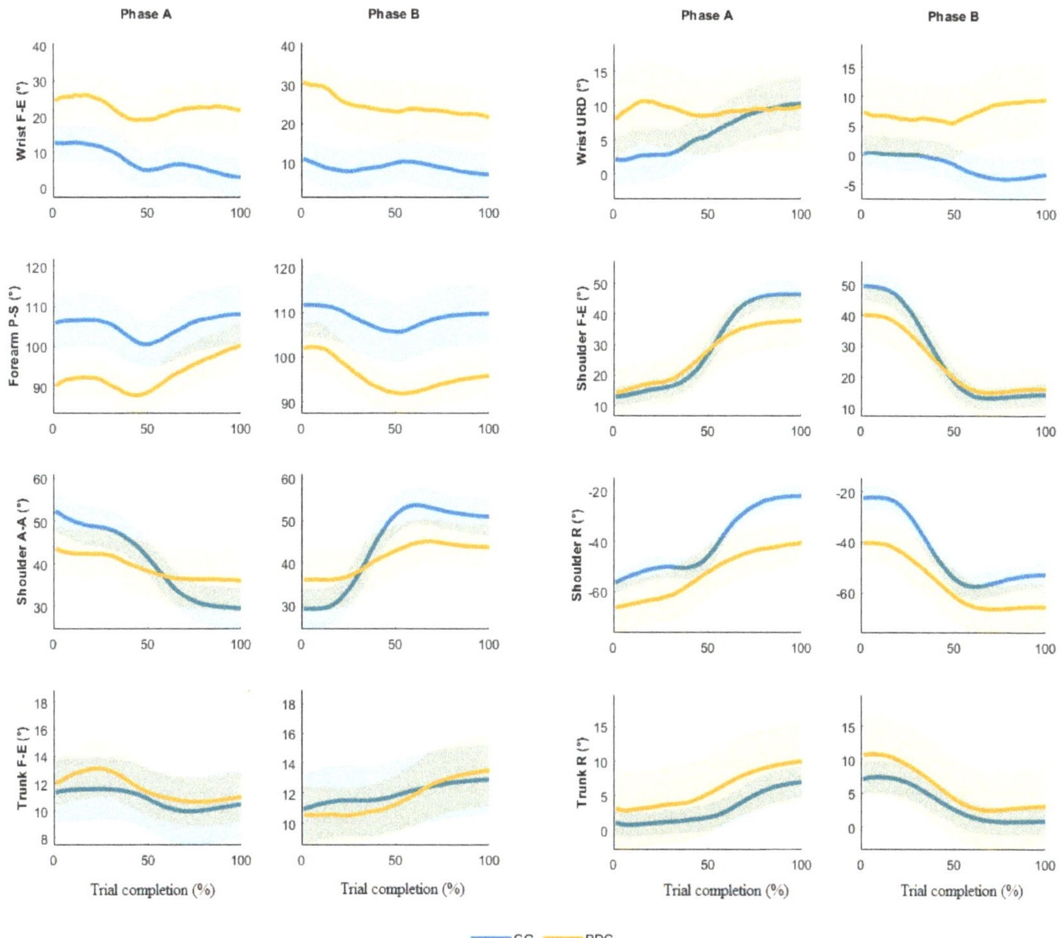

**Figure 2.** Time-normalized angle joints in phases A and B, dominant side.

In Figure 3, the averaged dominant hand trajectories are shown for both phase A and B, considering each block separately, highlighting differences especially in grasping and moving the more proximal blocks (from number 5 to number 10). The kinematic parameters calculated from the hand trajectories are depicted in Table 3; statistically significant inter-group differences have been found in all parameters. The mean velocity and the mean acceleration showed significantly lower values in PDG than CG. The DLJ index revealed that subjects with PD had lower movement smoothness than the able-bodied ones.

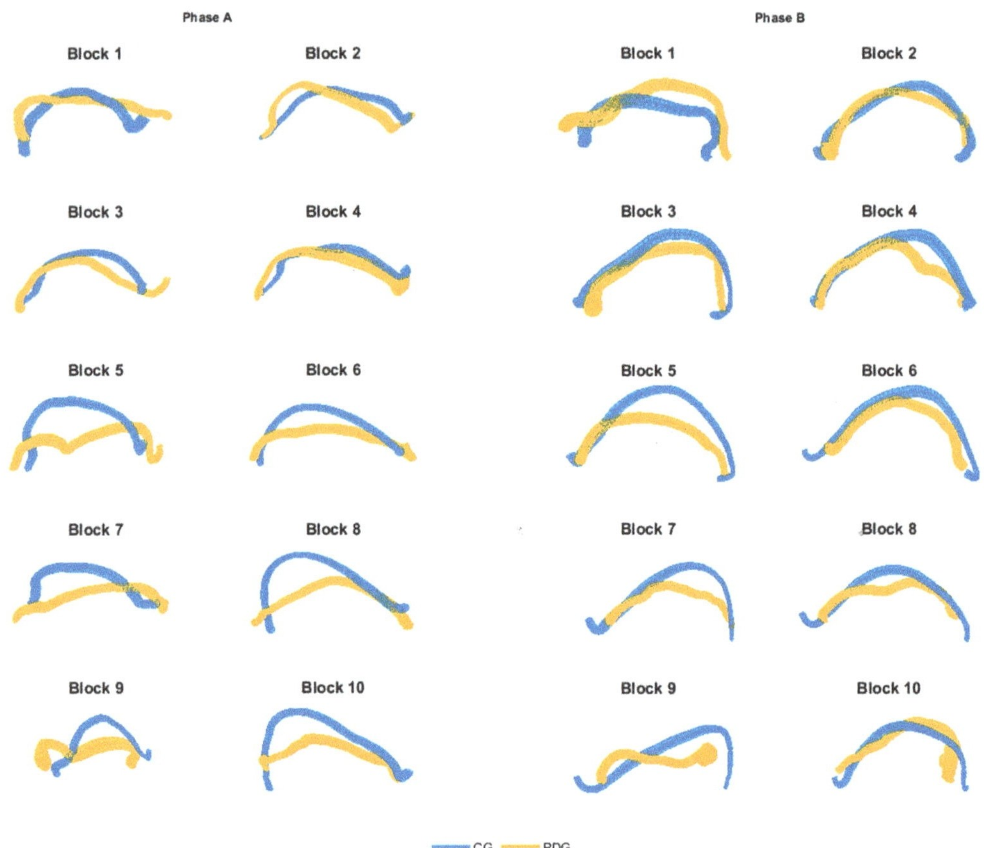



**Figure 3.** Averaged hand trajectories for both CG and PDG for each block of phase A and B (dominant side). Trajectories are obtained from the projection of 3D curves in the 2D coronal plane. The thickness of lines represents the dispersion of data around the mean trajectory.

**Table 3.** Averaged hand trajectory parameters calculated from the dominant and non-dominant arms in phases A and B.

| | | Hand Trajectory Parameters | | | | | | | | | | | |
| --- | --- | --- | --- | --- | --- | --- | --- | --- | --- | --- | --- | --- | --- |
| **Parameter** | **Group** | **Dominant** | | | | | | **Non-Dominant** | | | | | |
| | | Phase A | | | Phase B | | | Phase A | | | Phase B | | |
| $V_m$ (cm/s) | CG | 40.45 | ± | 2.66 | 42.13 | ± | 2.86 | 41.11 | ± | 2.78 | 38.47 | ± | 2.69 |
| | PDG | 31.79 | ± | 3.36 | 36.33 | ± | 3.73 | 37.22 | ± | 3.41 | 35.69 | ± | 3.50 |
| $A_m$ (cm/s$^2$) | CG | 687.45 | ± | 59.05 | 695.96 | ± | 65.33 | 665.95 | ± | 51.92 | 602.55 | ± | 48.88 |
| | PDG | 570.18 | ± | 65.98 | 622.77 | ± | 79.83 | 618.15 | ± | 63.82 | 548.90 | ± | 54.95 |
| DLJ | CG | −2.95 | ± | 0.78 | −2.40 | ± | 0.49 | −2.63 | ± | 0.69 | −2.97 | ± | 0.69 |
| | PDG | −8.85 | ± | 6.54 | −7.21 | ± | 4.78 | −4.91 | ± | 1.48 | −4.56 | ± | 1.59 |

*Abbreviations: $V_m$, mean velocity; $A_m$, mean acceleration; DLJ, DimensionLess Jerk index; CG, Control Group; PDG, Parkinson's Disease Group. Notes: Data are reported as mean ± standard deviation. The data marked in bold denote significant inter-group difference ($p < 0.05$).*

## 4. Discussion

This observational pilot study was conducted on 11 subjects with PD compared to 11 able-bodied subjects in order to assess the upper body kinematics during the targeted BBT with an ecological IMU-based system. To this extent, the IMU-based targeted BBT was analyzed for both the dominant and non-dominant upper limbs. All subjects were able to easily perform the requested motor tasks.

The analysis of the IMU data allowed us to calculate the joint angle kinematics. The outcomes from the able-bodied subjects were in accordance with the literature on similar studies based on stereophotogrammetry [16,17]. The results from subjects with PD allowed us to characterize the quality of the movements and the compensatory strategies typical of this disease [13,30,31]. Specifically, the wrist evidenced a significant higher mean flexion and ulnar deviation in the PDG compared to the CG, while the wrist F-E mean angular trajectories (Figure 2) were similar in both groups. The ROEs were significantly higher in the PDG than the CG in all wrist angles except for the URD dominant arm, Phase A. The forearm registered significantly higher supination values in the CG compared to the PDG; the variations of this angle over time were similar during the trial execution. The shoulder depicts significant lower ROEs in PDG than CG in F-E, A-A and R angles. Moreover, the mean shoulder F-E trajectories of the two groups were similar from 0% to 50% of trial completion, while when the block was carried over the partition the PDG evidenced a reduced shoulder flexion. This outcome was found in both dominant and non-dominant arms and in both phases A and B. The shoulder A-A is characterized by a significant smaller ROE in all motor tasks, thus revealing a limited angular excursion in PDG. Conversely, the PDG significantly rotated the shoulder more than the CG, showing significant inter-group differences in the mean angle and the ROE. Therefore, we can affirm that the PDG partially involved the shoulder during the execution of the motor tasks, except for the shoulder R, which seems to compensate for the limited ROE in FE and AA. The trunk exhibited higher ROEs in PDG than CG, thus confirming the employment of a compensatory strategy in subjects with PD [13,30,31].

The qualitative analysis of the hand trajectories (Figure 3) showed that the subjects with PD moved the end-effector like the able-bodied subjects in the movement of the first blocks (grasp and release of blocks 1–4), while they tended to decrease the range of motion and the precision in the subsequently blocks (grasp and release of blocks 5–10). Moreover, the PDG executed the movement with a significant lower mean velocity and mean acceleration of the hand in all considered motor tasks. The DLJ shows that the subjects with PD moved the end-effector with lower smoothness, in accordance with the literature on the quantitative analysis of bradykinesia and rigidity in PD [29,31].

The results of this study evidence that the proposed IMU-based targeted BBT is able to quantitatively and easily assess upper body kinematics during a test of manual dexterity. Moreover, the analysis of joint angle trajectories allows to characterize movements' quality and to find the compensatory strategies of subjects with PD. The analysis of such compensatory motor approaches could help the understanding of functional gain in a perspective of personalized punctual evaluation of patients with PD undergoing rehabilitative treatments.

The proposed IMU-based targeted BBT protocol is feasible, easy-to-do, low-cost and ecological. All recruited subjects participated in the experiments and executed the motor tasks without any difficulty. In a period in which motor rehabilitation increasingly needs an objectification of motor performance to personalize treatment, this system allows performing a quantitative movement analysis easily and accurately in the clinical setting.

The main limitations of the study are the restricted number of recruited subjects and the inclusion of PD subjects with a moderate impairment only. Future studies should consider a higher sample size to confirm our outcomes. Moreover, the analysis of subjects with different pathologies and motor impairment would allow us to discriminate different motor strategies.

## 5. Conclusions

An IMU-based targeted BBT allowed to analyze the upper body kinematics in subjects with PD and able-bodied persons. The analysis of joint angles and hand trajectories characterized the quality of the movements in the two groups and evidenced the compensatory strategies of subjects with PD. The obtained results suggest future studies on different pathologies since the IMU-based BBT could be a potential system for the standardized assessment of the upper limb in the clinical setting.

**Author Contributions:** Conceptualization, F.I., C.M. and M.G.; methodology, P.R., M.O, M.G. and F.I.; software, M.O., P.R. and C.M.; validation, A.G., M.F. and M.F.D.P.; formal analysis, M.O. and P.R.; investigation, M.G., S.P., M.O. and P.R.; resources, F.I. and M.F.; data curation, M.O. and P.R.; writing—original draft preparation, P.R., M.G., S.P. and M.O.; writing—review and editing, F.I., M.F., A.G., and M.F.D.P.; supervision, F.I., M.G.; project administration, P.R., M.O., M.G. and S.P.; funding acquisition, F.I. and M.F. All authors have read and agreed to the published version of the manuscript.

**Funding:** This project was partially funded by the Ministry of Health of Italy (Ricerca corrente).

**Institutional Review Board Statement:** The study was conducted according to the guidelines of the Declaration of Helsinki and approved by the Institutional Ethics Committee of IRCCS San Raffaele Roma (date: 22 January 2020; protocol code: RP19/34).

**Informed Consent Statement:** Informed consent was obtained from all subjects involved in the study.

**Data Availability Statement:** The data presented in this study are available on request from the corresponding author.

**Conflicts of Interest:** The authors declare no conflict of interest.

# References

1.  Murphy, M.A.; Resteghini, C.; Feys, P.; Lamers, I. An overview of systematic reviews on upper extremity outcome measures after stroke. *BMC Neurol.* **2015**, *15*, 1–15. [CrossRef]
2.  Platz, T.; Pinkowski, C.; Van Wijck, F.; Kim, I.-H.; Di Bella, P.; Johnson, G. Reliability and validity of arm function assessment with standardized guidelines for the Fugl-Meyer Test., Action Research Arm Test. and Box and Block Test.: A multicentre study. *Clin. Rehabil.* **2005**, *19*, 404–411. [CrossRef]
3.  Wang, S.; Hsu, C.J.; Trent, L.; Ryan, T.; Bs, N.T.K.; Civillico, E.F.; Kontson, K.L. Evaluation of Performance-Based Outcome Measures for the Upper Limb: A Comprehensive Narrative Review. *PM&R* **2018**, *10*, 951–962.
4.  Seccia, R.; Boresta, M.; Fusco, F.; Tronci, E.; Di Gemma, E.; Palagi, L.; Mangone, M.; Agostini, F.; Bernetti, A.; Santilli, V.; et al. Data of patients undergoing rehabilitation programs. *Data Brief* **2020**, *30*, 105419. [CrossRef] [PubMed]
5.  Franceschini, M.; Goffredo, M.; Pournajaf, S.; Paravati, S.; Agosti, M.; De Pisi, F.; Galafate, D.; Posteraro, F. Predictors of activities of daily living outcomes after upper limb robot-assisted therapy in subacute stroke patients. *PLoS ONE* **2018**, *13*, e0193235. [CrossRef]
6.  Goffredo, M.; Mazzoleni, S.; Gison, A.; Infarinato, F.; Pournajaf, S.; Galafate, D.; Agosti, M.; Posteraro, F.; Franceschini, M. Kinematic Parameters for Tracking Patient Progress during Upper Limb Robot.-Assisted Rehabilitation: An. Observational Study on Subacute Stroke Subjects. *Appl. Bionics Biomech.* **2019**, *2019*, 4251089. [CrossRef]
7.  McEwen, S.; Mayo, N.; Wood-Dauphinee, S. Inferring quality of life from performance-based assessments. *Disabil. Rehabil.* **2000**, *22*, 456–463. [CrossRef] [PubMed]
8.  Mathiowetz, V.; Volland, G.; Kashman, N.; Weber, K. Adult norms for the Box and Block Test. of manual dexterity. *Am. J. Occup. Ther.* **1985**, *39*, 386–391. [CrossRef] [PubMed]
9.  Chen, H.-M.; Chen, C.C.; Hsueh, I.-P.; Huang, S.-L.; Hsieh, C.-L. Test-retest reproducibility and smallest real difference of 5 hand function tests in patients with stroke. *Neurorehabilit. Neural Repair* **2009**, *23*, 435–440. [CrossRef] [PubMed]
10. Choi, Y.I.; Song, C.S.; Chun, B.Y. Activities of daily living and manual hand dexterity in persons with idiopathic parkinson disease. *J. Phys. Ther. Sci.* **2017**, *29*, 457–460. [CrossRef]
11. Hebert, J.S.; Lewicke, J. Case report of modified Box and Blocks test with motion capture to measure prosthetic function. *J. Rehabil. Res. Dev.* **2012**, *49*, 1163–1174. [CrossRef] [PubMed]
12. Collins, K.C.; Kennedy, N.C.; Clark, A.; Pomeroy, V.M. Getting a kinematic handle on reach-to-grasp: A meta-analysis. *Physiotherapy* **2018**, *104*, 153–166. [CrossRef] [PubMed]
13. Flash, T.; Inzelberg, R.; Schechtman, E.; Korczyn, A.D. Kinematic analysis of upper limb trajectories in Parkinson's disease. *Exp. Neurol.* **1992**, *118*, 215–226. [CrossRef]
14. Gates, D.H.; Walters, L.S.; Cowley, J.C.; Wilken, J.M.; Resnik, L. Range of Motion Requirements for Upper-Limb Activities of Daily Living. *Am. J. Occup. Ther.* **2015**, *70*, 7001350010. [CrossRef]
15. Valevicius, A.M.; Jun, P.; Hebert, J.S.; Vette, A.H. Use of optical motion capture for the analysis of normative upper body kinematics during functional upper limb tasks: A systematic review. *J. Electromyogr. Kinesiol.* **2018**, *40*, 1–15. [CrossRef]
16. Hebert, J.S.; Lewicke, J.; Williams, T.R.; Vette, A.H. Normative data for modified Box and Blocks test measuring upper-limb function via motion capture. *J. Rehabil. Res. Dev.* **2014**, *51*, 919–931. [CrossRef]
17. Kontson, K.; Marcus, I.; Myklebust, B.; Civillico, E. Targeted box and blocks test: Normative data and comparison to standard tests. *PLoS ONE* **2017**, *12*, e0177965. [CrossRef]

18. Zhang, Y.; Chen, Y.; Yu, H.; Lv, Z.; Shang, P.; Ouyang, Y.; Yang, X.; Lu, W. Wearable Sensors Based Automatic Box and Block Test System. In Proceedings of the 2019 IEEE SmartWorld, Ubiquitous Intelligence & Computing, Advanced & Trusted Computing, Scalable Computing & Communications, Cloud & Big Data Computing, Internet of People and Smart City Innovation (SmartWorld/SCALCOM/UIC/ATC/CBDCom/IOP/SCI), Leicester, UK, 19–23 August 2019.

19. Zhou, H.; Stone, T.; Hu, H.; Harris, N. Use of multiple wearable inertial sensors in upper limb motion tracking. *Med. Eng. Phys.* **2008**, *30*, 123–133. [CrossRef] [PubMed]

20. Filippeschi, A.; Schmitz, N.; Miezal, M.; Bleser, G.; Ruffaldi, E.; Stricker, D. Survey of Motion Tracking Methods Based on Inertial Sensors: A Focus on Upper Limb Human Motion. *Sensors* **2017**, *17*, 1257. [CrossRef]

21. Bai, L.; Pepper, M.G.; Yan, Y.; Spurgeon, S.K.; Sakel, M.; Phillips, M. Quantitative Assessment of Upper Limb Motion in Neurorehabilitation Utilizing Inertial Sensors. *IEEE Trans. Neural Syst. Rehabil. Eng.* **2015**, *23*, 232–243. [CrossRef] [PubMed]

22. Pérez, R.; Costa, Ú.; Torrent, M.; Solana, J.; Opisso, E.; Cáceres, C.; Tormos, J.M.; Medina, J.; Gómez, E.J. Upper Limb Portable Motion Analysis System Based on Inertial Technology for Neurorehabilitation Purposes. *Sensors* **2010**, *10*, 10733–10751. [CrossRef]

23. Carpinella, I.; Cattaneo, D.; Ferrarin, M. Quantitative assessment of upper limb motor function in Multiple Sclerosis using an instrumented Action Research Arm Test. *J. Neuroeng. Rehabil.* **2014**, *11*, 67. [CrossRef] [PubMed]

24. Patel, S.; Hughes, R.; Hester, T.; Stein, J.; Akay, M.; Dy, J.; Bonato, P. Tracking Motor Recovery in Stroke Survivors Undergoing Rehabilitation Using Wearable Technology. In Proceedings of the 2010 Annual International Conference of the Ieee Engineering in Medicine and Biology Society (Embc), Buenos Aires, Argentina, 31 August–4 September 2010; pp. 6858–6861.

25. Zhang, M.M.; Lange, B.; Chang, C.-Y.; Sawchuk, A.A.; Rizzo, A.A. Beyond the Standard Clinical Rating Scales: Fine-Grained Assessment of Post-Stroke Motor Functionality Using Wearable Inertial Sensors. In Proceedings of the 2012 Annual International Conference of the Ieee Engineering in Medicine and Biology Society (Embc), San Diego, CA, USA, 28 August–1 September 2012; pp. 6111–6115.

26. Hoffman, J.D.; McNames, J. Objective Measure of Upper Extremity Motor Impairment in Parkinson's Disease with Inertial Sensors. In Proceedings of the 2011 Annual International Conference of the Ieee Engineering in Medicine and Biology Society (Embc), Boston, MA, USA, 30 August–3 September 2011; pp. 4378–4381.

27. Goffredo, M.; Schmid, M.; Conforto, S.; D'Alessio, T. 3D Reaching in Visual Augmented Reality Using KinectTM: The Perception of Virtual Target. In *Converging Clinical and Engineering Research on Neurorehabilitation*; Springer: Berlin/Heidelberg, Germany, 2013.

28. Goffredo, M.; Carli, M.; Conforto, S.; Bibbo, D.; Neri, A.; D'Alessio, T. Evaluation of skin and muscular deformations in a non-rigid motion analysis. Medical Imaging. *SPIE* **2005**, *5746*, 535–541.

29. di Biase, L.; Summa, S.; Tosi, J.; Taffoni, F.; Marano, M.; Rizzo, A.C.; Vecchio, F.; Formica, D.; Di Lazzaro, V.; Di Pino, G. Quantitative Analysis of Bradykinesia and Rigidity in Parkinson's Disease. *Front. Neurol.* **2018**, *9*, 121. [CrossRef]

30. Lukos, J.R.; Snider, J.; Hernandez, M.E.; Tunik, E.; Hillyard, S.; Poizner, H. Parkinson's disease patients show impaired corrective grasp control and eye-hand coupling when reaching to grasp virtual objects. *Neuroscience* **2013**, *254*, 205–221. [CrossRef]

31. Tresilian, J.R.; Stelmach, G.E.; Adler, C.H. Stability of reach-to-grasp movement patterns in Parkinson's disease. *Brain* **1997**, *120 Pt 11*, 2093–2111. [CrossRef]

32. Verheyden, G.; Nieuwboer, A.; Van de Winckel, A.; De Weerdt, W. Clinical tools to measure trunk performance after stroke: A systematic review of the literature. *Clin. Rehabil.* **2007**, *21*, 387–394. [CrossRef]

33. Gill, D.J.; Freshman, A.; Blender, J.A.; Ravina, B. The Montreal cognitive assessment as a screening tool for cognitive impairment in Parkinson's disease. *Mov. Disord.* **2008**, *23*, 1043–1046. [CrossRef] [PubMed]

34. Shulman, L.M.; Gruber-Baldini, A.L.; Anderson, K.E.; Fishman, P.S.; Reich, S.G.; Weiner, W.J. The clinically important difference on the unified Parkinson's disease rating scale. *Arch. Neurol.* **2010**, *67*, 64–70. [CrossRef]

35. Repnik, E.; Puh, U.; Goljar, N.; Munih, M.; Mihelj, M. Using Inertial Measurement Units and Electromyography to Quantify Movement during Action Research Arm Test. Execution. *Sensors* **2018**, *18*, 2767. [CrossRef] [PubMed]

36. Engdahl, S.M.; Gates, D.H. Reliability of upper limb movement quality metrics during everyday tasks. *Gait Posture* **2019**, *71*, 253–260. [CrossRef] [PubMed]

37. Hogan, N.; Sternad, D. Sensitivity of smoothness measures to movement duration, amplitude, and arrests. *J. Mot. Behav.* **2009**, *41*, 529–534. [CrossRef] [PubMed]

38. Balasubramanian, S.; Melendez-Calderon, A.; Roby-Brami, A.; Burdet, E. On the analysis of movement smoothness. *J. Neuroeng. Rehabil.* **2015**, *12*, 112. [CrossRef] [PubMed]

*Communication*

# Rehabilitation and Return to Sport Assessment after Anterior Cruciate Ligament Injury: Quantifying Joint Kinematics during Complex High-Speed Tasks through Wearable Sensors

Stefano Di Paolo [1], Nicola Francesco Lopomo [2], Francesco Della Villa [3], Gabriele Paolini [4], Giulio Figari [4], Laura Bragonzoni [1], Alberto Grassi [5] and Stefano Zaffagnini [5,6,*]

[1] Department for Life Quality Studies, University of Bologna, 40136 Bologna, Italy; stefano.dipaolo@ior.it (S.D.P.); laura.bragonzoni4@unibo.it (L.B.)
[2] Department of Information Engineering, University of Brescia, 25123 Brescia, Italy; nicola.lopomo@unibs.it
[3] Isokinetic Medical Group, Educational & Research Department, FIFA Medical Centre of Excellence, 40132 Bologna, Italy; f.dellavilla@isokinetic.com
[4] GPEM srl, 07041 Alghero, Italy; gabriele.paolini@gpem.net (G.P.); giulio.figari@gpem.net (G.F.)
[5] Orthopaedic and Traumatologic Clinic II, IRCCS Istituto Ortopedico Rizzoli, 40136 Bologna, Italy; alberto.grassi@ior.it
[6] Department of Biomedical and Neuromotor Sciences, University of Bologna, 40136 Bologna, Italy
* Correspondence: stefano.zaffagnini@unibo.it; Tel.: +39-051-6507

**Citation:** Di Paolo, S.; Lopomo, N.F.; Della Villa, F.; Paolini, G.; Figari, G.; Bragonzoni, L.; Grassi, A.; Zaffagnini, S. Rehabilitation and Return to Sport Assessment after Anterior Cruciate Ligament Injury: Quantifying Joint Kinematics during Complex High-Speed Tasks through Wearable Sensors. *Sensors* **2021**, *21*, 2331. https://doi.org/10.3390/s21072331

Academic Editors: Kamiar Aminian, Mario Munoz-Organero, Paolo Capodaglio and Veronica Cimolin

Received: 3 March 2021
Accepted: 24 March 2021
Published: 26 March 2021



**Abstract:** The aim of the present study was to quantify joint kinematics through a wearable sensor system in multidirectional high-speed complex movements used in a protocol for rehabilitation and return to sport assessment after Anterior Cruciate Ligament (ACL) injury, and to validate it against a gold standard optoelectronic marker-based system. Thirty-four healthy athletes were evaluated through a full-body wearable sensor (MTw Awinda, Xsens) and a marker-based optoelectronic (Vicon Nexus, Vicon) system during the execution of three tasks: drop jump, forward sprint, and 90° change of direction. Clinically relevant joint angles of lower limbs and trunk were compared through Pearson's correlation coefficient (r), and the Coefficient of Multiple Correlation (CMC). An excellent agreement (r > 0.94, CMC > 0.96) was found for knee and hip sagittal plane kinematics in all the movements. A fair-to-excellent agreement was found for frontal (r 0.55–0.96, CMC 0.63–0.96) and transverse (r 0.45–0.84, CMC 0.59–0.90) plane kinematics. Movement complexity slightly affected the agreement between the systems. The system based on wearable sensors showed fair-to-excellent concurrent validity in the evaluation of the specific joint parameters commonly used in rehabilitation and return to sport assessment after ACL injury for complex movements. The ACL professionals could benefit from full-body wearable technology in the on-field rehabilitation of athletes.

**Keywords:** wearable inertial sensors; marker-based optoelectronic system; ACL; rehabilitation; motion capture validation; kinematics

## 1. Introduction

Biomechanical assessment of human movement represents a key tool to discriminate normal and pathological patterns in a wide variety of applications. In the sport-related context, the risky patterns lead to an increased risk for severe injuries such as the non-contact Anterior Cruciate Ligament (ACL) injury [1–5]. The ACL injury has detrimental consequences: a return to sport (RTS) at pre-injury level is not guaranteed, the re-injury rate is high (up to 30% [6,7]), and the risk of post-traumatic osteoarthritis increases by 4-fold [8]. All the main joints have a role in the ACL injury mechanism [1,9,10], and targeted neuromuscular training has been proposed to modify the specific risky patterns in the rehabilitation phase after injury [11–14]. Thus, the interest in tools for assessing multi-joint biomechanics in this context has increased more and more [15,16]. Such evaluation could help to understand the injury risk in high-speed and multidirectional movements, thus supporting the rehabilitation phase.

The optoelectronic marker-based (OMB) motion capture approach represents the gold standard for biomechanical evaluations. The OMB systems have been used for multiple applications in the analysis of both healthy and disease-related movement patterns [4,17]. Nevertheless, when it comes to an ACL scenario, the main and well documented-limitations of this technology are amplified: there is a need for a dedicated space, causing difficult on-field applications; there is a high cost in terms of money, time and technical skills; it is difficult to obtain quick reports [18,19]. The alternative solution is represented by wearable inertial sensors (WIS). This technology is indeed portable, less cumbersome on the body, and can produce real-time results. Different uses of the WIS technologies have been reported: single sensors settings have been mainly used in the assessment of athletes' performances at both individual and team level; multiple sensors settings have been mainly used for biomechanical assessments of both healthy movement patterns and to identify the effects of several diseases, including different contexts and applications [18–20].

The advantages brought by the WIS systems might have a strong impact on the ACL rehabilitation; ACL professionals might indeed benefit from a direct assessment of the biomechanical predictors of ACL injury risk, while performing the evaluation in players' actual environments, thus reducing the bias caused by controlled conditions and limited space.

Although previous studies have been carried out to validate the WIS against the OMB, demonstrating consistent results and good accuracy [18], extensive use of the WIS technology in the assessment of athletes' biomechanics during ACL rehabilitation is still not reported. A possible explanation is that no validation of WIS has been performed on clinically relevant parameters addressing the analysis of ACL injury risk in complex movements; in fact, movement complexity plays a significant role in motion capture system accuracy, adding extra sources of noise on the expected outcomes [19,21]. A feasibility analysis on clinically relevant parameters for ACL professionals is also crucial to focus on modifiable risk factors and athletes' progress. At present, little knowledge exists on the reliability of the WIS systems with respect to such specific requirements.

Therefore, the aim of the present study was to quantify joint kinematics through a full-body WIS system and to test its concurrent validity against a gold standard OMB system in a set of multidirectional high-speed complex movements included in a protocol for rehabilitation and RTS assessment after ACL reconstruction.

The main hypothesis was that the WIS system could be reliably used in the analysis of multi-joint kinematics when focusing on clinically relevant parameters and tasks.

## 2. Materials and Methods

The experimental session was conducted in the Green Room of the Isokinetic Medical Center of Bologna (Italy). An a priori power analysis was performed based on a previous similar study [22] to determine the correct sample size. At least 28 athletes were required to have a power of 0.9 with a minimum effect size of 0.5 (large) and a type I error of 0.05.

Overall, 34 recreational and elite athletes were recruited for the study (Table 1). Inclusion criteria were that they were aged between 18–50 years old, and had a Tegner level of at least 5. Exclusion criteria were musculoskeletal disorders or impairment, BMI > 35, previous surgery to lower limbs, and cardiopulmonary or cardiovascular disorders.

All the athletes signed informed consent before starting the acquisition protocol. The research study was approved by the Institutional Review Board (IRB approval: 555/2018/Sper/IOR of 12/09/2018) of Area Vasta Emilia Romagna Centro (AVEC, Bologna, Italy) and registered on ClinicalTrials.gov (Identifier: NCT03840551) (accessda on 15 march 2021).

**Table 1.** Demographic data, mean ± SD [range].

| | |
|---|---|
| Number of athletes | 34 |
| Age (years) | 22.8 ± 4.1 [18–31] |
| Gender (m/f) | 18/16 |
| Body Mass Index | 22.6 ± 2.6 [18–27] |
| Dominant limb (r/l) [1] | 30/4 |
| Tegner | 8.6 ± 1.0 [5–9] |

[1] Dominant limb is meant as the preferred one used to kick a ball.

### 2.1. Experimental Protocol

The analysis was performed in a specialized laboratory, equipped with artificial turf. Regarding the OMB motion capture, a set of 10 stereophotogrammetric cameras (Vicon Vero, Vicon Motion Systems Ltd., Oxford, UK) with a sampling frequency of 120 Hz were used. Three synchronized, calibrated, high-speed RGB cameras were also used to record the movement. The calibration of cameras and volume of the acquisition was performed at the beginning of the acquisition and repeated periodically during the session. A total of 42 retroreflective markers were placed on each athlete according to the full-body Plug-in Gait protocol (Appendix A, top row).

Regarding the WIS system, a set of 15 inertial sensors (MTw Awinda, Xsens Technologies, Enschede, Netherlands) with a sample frequency of 60 Hz were placed according to the Full body No Hands Xsens protocol (Appendix A, bottom row). The WIS has an internal sampling rate of 1000 Hz, an accelerometer range of 16 g, a gyroscope range of ±2000°/s, and a dynamic systematic uncertainty of 0.75°.

During data acquisition, the athletes were contemporary equipped with both systems. A single operator (G.P.) carried out the athletes' sensor and marker placement. After marker and sensor positioning, both static and dynamic subject calibrations were performed simultaneously for the two systems, and anthropometric measurements were collected. Data capture was triggered via hardware by using the OMB system as a master and the WIS system as a slave to directly compare the two systems' acquisitions.

Each athlete performed three motor tasks: a drop jump (DJ), a forward sprint (FS), and a change of direction at 90° (CD). The DJ consisted of a bilateral landing from a 41 cm-high box, with an immediate vertical jump at maximum force [2]. The FS consisted of a frontal sprint followed by a sudden stop on a single leg and by a further backward sprint, all performed at maximum speed [23]. The CD consisted of a frontal sprint followed by a sudden sidestep cutting maneuver at 90° and a further frontal sprint in the new direction, all performed at maximum speed [24] (Figure 1). These motor tasks are included in a protocol for the biomechanical assessment of return to sport after ACL reconstruction, developed and currently deployed at the Isokinetic Medical Group, a FIFA Medical Centre of Excellence.

Before the real test, the athletes were instructed and performed a few warm-up repetitions of each task in order to get confident with the movement. A sports medicine physician (F.D.V.) instructed each athlete on the movements performed and verified each trial's validity. All the athletes performed three valid repetitions of each task per leg (18 total valid trials per athlete). The tasks were performed consecutively, after a short rest (of a few seconds) per each athletes' fatigue.

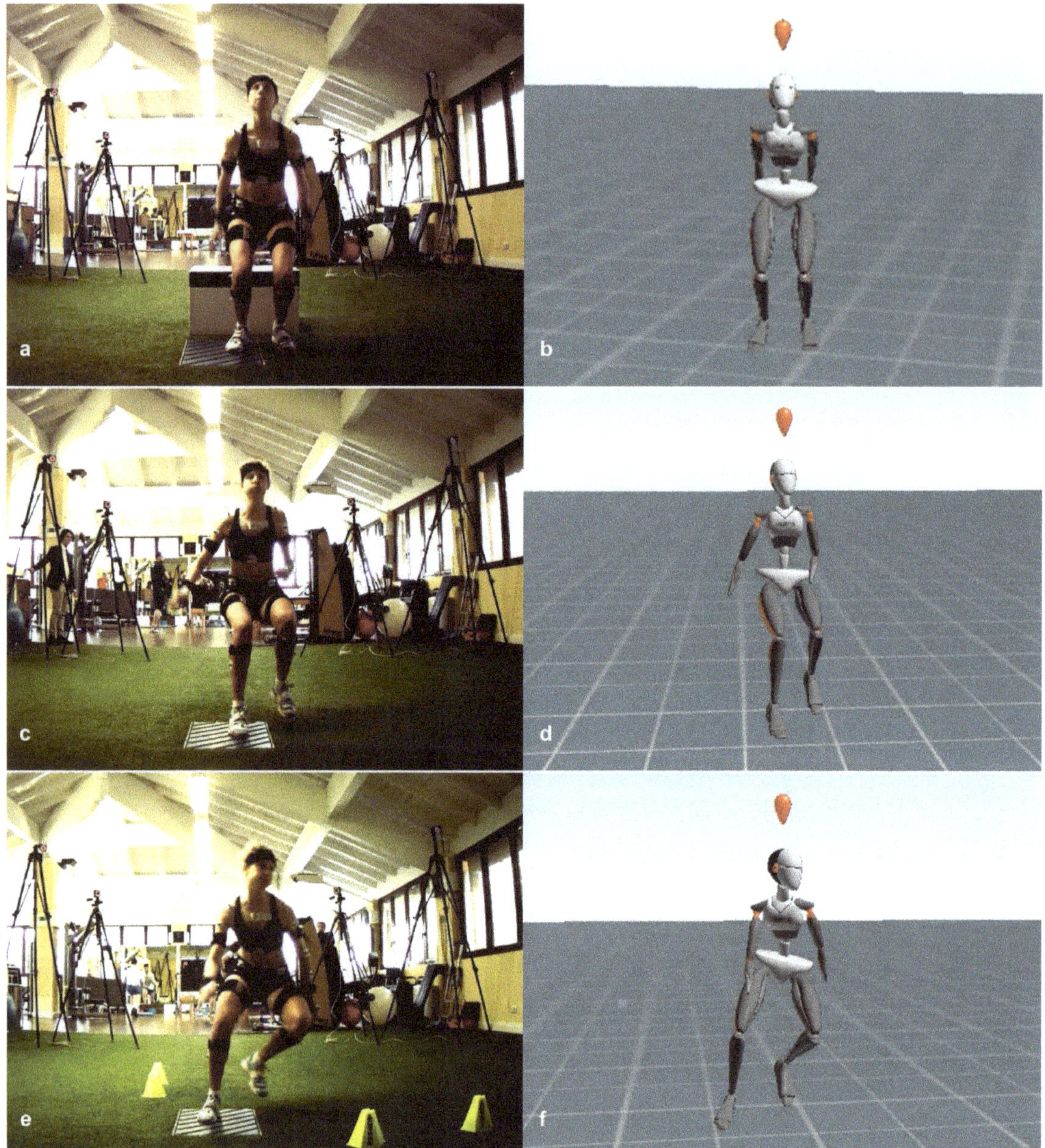

**Figure 1.** Representation of the three motor tasks performed by the athletes and used for comparative analysis alongside the real-time movement reconstruction in the wearable sensor software environment: (**a,b**) the drop jump (DJ); (**c,d**) the forward sprint (FS); (**e,f**) the change of direction at 90° (CD).

### 2.2. Features Selection and Data Processing

The kinematic parameters' selection was based on the current concepts of ACL injury mechanisms in the orthopedics and sports medicine community [1–3,11]. Since the main injury pattern, and the neuromuscular training to prevent it, involves both the lower limbs and the trunk complex, the ankle, knee, hip, and trunk joint angles were taken into account. Although a complete kinematical comparison (frontal, transverse, and sagittal plane for all the joints) between the systems was performed, the final analysis was focused on the following parameters: ankle transverse plane, knee frontal and sagittal plane, hip frontal, transverse and sagittal plane, trunk frontal and transverse plane (Figure 2). Data were resampled and normalized by considering the overall time length of each trial (0-100% of the task) to perform a direct comparison of each data frame. Joint angles were verified to match common conventions in all the three anatomical planes: abduction (+), adduction (-); internal (+), external (-) rotation; flexion (+), extension (-) [22].

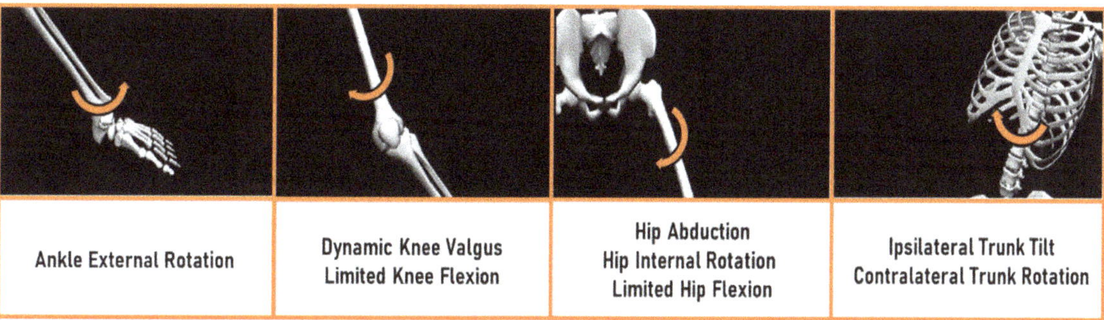


**Figure 2.** Common ACL injury mechanism according to the current literature. From the left: ankle, knee, hip, and trunk joint main kinematical mechanisms found in ACL injury. Arrows indicate the direction of joint motion. The limitation of these patterns is the goal of targeted neuromuscular training in rehabilitation after ACL injury. The figure is adapted from Della Villa et al. [1].

### 2.3. Statistical Analysis

The normal distribution of the data was verified through the Shapiro-Wilk test. Data were compared using different metrics [17,18,25,26]: Pearson's correlation coefficient (r); the Coefficient of Multiple Correlation (CMC); the offset defined as the difference between the means of the waveforms (ΔOFF); the Normalized Root Mean Square Error (NRMSE).

Pearson's r and the CMC coefficient indicated the agreement between the systems. For Pearson's r, the statistical significance of the correlation was also assessed with an alpha level of 5%. The CMC, calculated according to the definition given in Ferrari et al. [25], defined the similarity across the full movement, after offset removal. The validity was considered excellent if r and CMC > 0.75, fair if r and CMC 0.4–0.74, and poor if r and CMC < 0.39 [27].

The average offset error indicated the systematic error between the two waveforms and was expressed in degrees. In order to keep the information regarding which was the highest value between the two systems, positive values of ΔOFF indicated higher values for the OMB system, while negative values indicated higher values for the WIS system. The agreement was considered excellent for values of ΔOFF lower than ±5° [28,29]. The NRMSE indicated the dispersion of the data along with the waveforms and was normalized over the range of motion of the OMB system, thus ranging from 0 to 1. The validity was considered excellent for NRMSE < 0.2 [19]. All the statistical analyses were performed in MATLAB (The MathWorks, Natick, MA, USA).

### 3. Results

The athletes' age was 22.8 ± 4.1 years, and the Tegner level was 8.6 ± 1.0 (Table 1). Overall, a total of 469 valid trials (77% of the total)—168 for DJ, 137 for FS, 164 for CD—were kept and compared between the two systems. Trial exclusion was made due to technical reasons, i.e., error in data acquisition or export of either the OMB or the WIS system data. The waveforms for all the kinematical parameters investigated can be found in Appendix B.

### 3.1. Drop Jump

Excellent agreement was found for the knee and hip sagittal plane angles (r = 0.99, CMC = 0.98) (Table 2). A fair-to-excellent agreement was found for knee, hip, and trunk frontal plane angles (r 0.69–0.81, CMC 0.67–0.88). Fair agreement was found for ankle, hip, and trunk transverse plane angles (r 0.58–0.74, CMC 0.63–0.83). Errors ($\Delta$OFF) were always lower than $5°$, except for hip sagittal plane angles ($-6.91°$, NRMSE = 0.1, Table 3). Higher average values on lower limb frontal and transverse plane angles were found for the OMB system at the peak flexion angles (Appendix B).

**Table 2.** Agreement measurements (r, CMC) between the two motion capture systems for the three motor tasks, mean [range].

| | r | | | CMC | | |
| | DJ | FS | CD | DJ | FS | CD |
|---|---|---|---|---|---|---|
| **Ankle** | | | | | | |
| Transverse | 0.58 [0.49, 0.67] | 0.53 [0.43, 0.62] | 0.45 [0.35, 0.55] | 0.63 [0.57, 0.69] | 0.63 [0.56, 0.69] | 0.62 [0.55, 0.69] |
| **Knee** | | | | | | |
| Frontal | 0.69 [0.59, 0.79] | 0.55 [0.43, 0.67] | 0.60 [0.50, 0.70] | 0.67 [0.58, 0.76] | 0.65 [0.57, 0.73] | 0.63 [0.55, 0.72] |
| Sagittal | 0.99 [0.98, 0.99] | 0.95 [0.94, 0.97] | 0.95 [0.93, 0.96] | 0.99 [0.99, 0.99] | 0.98 [0.97, 0.98] | 0.97 [0.96, 0.98] |
| **Hip** | | | | | | |
| Frontal | 0.81 [0.76, 0.86] | 0.85 [0.81, 0.89] | 0.96 [0.95, 0.97] | 0.88 [0.85, 0.91] | 0.91 [0.88, 0.93] | 0.96 [0.96, 0.97] |
| Transverse | 0.64 [0.55, 0.73] | 0.47 [0.34, 0.59] | 0.63 [0.56, 0.71] | 0.74 [0.68, 0.81] | 0.59 [0.48, 0.70] | 0.72 [0.67, 0.77] |
| Sagittal | 0.99 [0.98, 0.99] | 0.97 [0.97, 0.98] | 0.96 [0.94, 0.97] | 0.99 [0.99, 0.99] | 0.98 [0.98, 0.99] | 0.97 [0.96, 0.97] |
| **Trunk** | | | | | | |
| Frontal | 0.77 [0.72, 0.82] | 0.67 [0.55, 0.79] | 0.87 [0.84, 0.90] | 0.85 [0.82, 0.89] | 0.81 [0.75, 0.88] | 0.91 [0.89, 0.94] |
| Transverse | 0.74 [0.67, 0.81] | 0.84 [0.79, 0.89] | 0.81 [0.76, 0.86] | 0.83 [0.78, 0.88] | 0.90 [0.87, 0.93] | 0.89 [0.87, 0.92] |

Note: DJ = Drop Jump; FS = Frontal Sprint; CD = Change of Direction at $90°$; CMC = Coefficient of Multiple Correlation.

**Table 3.** Error measurements ($\Delta$OFF, NRMSE) between the two motion capture systems for the three motor tasks, mean [range].

| | $\Delta$OFF ($°$) | | | NRMSE (%) | | |
| | DJ | FS | CD | DJ | FS | CD |
|---|---|---|---|---|---|---|
| **Ankle** | | | | | | |
| Transverse | 3.31 [0.38, 6.24] | $-8.51$ [$-11.39$, $-5.64$] | $-7.91$ [$-11.16$, $-4.66$] | 0.26 [0.23, 0.28] | 0.31 [0.27, 0.35] | 0.37 [0.32, 0.42] |
| **Knee** | | | | | | |
| Frontal | $-4.14$ [$-5.90$, $-2.37$] | $-9.93$ [$-13.63$, $-6.22$] | $-10.93$ [$-14.67$, $-7.19$] | 0.27 [0.22, 0.32] | 0.43 [0.35, 0.51] | 0.40 [0.33, 0.47] |
| Sagittal | $-4.67$ [$-6.63$, $-2.71$] | $-2.45$ [$-5.24$, 0.35] | $-3.86$ [$-6.28$, $-1.43$] | 0.08 [0.06, 0.09] | 0.12 [0.09, 0.14] | 0.12 [0.10, 0.14] |
| **Hip** | | | | | | |
| Frontal | 3.91 [2.55, 5.27] | 4.82 [3.19, 6.45] | 5.18 [3.38, 6.99] | 0.36 [0.28, 0.44] | 0.29 [0.23, 0.35] | 0.21 [0.17, 0.25] |
| Transverse | $-1.05$ [$-4.69$, 2.6] | 6.57 [1.83, 11.31] | 5.49 [1.21, 9.76] | 0.26 [0.22, 0.29] | 0.38 [0.31, 0.45] | 0.30 [0.25, 0.35] |
| Sagittal | $-6.91$ [$-9.13$, $-4.68$] | $-2.94$ [$-5.62$, $-0.27$] | $-4.99$ [$-7.54$, $-2.44$] | 0.10 [0.09, 0.12] | 0.13 [0.10, 0.16] | 0.14 [0.12, 0.17] |
| **Trunk** | | | | | | |
| Frontal | $-0.33$ [$-1.30$, 0.64] | $-0.69$ [$-2.0$, 0.61] | $-1.05$ [$-2.40$, 0.31] | 0.30 [0.25, 0.35] | 0.31 [0.25, 0.37] | 0.20 [0.17, 0.23] |
| Transverse | 0.38 [$-1.06$, 1.83] | $-0.85$ [$-2.66$, 0.97] | $-1.14$ [$-3.26$, 0.98] | 0.37 [0.30, 0.45] | 0.21 [0.17, 0.25] | 0.26 [0.21, 0.30] |

Note: DJ = Drop Jump; FS = Frontal Sprint; CD = Change of Direction at $90°$; $\Delta$OFF = difference between means of the waveforms; NRMSE = Normalized Root Mean Square Error.

### 3.2. Frontal Sprint

Excellent agreement was found for the knee and hip sagittal plane angles (r > 0.94, CMC = 0.98) (Table 2). A fair-to-excellent agreement was found for frontal and transverse plane joint angles (r 0.47–0.85, CMC 0.59–0.91). The $\Delta$OFF were lower than $1°$ and $7°$ for trunk and hip angles, respectively (Table 3). The highest offset errors were found for knee frontal ($-9.93°$, NRMSE = 0.43) plane angles, with higher average values for the OMB system (Appendix B).

*3.3. 90° Change of Direction*

Excellent agreement was found for the knee and hip sagittal plane angles, and for hip and trunk frontal plane angles (r > 0.87, CMC > 0.91) (Table 2). A fair-to-excellent agreement was found for the remaining frontal and transverse plane joint angles (r 0.45–0.87, CMC 0.66–0.91). The ΔOFF were always lower than 6°, except for knee frontal and ankle transverse plane angles (Table 3). The OMB system showed higher average values for knee frontal and transverse angles (Appendix B).

## 4. Discussion

The aim of the present study was to quantify joint kinematics and validate a full-body WIS motion capture system against a gold standard OMB in complex movements specifically used for the clinical evaluation of the ACL injury risk and return to sport. The analysis was carried out on a consistent number of healthy athletes, mainly coming from competitive sports (Tegner level 9 in 85% of athletes).

The main finding of the present study was that the WIS motion capture system showed overall fair-to-excellent correlation with respect to the OMB system, and acceptable measurement errors in all the movements assessed. This finding confirmed the concurrent validity of the WIS system already underlined in previous studies [18,28–30], and further extended its usability to clinical applications in the ACL rehabilitation context. For the first time, a stronger focus was put on clinically relevant biomechanical parameters used by the ACL professionals in the rehabilitation protocols after ACL injury in sport-specific, high-speed, and multidirectional movements. Indeed, the parameters investigated are the common targets of neuromuscular training used by sports physicians and orthopedic surgeons in the rehabilitation phase, and to clear patients for RTS [12,13,16,31–33] (Figure 2). The latter finding represents a great step forward in the use of full-body wearable technology by health professionals for ACL rehabilitation and RTS, both in-lab and on-field.

The differences found between the two systems were tolerable according to the literature requirements [18,19,27–29]. Knee and hip sagittal plane angles showed the highest agreement between the two systems (minimum CMC = 0.95). Similar levels of agreement and errors were reported in literature considering the very same WIS system used in the present study and either the same [22] or different OMB systems [28,29,34] addressing walking, stair climbing, and landing tasks. The results of the present study confirmed that sagittal plane angles could be accurately assessed in sprints and counter-movements. A trustful evaluation of sagittal plane joint angles is of primary importance for ACL injury risk and RTS. Many rehabilitation programs focus on reaching a good joint range of motion in dynamic tasks, as this reduces the stress on lower limb joints and the trunk [5,35,36]. Landing strategies favoring the hip (i.e., higher hip than knee flexion, namely "hip strategy") or knee (i.e., higher knee than hip flexion, namely "knee strategy") are also widely assessed in rehabilitation programs since they were shown to correlate with lower and higher knee abduction moment, respectively [37].

The agreement between the two systems on the frontal plane was fair-to-excellent for the hip and trunk, and fair for the knee. Measurement errors and agreement were only slightly affected by the different complexity of the three movements evaluated. This aspect partially extended the validity of the WIS system on frontal plane angles to include complex movements. Furthermore, the average and range of motion data were generally higher for the OMB system compared to the WIS system, particularly for the knee joint. These findings are in line with previous studies which focused on either gait or counter-movements [22,38]. The evaluation of frontal plane angles is crucial in the assessment of ACL injury risk and RTS [39,40]. Primary attention is paid to knee and hip frontal plane kinematics. The limitation of the dynamic valgus pattern represents a milestone of every ACL rehabilitation program since literature extensively underlines how this pattern increases the knee abduction moment and it is present in almost all the ACL rupture mechanisms in the athletes [1,2]. Trunk kinematics is also largely evaluated since excessive homolateral lean also increases knee abduction moment [41].

Fair-to-excellent agreement was found for the hip and trunk transverse plane angles. This aspect extends the possibilities of the ACL professionals during the rehabilitation: without 3D technologies, the correct assessment of such angles is critical and often neglected. The importance of rotational patterns has been already highlighted: the coupling of hip abduction and internal rotation leads to the dynamic knee valgus [4,24,42,43], and the excessive trunk rotation magnifies the knee's external moments and the loss of core stability [41,42,44]. The lowest agreement was found for ankle transverse plane angles. The errors also increased with movement complexity (highest for the DJ task, lowest for the CD task). As for the knee frontal plane angles, absolute average values and range of motion were higher for the OMB system compared to the WIS in all the tasks. All the previous studies investigating the validity of WIS systems against the OMB systems reported a low agreement for transverse plane kinematics [17,18,20,28,29,37]. The assessment of such angles is critically related to the definition of the biomechanical models, the markers/sensors' positioning, and the limited range of motion [18,28]. Ankle joint definition is probably the most critical for both the OMB and the WIS systems [26,45]. For the former, the relatively large shape of medial and lateral malleoli introduces complexity in landmark palpation [28] and the presence of an expert operator is mandatory. For the latter, the malleoli are automatically placed at the same height in the biomechanical model, thus probably introducing an offset on frontal orientation with respect to the OMB reference system. Moreover, the limited range of motion of such angles has been associated with an intrinsic decrease of measurement agreement like the CMC [46].

Regarding the clinical interpretation of the results obtained, the high peaks and range of motion evaluated through OMB seem to be highly non-physiological, even for countermovements. This aspect was likely due to cross-talk between the sagittal and frontal plane in the OMB data analysis, thus mainly related to marker placement. Although this aspect could be of limited interest in gait analysis, it could cause severe flaws in an ACL scenario, where such high varus/valgus and internal/external rotation values could cause unreasonable alerts in the data interpretation [26]. The same joint angles evaluated with the WIS systems are much smaller. Despite this could be symptomatic of an "over-constrained" biomechanical model, these results seem more appropriate for a healthy athlete' population. In clinical scenarios, attention should be paid when interpreting specific angle values. The strength of such WIS system relies on the kinematical assessment of multiple joints, which offers an overall consideration of the movement. ACL professionals should, therefore, take into account multiple variables when drawing clinical conclusions from such analyses.

The novelty of the present work relies on the clinically-oriented analysis carried out in terms of both motor tasks and parameters evaluated. Compared to the previous literature, the set of motor tasks evaluated in the present study is one of the most demanding in terms of complexity and has, most of all, practical clinical use in rehabilitation after ACL injury and the RTS in terms of parameters assessed. Therefore, the WIS system used in the present study (MTw Awinda, Xsens) resulted in a suitable solution for motion capture in the sports environment for the biomechanical assessment of ACL rehabilitation. The assessment of wearables' accuracy before experimental applications has been recently advocated in specific applications such as outdoor sport and the military [19,47]. The present study, alongside all the limitations of the OMB system in outdoor use, endorses the use of WIS in ACL rehabilitation during on-field assessments. Such kinematical analyses could be of crucial importance to deeply evaluate movement quality directly on-field, thus improving and personalizing the rehabilitation strategies [20,48–50].

The present study has some limitations. First, the athletes' velocity was not controlled. Although each athlete performed the movement at his/her best, intra-subject differences could have influenced the obtained outcome. Such analysis could contribute to understanding whether differences between the two systems increase over a defined level of velocity. Second, the systems' comparison was based only on healthy athletes' kinematic data. The analysis of those with ACL injuries or ACL reconstructed athletes could offer stronger insights on the sensibility of the two systems and should be objectives of future investigations. Third, the systems' comparison was based on kinematic data only. Recent studies underlined the correlation between the angular velocity evaluated through WIS and the knee abduction moment evaluated through OMB and force platform in single-leg landing [51]. The assessment of joint moments, powers and inter-segmental forces is an intrinsic limitation of the wearable technology compared to optical tracking, alongside a lower applicability in specific motion capture fields, i.e., facial, hands, etc. Furthermore, the accuracy of the WIS system on knee and ankle frontal and transverse plane angles was flawed by the high—and probably non-physiological—values of the OMB systems. This is reported in the literature as an intrinsic limitation of the OMB biomechanical model and protocol selected for the study, which, on the other hand, were selected for their ease of use [18,47]. Lastly, a single full-body configuration was adopted for the WIS system. The possibility to reduce the number of inertial units while achieving a comparable accuracy [52] should be an objective of future investigations.

## 5. Conclusions

The full-body WIS motion capture system showed a fair-to-excellent concurrent validity in the evaluation of complex movements commonly used in rehabilitation after ACL injury. The ACL professionals could benefit from full-body wearable technology in the on-field rehabilitation of athletes.

**Author Contributions:** The All the authors actively contribute to the development of the present work. In particular: conceptualization, S.D.P., S.Z., and N.F.L.; methodology, S.D.P., N.F.L., G.P., and F.D.V.; software, S.D.P., G.P., and G.F.; validation, S.D.P., G.P., and G.F.; formal analysis, S.D.P., G.P., and G.F.; investigation, S.D.P., G.P., G.F., and F.D.V.; resources, F.D.V., N.F.L.; data curation, S.D.P., G.P., G.F.; writing—original draft preparation, S.D.P.; writing—review and editing, N.F.L., A.G., S.Z.; visualization, S.D.P., L.B.; supervision, S.Z., A.G., L.B.; project administration, S.Z.; funding acquisition, none. All authors have read and agreed to the published version of the manuscript.

**Funding:** This research received no external funding.

**Institutional Review Board Statement:** The study was conducted according to the guidelines of the Declaration of Helsinki, and approved by the Institutional Review Board of Area Vasta Emilia Romagna Centro (AVEC, Bologna, Italy) (protocol code 555/2018/Sper/IOR of 12/09/2018), and registered on ClinicalTrials.gov (Identifier: NCT03840551) (accessda on 15 march 2021).

**Informed Consent Statement:** Informed consent was obtained from all subjects involved in the study. Written informed consent has been obtained from the subject in Figure 1 to publish this paper.

**Data Availability Statement:** Data are available on reasonable request and due to restrictions, e.g., privacy or ethical.

**Acknowledgments:** The authors want to thank Edoardo Della Croce, Francesco Agostinis, and Eng Andrea Taddio for the support in the data acquisition process.

**Conflicts of Interest:** The authors declare no conflict of interest.

## Appendix A

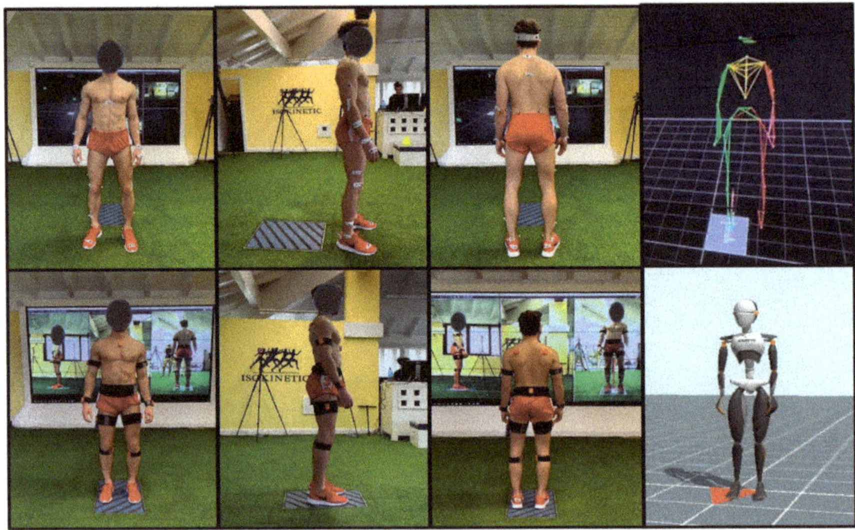

**Figure A1.** Motion capture systems used in the study: (top row) Plug-in-Gait VICON marker-based protocol in (from left to right) front, side, back, and 3D-environment view; (bottom row) Full Body Xsens inertial sensors protocol in (from left to right) front, side, back, and 3D-environment view.

## Appendix B

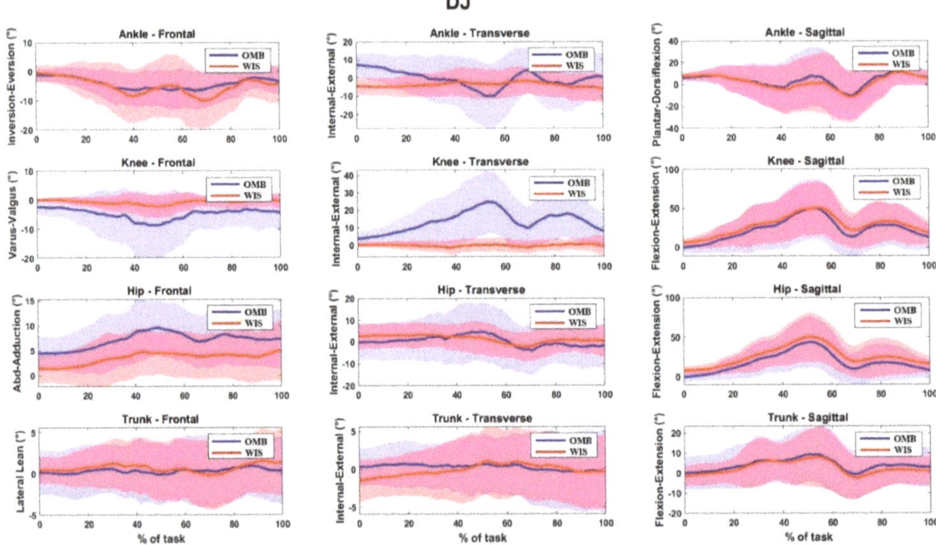


**Figure A2.** Joint kinematics of marker-based (OMB, blue) and inertial sensors (WIS, red) for the drop jump (DJ) task. Data are normalized by the length of the motor task. The waveforms are shown as average (solid lines) and standard deviations (shaded lines).

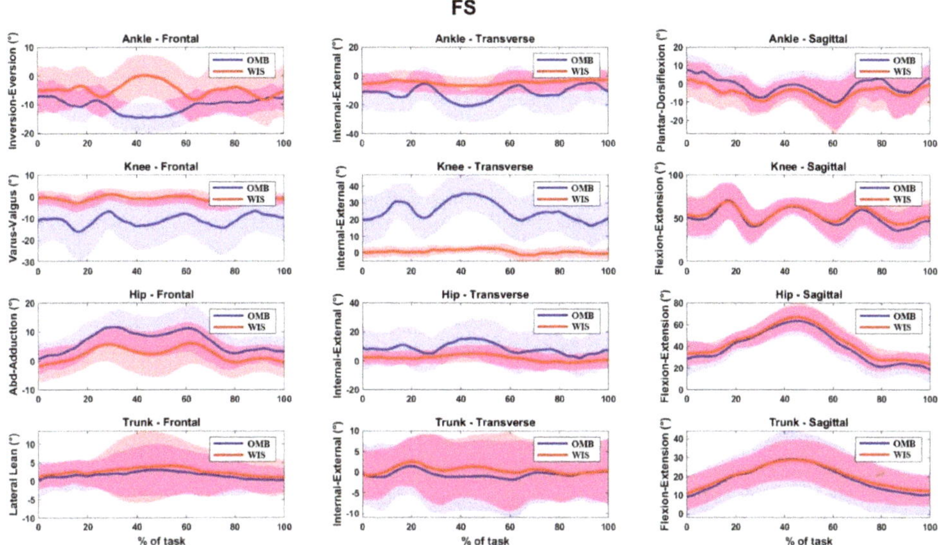

**Figure A3.** Joint kinematics of marker-based (OMB, blue) and inertial sensors (WIS, red) for the forward sprint (FS) task. Data are normalized by the length of the motor task. The waveforms are shown as average (solid lines) and standard deviations (shaded lines).

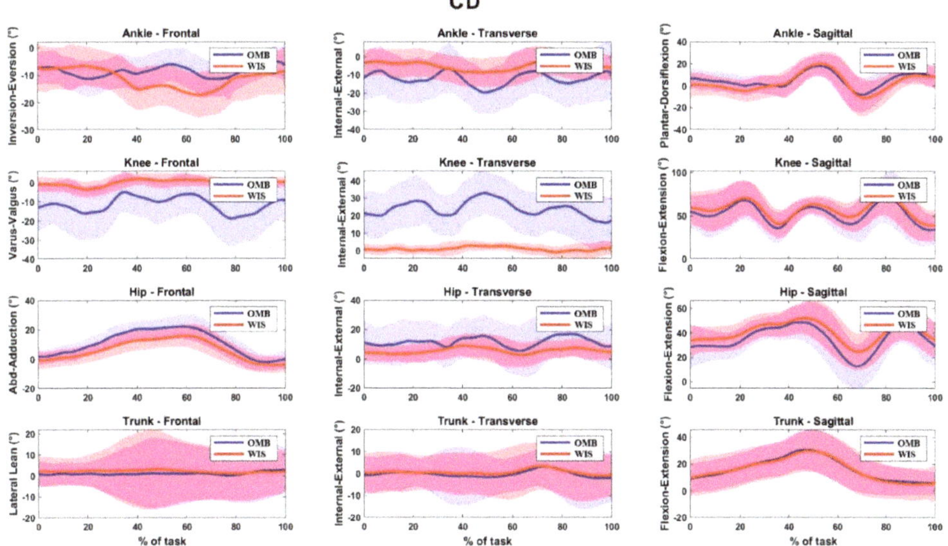

**Figure A4.** Joint kinematics of marker-based (OMB, blue) and inertial sensors (WIS, red) for the change of direction (CD) task. Data are normalized by the length of the motor task. The waveforms are shown as average (solid lines) and standard deviations (shaded lines).

## References

1. Della Villa, F.; Buckthorpe, M.; Grassi, A.; Nabiuzzi, A.; Tosarelli, F.; Zaffagnini, S.; Della Villa, S. Systematic Video Analysis of ACL Injuries in Professional Male Football (Soccer): Injury Mechanisms, Situational Patterns and Biomechanics Study on 134 Consecutive Cases. *Br. J. Sports Med.* **2020**, *54*, 1423–1432. [CrossRef]
2. Hewett, T.E.; Myer, G.D.; Ford, K.R.; Heidt, R.S.; Colosimo, A.J.; McLean, S.G.; van den Bogert, A.J.; Paterno, M.V.; Succop, P. Biomechanical Measures of Neuromuscular Control and Valgus Loading of the Knee Predict Anterior Cruciate Ligament Injury Risk in Female Athletes: A Prospective Study. *Am. J. Sports Med.* **2005**, *33*, 492–501. [CrossRef]
3. Alentorn-Geli, E.; Myer, G.D.; Silvers, H.J.; Samitier, G.; Romero, D.; Lázaro-Haro, C.; Cugat, R. Prevention of Non-Contact Anterior Cruciate Ligament Injuries in Soccer Players. Part 1: Mechanisms of Injury and Underlying Risk Factors. *Knee Surg. Sports Traumatol. Arthrosc.* **2009**, *17*, 705–729. [CrossRef]
4. Leppänen, M.; Pasanen, K.; Krosshaug, T.; Kannus, P.; Vasankari, T.; Kujala, U.M.; Bahr, R.; Perttunen, J.; Parkkari, J. Sagittal Plane Hip, Knee, and Ankle Biomechanics and the Risk of Anterior Cruciate Ligament Injury: A Prospective Study. *Orthop. J. Sports Med.* **2017**, *5*, 2325967117745487. [CrossRef]
5. Leppänen, M.; Pasanen, K.; Kujala, U.M.; Vasankari, T.; Kannus, P.; Äyrämö, S.; Krosshaug, T.; Bahr, R.; Avela, J.; Perttunen, J.; et al. Stiff Landings Are Associated with Increased ACL Injury Risk in Young Female Basketball and Floorball Players. *Am. J. Sports Med.* **2017**, *45*, 386–393. [CrossRef]
6. Webster, K.E.; Feller, J.A. Exploring the High Reinjury Rate in Younger Patients Undergoing Anterior Cruciate Ligament Reconstruction. *Am. J. Sports Med.* **2016**, *44*, 2827–2832. [CrossRef]
7. Wiggins, A.J.; Grandhi, R.K.; Schneider, D.K.; Stanfield, D.; Webster, K.E.; Myer, G.D. Risk of Secondary Injury in Younger Athletes After Anterior Cruciate Ligament Reconstruction: A Systematic Review and Meta-Analysis. *Am. J. Sports Med.* **2016**, *44*, 1861–1876. [CrossRef]
8. Poulsen, E.; Goncalves, G.H.; Bricca, A.; Roos, E.M.; Thorlund, J.B.; Juhl, C.B. Knee Osteoarthritis Risk Is Increased 4-6 Fold after Knee Injury—A Systematic Review and Meta-Analysis. *Br. J. Sports Med.* **2019**, *53*, 1454–1463. [CrossRef] [PubMed]
9. Benjaminse, A.; Holden, S.; Myer, G.D. ACL Rupture Is a Single Leg Injury but a Double Leg Problem: Too Much Focus on "symmetry" Alone and That's Not Enough! *Br. J. Sports Med.* **2018**, *52*, 1029–1030. [CrossRef]
10. Bencke, J.; Aagaard, P.; Zebis, M.K. Muscle Activation During ACL Injury Risk Movements in Young Female Athletes: A Narrative Review. *Front. Physiol.* **2018**, *9*, 445. [CrossRef] [PubMed]
11. Alentorn-Geli, E.; Myer, G.D.; Silvers, H.J.; Samitier, G.; Romero, D.; Lázaro-Haro, C.; Cugat, R. Prevention of Non-Contact Anterior Cruciate Ligament Injuries in Soccer Players. Part 2: A Review of Prevention Programs Aimed to Modify Risk Factors and to Reduce Injury Rates. *Knee Surg. Sports Traumatol. Arthrosc.* **2009**, *17*, 859–879. [CrossRef]
12. Hewett, T.E.; Ford, K.R.; Xu, Y.Y.; Khoury, J.; Myer, G.D. Effectiveness of Neuromuscular Training Based on the Neuromuscular Risk Profile. *Am. J. Sports Med.* **2017**, *45*, 2142–2147. [CrossRef] [PubMed]
13. Mandelbaum, B.R.; Silvers, H.J.; Watanabe, D.S.; Knarr, J.F.; Thomas, S.D.; Griffin, L.Y.; Kirkendall, D.T.; Garrett, W. Effectiveness of a Neuromuscular and Proprioceptive Training Program in Preventing Anterior Cruciate Ligament Injuries in Female Athletes: 2-Year Follow-Up. *Am. J. Sports Med.* **2005**, *33*, 1003–1010. [CrossRef]
14. Hewett, T.E.; Bates, N.A. Preventive Biomechanics: A Paradigm Shift with a Translational Approach to Injury Prevention. *Am. J. Sports Med.* **2017**, *45*, 2654–2664. [CrossRef]
15. Buckthorpe, M.; Della Villa, F. Optimising the "Mid-Stage" Training and Testing Process After ACL Reconstruction. *Sports Med.* **2020**, *50*, 657–678. [CrossRef]
16. King, E.; Richter, C.; Daniels, K.A.J.; Franklyn-Miller, A.; Falvey, E.; Myer, G.D.; Jackson, M.; Moran, R.; Strike, S. Biomechanical but Not Strength or Performance Measures Differentiate Male Athletes Who Experience ACL Reinjury on Return to Level 1 Sports. *Am. J. Sports Med.* **2021**, *49*, 918–927. [CrossRef] [PubMed]
17. Kaufman, K.; Miller, E.; Kingsbury, T.; Russell Esposito, E.; Wolf, E.; Wilken, J.; Wyatt, M. Reliability of 3D Gait Data across Multiple Laboratories. *Gait Posture* **2016**, *49*, 375–381. [CrossRef] [PubMed]
18. Poitras, I.; Dupuis, F.; Bielmann, M.; Campeau-Lecours, A.; Mercier, C.; Bouyer, L.; Roy, J.-S. Validity and Reliability of Wearable Sensors for Joint Angle Estimation: A Systematic Review. *Sensors* **2019**, *19*, 1555. [CrossRef]
19. Van der Kruk, E.; Reijne, M.M. Accuracy of Human Motion Capture Systems for Sport Applications; State-of-the-Art Review. *Eur. J. Sport Sci.* **2018**, *18*, 806–819. [CrossRef]
20. Camomilla, V.; Bergamini, E.; Fantozzi, S.; Vannozzi, G. Trends Supporting the In-Field Use of Wearable Inertial Sensors for Sport Performance Evaluation: A Systematic Review. *Sensors* **2018**, *18*, 873. [CrossRef]
21. Godwin, A.; Agnew, M.; Stevenson, J. Accuracy of Inertial Motion Sensors in Static, Quasistatic, and Complex Dynamic Motion. *J. Biomech. Eng.* **2009**, *131*, 114501. [CrossRef]
22. Al-Amri, M.; Nicholas, K.; Button, K.; Sparkes, V.; Sheeran, L.; Davies, J. Inertial Measurement Units for Clinical Movement Analysis: Reliability and Concurrent Validity. *Sensors* **2018**, *18*, 719. [CrossRef] [PubMed]
23. Jones, P.A.; Herrington, L.C.; Graham-Smith, P. Technique Determinants of Knee Abduction Moments during Pivoting in Female Soccer Players. *Clin. Biomech.* **2016**, *31*, 107–112. [CrossRef]
24. Dempsey, A.R.; Lloyd, D.G.; Elliott, B.C.; Steele, J.R.; Munro, B.J. Changing Sidestep Cutting Technique Reduces Knee Valgus Loading. *Am. J. Sports Med.* **2009**, *37*, 2194–2200. [CrossRef]

25. Ferrari, A.; Cutti, A.G.; Cappello, A. A New Formulation of the Coefficient of Multiple Correlation to Assess the Similarity of Waveforms Measured Synchronously by Different Motion Analysis Protocols. *Gait Posture* **2010**, *31*, 540–542. [CrossRef]
26. Ferrari, A.; Benedetti, M.G.; Pavan, E.; Frigo, C.; Bettinelli, D.; Rabuffetti, M.; Crenna, P.; Leardini, A. Quantitative Comparison of Five Current Protocols in Gait Analysis. *Gait Posture* **2008**, *28*, 207–216. [CrossRef] [PubMed]
27. Kadaba, M.P.; Ramakrishnan, H.K.; Wootten, M.E.; Gainey, J.; Gorton, G.; Cochran, G.V. Repeatability of Kinematic, Kinetic, and Electromyographic Data in Normal Adult Gait. *J. Orthop. Res.* **1989**, *7*, 849–860. [CrossRef]
28. Robert-Lachaine, X.; Mecheri, H.; Larue, C.; Plamondon, A. Validation of Inertial Measurement Units with an Optoelectronic System for Whole-Body Motion Analysis. *Med. Biol. Eng. Comput.* **2017**, *55*, 609–619. [CrossRef]
29. Zhang, J.-T.; Novak, A.C.; Brouwer, B.; Li, Q. Concurrent Validation of Xsens MVN Measurement of Lower Limb Joint Angular Kinematics. *Physiol. Meas.* **2013**, *34*, N63–N69. [CrossRef]
30. Richter, C.; Daniels, K.A.J.; King, E.; Franklyn-Miller, A. Agreement between Inertia and Optical Based Motion Capture during the VU-Return-to-Play- Field-Test. *Sensors* **2020**, *20*, 831. [CrossRef]
31. Cannon, J.; Cambridge, E.D.J.; McGill, S.M. Anterior Cruciate Ligament Injury Mechanisms and the Kinetic Chain Linkage: The Effect of Proximal Joint Stiffness on Distal Knee Control during Bilateral Landings. *J. Orthop. Sports Phys. Ther.* **2019**, *49*, 601–610. [CrossRef]
32. Sabet, S.; Letafatkar, A.; Eftekhari, F.; Khosrokiani, Z.; Gokeler, A. Trunk and Hip Control Neuromuscular Training to Target Inter Limb Asymmetry Deficits Associated with Anterior Cruciate Ligament Injury. *Phys. Ther. Sport* **2019**, *38*, 71–79. [CrossRef]
33. King, E.; Richter, C.; Daniels, K.A.J.; Franklyn-Miller, A.; Falvey, E.; Myer, G.D.; Jackson, M.; Moran, R.; Strike, S. Can Biomechanical Testing After Anterior Cruciate Ligament Reconstruction Identify Athletes at Risk for Subsequent ACL Injury to the Contralateral Uninjured Limb? *Am. J. Sports Med.* **2021**, *49*, 609–619. [CrossRef]
34. van der Straaten, R.; Bruijnes, A.K.B.D.; Vanwanseele, B.; Jonkers, I.; De Baets, L.; Timmermans, A. Reliability and Agreement of 3D Trunk and Lower Extremity Movement Analysis by Means of Inertial Sensor Technology for Unipodal and Bipodal Tasks. *Sensors* **2019**, *19*, 141. [CrossRef]
35. Pollard, C.D.; Sigward, S.M.; Powers, C.M. Limited Hip and Knee Flexion during Landing Is Associated with Increased Frontal Plane Knee Motion and Moments. *Clin. Biomech.* **2010**, *25*, 142–146. [CrossRef] [PubMed]
36. Kotsifaki, A.; Korakakis, V.; Whiteley, R.; Van Rossom, S.; Jonkers, I. Measuring Only Hop Distance during Single Leg Hop Testing Is Insufficient to Detect Deficits in Knee Function after ACL Reconstruction: A Systematic Review and Meta-Analysis. *Br. J. Sports Med.* **2020**, *54*, 139–153. [CrossRef] [PubMed]
37. Nguyen, A.-D.; Taylor, J.B.; Wimbish, T.G.; Keith, J.L.; Ford, K.R. Preferred Hip Strategy During Landing Reduces Knee Abduction Moment in Collegiate Female Soccer Players. *J. Sport Rehabil.* **2018**, *27*, 213–217. [CrossRef]
38. Teufl, W.; Miezal, M.; Taetz, B.; Fröhlich, M.; Bleser, G. Validity of Inertial Sensor Based 3D Joint Kinematics of Static and Dynamic Sport and Physiotherapy Specific Movements. *PLoS ONE* **2019**, *14*, e0213064. [CrossRef]
39. Bates, N.A.; Myer, G.D.; Hale, R.F.; Schilaty, N.D.; Hewett, T.E. Prospective Frontal Plane Angles Used to Predict ACL Strain and Identify Those at High Risk for Sports-Related ACL Injury. *Orthop. J. Sports Med.* **2020**, *8*, 2325967120957646. [CrossRef]
40. Räisänen, A.M.; Pasanen, K.; Krosshaug, T.; Vasankari, T.; Kannus, P.; Heinonen, A.; Kujala, U.M.; Avela, J.; Perttunen, J.; Parkkari, J. Association between Frontal Plane Knee Control and Lower Extremity Injuries: A Prospective Study on Young Team Sport Athletes. *BMJ Open Sport Exerc. Med.* **2018**, *4*, e000311. [CrossRef]
41. Jamison, S.T.; Pan, X.; Chaudhari, A.M.W. Knee Moments during Run-to-Cut Maneuvers Are Associated with Lateral Trunk Positioning. *J. Biomech.* **2012**, *45*, 1881–1885. [CrossRef]
42. Markström, J.L.; Tengman, E.; Häger, C.K. ACL-Reconstructed and ACL-Deficient Individuals Show Differentiated Trunk, Hip, and Knee Kinematics during Vertical Hops More than 20 Years Post-Injury. *Knee Surg. Sports Traumatol. Arthrosc.* **2018**, *26*, 358–367. [CrossRef]
43. Della Villa, F.; Ricci, M.; Perdisa, F.; Filardo, G.; Gamberini, J.; Caminati, D.; Della Villa, S. Anterior Cruciate Ligament Reconstruction and Rehabilitation: Predictors of Functional Outcome. *Joints* **2015**, *3*, 179–185. [CrossRef] [PubMed]
44. Hewett, T.E.; Torg, J.S.; Boden, B.P. Video Analysis of Trunk and Knee Motion during Non-Contact Anterior Cruciate Ligament Injury in Female Athletes: Lateral Trunk and Knee Abduction Motion Are Combined Components of the Injury Mechanism. *Br. J. Sports Med.* **2009**, *43*, 417–422. [CrossRef]
45. Piazza, S.J.; Cavanagh, P.R. Measurement of the Screw-Home Motion of the Knee Is Sensitive to Errors in Axis Alignment. *J. Biomech.* **2000**, *33*, 1029–1034. [CrossRef]
46. Røislien, J.; Skare, O.; Opheim, A.; Rennie, L. Evaluating the Properties of the Coefficient of Multiple Correlation (CMC) for Kinematic Gait Data. *J. Biomech.* **2012**, *45*, 2014–2018. [CrossRef]
47. Mavor, M.P.; Ross, G.B.; Clouthier, A.L.; Karakolis, T.; Graham, R.B. Validation of an IMU Suit for Military-Based Tasks. *Sensors* **2020**, *20*, 4280. [CrossRef]
48. Buckthorpe, M.; Della Villa, F.; Della Villa, S.; Roi, G.S. On-Field Rehabilitation Part 1: 4 Pillars of High-Quality On-Field Rehabilitation Are Restoring Movement Quality, Physical Conditioning, Restoring Sport-Specific Skills, and Progressively Developing Chronic Training Load. *J. Orthop. Sports Phys. Ther.* **2019**, *49*, 565–569. [CrossRef]
49. Buckthorpe, M.; Della Villa, F.; Della Villa, S.; Roi, G.S. On-Field Rehabilitation Part 2: A 5-Stage Program for the Soccer Player Focused on Linear Movements, Multidirectional Movements, Soccer-Specific Skills, Soccer-Specific Movements, and Modified Practice. *J. Orthop. Sports Phys. Ther.* **2019**, *49*, 570–575. [CrossRef] [PubMed]

50. Verheul, J.; Nedergaard, N.J.; Vanrenterghem, J.; Robinson, M.A. *Measuring Biomechanical Loads in Team Sports—From Lab to Field*; SportRxiv; Taylor & Francis: Oxfordshire, UK, 2019.
51. Pratt, K.A.; Sigward, S.M. Detection of Knee Power Deficits Following Anterior Cruciate Ligament Reconstruction Using Wearable Sensors. *J. Orthop. Sports Phys. Ther.* **2018**, *48*, 895–902. [CrossRef]
52. Preatoni, E.; Nodari, S.; Lopomo, N.F. Supervised Machine Learning Applied to Wearable Sensor Data Can Accurately Classify Functional Fitness Exercises Within a Continuous Workout. *Front. Bioeng. Biotechnol.* **2020**, *8*, 664. [CrossRef] [PubMed]

*Communication*

# Functional Electrical Stimulation for Foot Drop in Post-Stroke People: Quantitative Effects on Step-to-Step Symmetry of Gait Using a Wearable Inertial Sensor

**Giulia Schifino** [1,2]**, Veronica Cimolin** [3,]*****, Massimiliano Pau** [4]**, Maira Jaqueline da Cunha** [1,2]**, Bruno Leban** [4]**, Micaela Porta** [4]**, Manuela Galli** [3] **and Aline Souza Pagnussat** [1,2,5]

[1]  Rehabilitation Sciences Graduate Program, Universidade Federal de Ciências da Saúde de Porto Alegre (UFCSPA), Porto Alegre 90050-170, Brazil; giuliaschifino@hotmail.com (G.S.); maira.estudo@gmail.com (M.J.d.C.); alinespagnussat@gmail.com (A.S.P.)

[2]  Movement Analysis and Rehabilitation Laboratory, Universidade Federal de Ciências da Saúde de Porto Alegre (UFCSPA), Porto Alegre 90050-170, Brazil

[3]  Department of Electronics, Information and Bioengineering, Politecnico di Milano, Piazza Leonardo da Vinci 32, 20133 Milano, Italy; manuela.galli@polimi.it

[4]  Department of Mechanical, Chemical and Materials Engineering, University of Cagliari, Piazza d'Armi, 09123 Cagliari, Italy; massimiliano.pau@dimcm.unica.it (M.P.); bruno.leban@dimcm.unica.it (B.L.); m.porta@dimcm.unica.it (M.P.)

[5]  Department of Physiotherapy, Universidade Federal de Ciências da Saúde de Porto Alegre (UFCSPA), Porto Alegre 900050-170, Brazil

*  Correspondence: veronica.cimolin@polimi.it; Tel.: +39-02-2399-3359; Fax: +39-02-2399-3360

**Abstract:** The main purpose of the present study was to assess the effects of foot drop stimulators (FDS) in individuals with stroke by means of spatio-temporal and step-to-step symmetry, harmonic ratio (HR), parameters obtained from trunk accelerations acquired using a wearable inertial sensor. Thirty-two patients (age: 56.84 ± 9.10 years; 68.8% male) underwent an instrumental gait analysis, performed using a wearable inertial sensor before and a day after the 10-session treatment (PRE and POST sessions). The treatment consisted of 10 sessions of 20 min of walking on a treadmill while using the FDS device. The spatio-temporal parameters and the HR in the anteroposterior (AP), vertical (V), and mediolateral (ML) directions were computed from trunk acceleration data. The results showed that time had a significant effect on the spatio-temporal parameters; in particular, a significant increase in gait speed was detected. Regarding the HRs, the HR in the ML direction was found to have significantly increased (+20%), while those in the AP and V directions decreased (approximately 13%). Even if further studies are necessary, from these results, the HR seems to provide additional information on gait patterns with respect to the traditional spatio-temporal parameters, advancing the assessment of the effects of FDS devices in stroke patients.

**Keywords:** foot drop stimulation; gait; symmetry; stroke; inertial measurement sensor

**Citation:** Schifino, G.; Cimolin, V.; Pau, M.; da Cunha, M.J.; Leban, B.; Porta, M.; Galli, M.; Souza Pagnussat, A. Functional Electrical Stimulation for Foot Drop in Post-Stroke People: Quantitative Effects on Step-to-Step Symmetry of Gait Using a Wearable Inertial Sensor. *Sensors* **2021**, *21*, 921. https://doi.org/10.3390/s21030921

Academic Editor: Marco Iosa
Received: 18 December 2020
Accepted: 25 January 2021
Published: 29 January 2021

**Publisher's Note:** MDPI stays neutral with regard to jurisdictional claims in published maps and institutional affiliations.

## 1. Introduction

Stroke and cerebrovascular disease are leading causes of morbidity, mortality, and disability and represent the most common reason for long-term care not only in developed countries, but also in low- and middle-income countries where stroke is the fourth-leading cause of disability among people older than age 65 [1]. Stroke may severely affect a wide range of motor skills at different levels, including upper and lower limb functioning, particularly due to muscular weakness or partial paralysis (often restricted to one side of the body), which are present in more than 80% of individuals [2]. Hemiparetic individuals often suffer from limitations in mobility and the most common post-stroke impairment that affects gait is foot drop [3]. This motor impairment is associated with the weakness or lack of voluntary control in ankle dorsiflexors and/or the increased spasticity of plantar flexor muscles [4–6]. Foot drop interferes with ankle dorsiflexion during the swing phase of gait and contributes to the disruption in weight acceptance and weight transfer in the initial foot contact and stance phases [7]. The most evident alterations in gait, besides a marked asymmetry, include walking speed reduction, longer durations of double-stance and paretic swing phases, reduced paretic single-stance phase duration, cadence, and stride length [3,8], asymmetric postural behavior during walking and standing [9], altered kinematics, and reduced ankle push-off ability in terminal stance [10,11]. In this context, gait impairments cause difficulties with performing the activities of daily living and mobility, thus reducing independence and quality of life [12]. The different degrees of impairment that characterize the affected and not-affected side suggest that the study of gait symmetry represents a crucial feature in characterizing and quantifying locomotion in hemiparetic individuals, also considering that an asymmetric gait pattern is generally characterized by poor efficiency and requires high energy expenditure. In addition, the restoration of gait symmetry is not only an indicator of functional recovery but also an important aim for clinical rehabilitation practice.

Generally, the gait asymmetry metrics of both healthy [13] and injured individuals [14–16] are based on the assessments of right vs. left spatio-temporal parameters, kinematics (e.g., joint angles), and kinetics (e.g., ground reaction forces, GRFs) carried out using three-dimensional motion analysis systems. However, due to a number of issues associated with high cost, operator skills, and space requirements, their application is often limited to research settings [17]. Therefore, it appears important to have available reliable quantitative tools that are suitable for clinical daily use and are able to effectively quantify gait asymmetry. For this purpose, wearable accelerometers represent a very appealing option due to their relatively low cost and ease of use, and are becoming widespread in investigating several aspects of human movement in a variety of contexts [18]. Usually, gait asymmetry is quantified on the basis of conventional spatio-temporal parameters, including step time asymmetry, stance time asymmetry, swing time asymmetry, and step length asymmetry; these are calculated as the absolute difference between consecutive left and right steps. However, further asymmetry variables derived from the cyclical acceleration signals during gait have been found to be effective in detecting gait alterations. In particular, trunk accelerations acquired during gait, using a single sensor placed on the lower back, allows us to obtain information about the so-called "smoothness" of gait (also defined as step-to-step symmetry) by means of a parameter called the Harmonic Ratio (HR) [19–21]. The HR is based on a spectral analysis of the acceleration signals and is related to the bilateral rhythmicity of movement, based on the measure of trunk acceleration during a stride that is expected to be formed by two alternating symmetric steps; it provides different kinds of information with respect to the traditional spatio-temporal parameters, which are focused on the lower limb symmetry at the distal level [22]. Recent studies demonstrated that the HR parameter is worthwhile in quantifying gait alterations associated with neurologic and orthopedic conditions, such as older people [23], Parkinson's disease patients [22], multiple sclerosis [24], normal weight and obese children/adolescents [25,26], Prader–Willi patients [27,28], and cognitively impaired individuals [29]; in several cases, it is able to reveal subtle changes in gait that might occur well before they become detectable in terms of conventional spatio-temporal parameters [22,24–26,30]. Furthermore, it must be emphasized that trunk accelerations could be easily recorded by a single sensor in clinical settings or in other ecological contexts, without the limitations of a movement analysis laboratory, which requires expensive equipment, long setup times, and time-consuming post-processing procedures [26,28,30,31].

Concerning its application on individuals with stroke, to date, several studies reported that the HR could be considered a robust outcome in quantifying the step-to-step asymmetry during gait [18,32–37]. However, to the best of our knowledge, the HR has not been used so far as an indicator of the effectiveness of rehabilitative treatment targeting the improvement of gait in individuals with stroke. Thus, in the present study, we employed the HR to quantify the possible changes originating from the use of the foot drop stimulator

(FDS) on gait asymmetry in chronic post-stroke subjects while walking in an outdoor environment. FDSs are based on the functional electrical stimulation (FES) of the peroneal nerve to elicit ankle dorsiflexion during the swing phase of the step cycle [38–40]. Such devices have been proven to be effective in enhancing gait speed in short- and long-term studies [41–43], but there is no evidence about the reduction of step-to-step symmetry.

The FDS device seems to have positive effects on gait in stroke patients [40,44]; thus, it appears of interest to assess the feasibility of the HR and to compare it to conventional spatio-temporal measures. Our hypothesis is that changes in the spatio-temporal parameters of gait previously reported [41,43] could also be accompanied by modifications of step-to-step symmetry in stroke patients.

## 2. Materials and Methods

### 2.1. Study Design

This quasi-experimental clinical trial was registered at ClinicalTrial.gov (NCT04266899) and approved by the Ethics and Research Committee of the Santa Casa de Misericórdia Hospital of Porto Alegre (CAAE 64819617.0.0000.5335). Procedures were conducted according to the Template for Intervention Description and Replication (TIDeR) checklist [45].

### 2.2. Participants

Participants were recruited through a database of the Santa Casa de Misericórdia Neurology service in Porto Alegre and social networks, and then selected according to eligibility criteria. We included individuals with ischemic or hemorrhagic chronic stroke confirmed by head Computerized Tomography (CT) or Magnetic Resonance Imaging (MRI) at least 6 months before recruitment, aged 20 to 80 years, with mild (29–34/34), moderate (20–28/34), or severe hemiparesis (0–19/34) according to the Fugl–Meyer score's lower limb subdivision [46]. Patients had to have minimal cognitive ability on the Mini-Mental State Examination (>20 points (illiterate) or >24 (literate)) [47], and no history of seizures or recent episodes of a fall (at least 3 months before study engagement). In addition, participants were required to be able to walk at least 30 m autonomously without assistive devices. Individuals who presented any contraindication for electrical stimulation were excluded (any electric or metallic implant; skin problems or lesion in proximity to the site of FDS stimulation; pregnancy). Furthermore, subjects with any lower limb musculoskeletal disorder, significant visual impairment, low response to FDS electrical stimulation (no response to the highest stimulation intensity provided by the FDS device, namely 200 mA intensity), or relevant ankle restriction (fixed ankle contracture at $\geq 10$ degrees of plantar flexion in the hemiplegic leg with the knee extended) were also excluded.

A group of healthy individuals (Control Group: CG) matched by age and sex were also tested. Exclusion criteria for the CG were the existence of cardiorespiratory, neurological, or musculoskeletal disorders. All of them exhibited normal flexibility and muscle strength, had no evident gait abnormalities, and were able to walk independently.

The experimental protocol was carried out in accordance with the ethical standards of the institute and the 1964 Declaration of Helsinki and its later amendments. All participants signed a free and informed consent form before enrolment.

### 2.3. Procedures

This study was conducted at the Movement Analysis and Rehabilitation Laboratory of the Federal University of Health Sciences of Porto Alegre (UFCSPA) between January 2018 and May 2019. Each participant participated in a clinical and documented evaluation session. Indirect assessment of spasticity was done by the Modified Ashworth Scale (MAS) [48], which consists of five ordinal values ranging from 0 (no tonus increase) to 4 (stiffness) [48]. Participants were evaluated while lying in a supine position and were instructed to remain relaxed during the test. The spasticity of plantar flexors, knee extensors, and hip adductors was tested. All clinical assessments were performed by the same researcher at baseline (pre-intervention; 2 days before the first session) and post-intervention (1 day after the last session).

### 2.4. Intervention

Subjects underwent 10 face-to-face sessions of 20 min of walking on a treadmill (Athletic advanced 720EE, Buenos Aires, Argentina) with a self-selected comfortable velocity while using the FDS device, configured according to each subject's need (Figure 1).

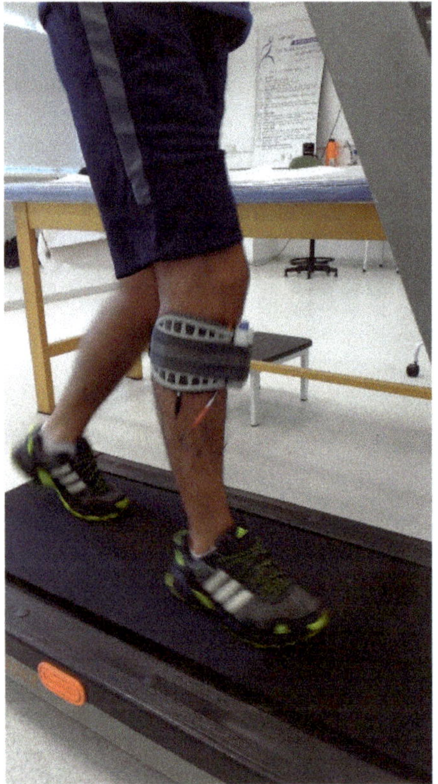

**Figure 1.** The experimental setup.

The WalkAide device (Innovative Neurotronics, Austin, TX, USA) was used to stimulate the peroneal nerve on the affected side through a tilt sensor that detects the affected leg tilt when the foot contact on the ground changes from posterior to anterior (pre-swing phase). The stimulus stops when the leg is tilted forward on foot strike [49,50]. Before treatment, subjects underwent a 3-day adaptation period with the FDS device that included walking overground on a flat surface, walking up and down stairs, and walking on a treadmill. The FDS device was set at enough intensity to achieve the movement, but, at the same time, was required to be comfortable. Before each session, subjects underwent 15 min of lower limb stretching and vital sign measurements. The training sessions, which were interspersed by 5 min rest periods, consisted of 20 min of walking on a treadmill. Each session was administered by the same physical therapist that annotated the overall covered distance and also periodically checked subjects' heart rate and blood pressure. Subjects were allowed to stop the trial at any time, if necessary.

### 2.5. Data Acquisition

A single miniaturized inertial sensor (G-Sensor®, BTS Bioengineering, Milano, Italy), previously validated for investigations on gait in unaffected individuals and people with neurologic conditions [25–27,51–53], was placed on the participants' lower back, approximately at the L4–L5 vertebrae position. The sensor, which is sized 70 mm $\times$ 40 mm $\times$ 18 mm and weighs 37 g, is composed of a three-axis accelerometer, a three-axis gyroscope, and a three-axis magnetometer. After a brief familiarization phase, participants were asked to walk at a self-selected speed on a 30-m flat pathway of the university outdoors. The gait test was performed before and a day after the 10-session treatment and named as PRE and POST session, respectively. Each trial was repeated three times and a mean of them was calculated. Participants were equipped with the FDS device during the gait assessment of the PRE and POST session.

The values of the linear accelerations along the antero-posterior (AP), medio-lateral (ML), and vertical (V) directions were acquired by means of the inertial sensor at a frequency of 100 Hz. The elaboration and parameter computations were performed with a custom Matlab® routine. The first 5 s of the acquisition (during which the subject was requested to stand still) were used to verify the orientation of the sensor, and this information was then used to correct the acceleration vectors' data during the gait trial.

Based on the raw acceleration data, the main spatio-temporal parameters (gait speed, stride length, cadence, and duration of stance and double support phase) were calculated following the approaches described in the literature [27,54,55]. The HRs for the AP, ML, and V directions were computed according to the procedure proposed by Pasciuto et al. [20].

### 2.6. Statistical Analysis

Sample size was determined by G-Power 3.0 software (version 3.1.9.4.; Faul & Buchner, Germany) based on a previous study [18] considering the minimum effect size of 0.56% to detect a minimum clinical difference in the HR of the ML direction. Sample size was calculated by adopting 90% power and an alpha value of 0.05. A total of 29 participants was calculated as necessary to perform this study. The parametric Student's t-test, non-parametric Mann–Whitney U test, and Chi-square test were used to compare the demographic characteristics between the stroke and the control groups. After verifying their normality (using the Shapiro–Wilk test) and homogeneity of variances (Levene's test), a one-way analysis of variance for repeated measure (RM-ANOVA) was conducted using SPSS software (v.20, IBM, Armonk, NY, USA) to verify the effect of the use of the foot drop stimulator (FDS) on spatio-temporal parameters and the HR PRE and POST training. Time (PRE/POST) was set as an independent variable, while the five gait parameters previously listed and the three HRs represented the dependent variables. After the Bonferroni correction was performed considering the three main outcomes (HR in the AP, ML, and V direction), the statistical significance was fixed at $p = 0.017$. The Student's t-test assessed the differences between PRE evaluations and the controls (Control Group).

## 3. Results

Participants were recruited from March 2017 to August 2019, while the final measurements were carried out in August 2019. Forty-one stroke survivors were contacted and the final tested sample included 32 individuals. Their baseline demographic and clinical characteristics are reported in Table 1.

**Table 1.** Participant characteristics.

|  | Stroke ($n = 32$) | Control Group ($n = 32$) | $p$ Value |
|---|---|---|---|
| Gender, $n$ (%) |  |  |  |
| Male | 22 (68.8) | 22 (68.8) | 0.444 [¥] |
| Age (years) | 56.84 ± 9.10 | 56.81 ± 8.88 | 0.989 [#] |
| Height (m) | 1.68 ± 9.73 | 1.72 ± 8.23 | 0.221 [#] |
| Body mass (kg) | 75.13 ±11.90 | 78.26 ± 12.97 | 0.318 [*] |
| Time since stroke (months) (min–max) | 39.41 (6–96) |  |  |
| Stroke type, $n$ (%) |  |  |  |
| Ischemic | 24 (75) |  |  |
| Hemorrhagic | 8 (25) |  |  |
| Affected hemisphere, $n$ (%) |  |  |  |
| Right | 19 (59.4) |  |  |
| Left | 13 (40.6) |  |  |
| FMA–LL (0–34), (min–max) | 19.63 (11–32) |  |  |
| MAS, frequency (0/1/1 + /2/3/4) |  |  |  |
| Plantiflexors | 0/3/2/4/11/12 |  |  |
| Knee extensors | 5/7/6/4/8/2 |  |  |
| Adductors | 5/4/4/12/7/0 |  |  |

FMA—LL, Fugl–Meyer Assessment—Lower Limb; MAS, Modified Ashworth Scale; max, maximum; min, minimum; $n$, number of participants; SD, standard deviation. [#] $p$-value parametric Student's t-tests; [*] $p$-value non-parametric Mann–Whitney and [¥] $p$-value Chi-square tests were used to compare the demographic characteristics between the stroke and control groups.

Table 2 presents the values of the spatio-temporal parameters and the HR features for stroke patients in the PRE and POST session and for the Control Group. In the comparison of the PRE session of stroke individuals vs. the Control Group, all parameters exhibited significant differences, with the exception of the double support phase.

**Table 2.** The spatio-temporal parameters of gait and the Harmonic Ratio values of the participants.

|  | Stroke | | | | Control Group | |
|---|---|---|---|---|---|---|
|  | PRE | POST | F | $p$ Value |  | $p$ Value |
| **Spatio-temporal parameters** | | | | | | |
| Gait speed (m/s) | 0.62 ± 0.47 | 0.66 ± 0.25 [*] | 4.615 | 0.040 | 1.22 ± 0.23 | <0.001 [#] |
| Stride length (m) | 1.28 ± 0.47 | 1.23 ± 0.48 | 3.23 | 0.082 | 1.47± 0.12 | 0.044 [#] |
| Cadence (steps/min) | 85.06 ± 26.78 | 88.64 ± 25.81 | 4.49 | 0.043 | 116.99 ± 9.60 | <0.001 [#] |
| Stance phase (% Gait Cycle) | 55.66 ± 7.98 | 56.17 ± 6.92 | 0.845 | 0.365 | 59.46 ± 1.40 | 0.010 [#] |
| Double support phase (% Gait Cycle) | 9.72 ± 4.26 | 10.13± 5.03 | 0.80 | 0.379 | 9.82 ± 1.50 | 0.102 |
| **Harmonic Ratio** | | | | | | |
| AP direction | 80.87 ± 11.24 | 71.44 ± 18.00 [*] | 10.47 | 0.003 | 95.12 ± 2.33 | <0.001 [#] |
| ML direction | 38.02 ± 19.92 | 47.65 ± 21.44 [*] | 6.05 | 0.020 | 85.61 ± 8.03 | <0.001 [#] |
| V direction | 72.38 ± 11.89 | 64.13 ± 19.97 [*] | 8.39 | 0.007 | 95.18 ± 2.02 | <0.001 [#] |

Values are expressed as mean ± SD. PRE = pre-training (after habituation with FDS—3 days); POST = post-training (after 10 sessions of intensive training with the Foot Drop Stimulator); AP = Antero-posterior; ML = medio-lateral; V = vertical; [*] significant difference between PRE and POST training: One-way Repeated Measure Analysis of variance (spatio-temporal parameters $p = 0.05$; Harmonic Ratio parameters' statistical significance after Bonferroni correction ($p < 0.017$); [#] significant difference when comparing the PRE intervention and control group: Student's t-test ($p < 0.05$).

As for the assessment of time effects (PRE vs. POST session in stroke patients), a significant effect of time was found and, in particular, the post-hoc analysis detected a significant increase in gait speed ($p = 0.028$). Regarding the symmetry parameters (HR features), the statistical analysis detected a significant effect of time for the HR in all three directions. In particular, the HR in the ML direction increased (+20%, $p = 0.02$), while those in the AP and V directions decreased after training (both approximately 13%, AP direction: $p = 0.003$, V direction: $p = 0.007$).

## 4. Discussion

The purpose of the present study was to assess the effects of FDS use in individuals with stroke by means of spatio-temporal and step-to-step symmetry parameters obtained from trunk accelerations acquired using a wearable inertial sensor.

In terms of the spatio-temporal parameters, the only significant change observed post-treatment involved gait speed. This result is consistent with previous, similar studies [41–43]. However, it is worth noting that such improvement (+0.04 m/s), although statistically significant, was lower than the value indicated in the literature as clinically meaningful; several authors reported that the minimally clinical important difference in individuals who undergo inpatient rehabilitation after stroke lies between 0.10 and 0.18 m/s [56,57]. Concerning the HR parameters, mixed effects were observed in the treatment. Generally speaking, higher HR values denote better gait stability [30] and improved symmetry, smoothness, and rhythmicity. Our results show that the HR in the ML direction significantly increased, and this result may be explained by the fact that, during gait, the central nervous system controls the ML displacements related to the weight acceptance of each step [58,59]. However, in hemiplegic subjects the lack of lower muscle strength and increased instability observed in the affected side often acts by disrupting this strategy [60–67]. In this sense, the FDS walking training may have increased the ML gait symmetry by generating better foot contact [68] while shifting the body weight to a medial position, resulting in an improved ML stability during the walking movement and possibly, in the medium term, improving muscle strength and spasticity [69]. The important role played by the HR in the ML direction as a determinant of stability is also confirmed by previous studies reporting that a good lateral harmonic stability in gait may be important for minimizing fall risk in older people [70,71]. In addition, however, the $p$ value observed in the HR of the ML direction is slightly higher than the post-Bonferroni correction fixed statistical significance; the improvement of the HR in the ML direction is highly clinically relevant (+20%) and must be taken into consideration. Furthermore, it is important to note that the improvement of the HR appears in the more critical direction, the ML direction, which exhibited, in the PRE session, a much lower value than the control group value with respect to the V and AP directions.

In contrast, we also detected a reduction of the AP and V components of the HR, even though it is worth noting that the magnitude of such changes is approximately half compared with those related to the ML direction (−13% vs. +20%). This result suggests that the number of training sessions may be insufficient to let participants completely adapt to the new gait strategy. Further studies are thus necessary to verify whether a longer training period may trigger a complete readaptation of gait, from the point of view of symmetry, which involves all directions. This hypothesis is somewhat supported by previous studies that report how neuroprostheses are effective in enhancing balance control during walking (and thus effectively manage foot drop) after 8 weeks [41].

In addition, it is important to consider that our study's participants are mainly severely impaired. In this view, they probably need longer intervention periods to exhibit substantial changes in walking symmetry, which may contribute to the lack of improvement in the AP and V components of the HR.

While walking, maintaining balance requires continuous integrative control, especially in the ML direction, in order to cope with instability during single limb support [72]; thus, we can hypothesize that, after the treatment, patients are trying to adapt to the new gait strategy, which is characterized by higher velocity too, and the importance is given to the ML direction. Furthermore, it has been demonstrated that the HR is speed-dependent and it is especially affected in the very slow condition [73]. It is important to underline that, in this study, only the HR and spatio-temporal parameters were investigated; further research should be conducted, integrating these parameters with kinematic and kinetic data, to evaluate the most sensitive measures to changes in walking due to the FDS. Kinematics and kinetics may clarify where the symmetry deviations occur. In addition, in this study, no placebo control group was included. Further research that includes a placebo control group should be conducted in order to distinguish, more clearly, whether the procedure is effective or if the changes could be associated with a placebo effect.

Future studies may also wish to assess the effects of the FDS at a longer follow-up to more fully understand if the gait changes are maintained over time. In addition, this study contains a disproportionate number of men, who make up 68.8% of the study participants. Thus, by increasing the number of female patients, further work could be conducted to better understand the possible sex-related differences in trunk movement asymmetry. In addition, a large range of time since stroke (from 6 to 96 months) could have influenced the results. Further studies with a larger sample size and with a restricted range of time since stroke should investigate whether the results differ in stroke patients with dissimilar levels of impairment (i.e., mild–moderate vs. severe impairment).

Even though this study presents some limitations, it presents two original aspects: (1) the assessments were conducted using an inertial wearable sensor to document the effects of the FDS on gait in stroke patients; and (2) the analysis was conducted considering not only the traditional spatio-temporal parameters but also the HR, which has never been used to quantify the modifications of step-to-step symmetry in stroke patients induced by FDS treatment.

**Author Contributions:** G.S. and M.J.d.C.: conceptualization, methodology, clinical and instrumental assessment of the participants, and writing—original draft; V.C.: writing—original draft, formal analysis, and data curation; M.P. (Massimiliano Pau), B.L., and M.P. (Micaela Porta): formal analysis and software; M.G.: review and editing; A.S.P.: conceptualization, methodology, and review. All authors have read and agreed to the published version of the manuscript.

**Funding:** This research received no external funding.

**Institutional Review Board Statement:** The study was conducted according to the guidelines of the Declaration of Helsinki, and approved by the Ethics and Research Committee of the Santa Casa de Misericórdia Hospital of Porto Alegre (CAAE 64819617.0.0000.5335).

**Informed Consent Statement:** Informed consent was obtained from all subjects involved in the study.

**Data Availability Statement:** The data presented in this study are available on request from the corresponding author.

**Conflicts of Interest:** The authors declare no conflict of interest.

# References


1. Sousa, R.M.; Ferri, C.P.; Acosta, D.; Albanese, E.; Guerra, M.; Huang, Y.; Jacob, K.S.; Jotheeswaran, A.T.; Rodriguez, J.J.; Pichardo, G.R.; et al. Contribution of chronic diseases to disability in elderly people in countries with low and middle incomes: A 10/66 Dementia Research Group population-based survey. *Lancet* **2009**, *374*, 1821–1830. [CrossRef]
2. Aqueveque, P.; Ortega, P.; Pino, E.; Saavedra, F.; Germany, E.; Gómez, B. After Stroke Movement Impairments: A Review of Current Technologies for Rehabilitation. In *Physical Disabilities—Therapeutic Implications*; Tan, U., Ed.; IntechOpen Limited: London, UK, 2017.
3. Sheffler, L.R.; Chae, J. Hemiparetic Gait. *Phys. Med. Rehabil. Clin. N. Am.* **2015**, *26*, 611–623. [CrossRef] [PubMed]
4. Chisholm, A.E.; Perry, S.D.; McIlroy, W.E. Correlations between ankle–foot impairments and dropped foot gait deviations among stroke survivors. *Clin. Biomech.* **2013**, *28*, 1049–1054. [CrossRef] [PubMed]
5. Stewart, J.D. Foot drop: Where, why and what to do? *Pr. Neurol.* **2008**, *8*, 158–169. [CrossRef] [PubMed]
6. Pittock, S.J.; Moore, A.; Hardiman, O.; Ehler, E.; Kovac, M.; Bojakowski, J.; Al Khawaja, I.; Brozman, M.; Kaňovský, P.; Skorometz, A.; et al. A Double-Blind Randomised Placebo-Controlled Evaluation of Three Doses of Botulinum Toxin Type A (Dysport®) in the Treatment of Spastic Equinovarus Deformity after Stroke. *Cerebrovasc. Dis.* **2003**, *15*, 289–300. [CrossRef] [PubMed]
7. Duncan, P.W. Stroke Disability. *Phys. Ther.* **1994**, *74*, 399–407. [CrossRef] [PubMed]
8. Titianova, E.B.; Pitkänen, K.; Pääkkönen, A.; Sivenius, J.; Tarkka, I.M. Gait Characteristics and Functional Ambulation Profile in Patients with Chronic Unilateral Stroke. *Am. J. Phys. Med. Rehabil.* **2003**, *82*, 778–786, quiz 787–779, 823. [CrossRef]
9. Beyaert, C.; Vasa, R.; Frykberg, G. Gait post-stroke: Pathophysiology and rehabilitation strategies. *Neurophysiol. Clin. Neurophysiol.* **2015**, *45*, 335–355. [CrossRef]
10. Wonsetler, E.C.; Bowden, M.G. A systematic review of mechanisms of gait speed change post-stroke. Part 1: Spatiotemporal parameters and asymmetry ratios. *Top. Stroke Rehabil.* **2017**, *24*, 435–446. [CrossRef]
11. Wonsetler, E.C.; Bowden, M.G. A systematic review of mechanisms of gait speed change post-stroke. Part 2: Exercise capacity, muscle activation, kinetics, and kinematics. *Top. Stroke Rehabil.* **2017**, *24*, 394–403. [CrossRef]

12. Li, S.; Francisco, G.E.; Zhou, P. Post-stroke Hemiplegic Gait: New Perspective and Insights. *Front. Physiol.* **2018**, *9*, 1021. [CrossRef] [PubMed]

13. Herzog, W.; Nigg, B.M.; Read, L.J.; Olsson, E. Asymmetries in ground reaction force patterns in normal human gait. *Med. Sci. Sports Exerc.* **1989**, *21*, 110–114. [CrossRef] [PubMed]

14. McCrory, J.L.; White, S.C.; Lifeso, R.M. Vertical ground reaction forces: Objective measures of gait following hip arthroplasty. *Gait Posture* **2001**, *14*, 104–109. [CrossRef]

15. Queen, R.M.; Attarian, D.; Bolognesi, M.P.; Butler, R.J. Bilateral symmetry in lower extremity mechanics during stair ascent and descent following a total hip arthroplasty: A one-year longitudinal study. *Clin. Biomech.* **2015**, *30*, 53–58. [CrossRef]

16. Wiik, A.V.; Aqil, A.; Brevadt, M.; Jones, G.G.; Cobb, J. Abnormal ground reaction forces lead to a general decline in gait speed in knee osteoarthritis patients. *World J. Orthop.* **2017**, *8*, 322–328. [CrossRef]

17. Buckley, C.; Alcock, L.; McArdle, R.; Rehman, R.Z.U.; Del Din, S.; Mazza, C.; Yarnall, A.J.; Rochester, L. The Role of Movement Analysis in Diagnosing and Monitoring Neurodegenerative Conditions: Insights from Gait and Postural Control. *Brain Sci.* **2019**, *9*, 34. [CrossRef]

18. Iosa, M.; Bini, F.; Marinozzi, F.; Fusco, A.; Morone, G.; Koch, G.; Cinnera, A.M.; Bonnì, S.; Paolucci, S. Stability and Harmony of Gait in Patients with Subacute Stroke. *J. Med. Biol. Eng.* **2016**, *36*, 635–643. [CrossRef]

19. Menz, H.B.; Lord, S.R.; Fitzpatrick, R.C. Acceleration patterns of the head and pelvis when walking on level and irregular surfaces. *Gait Posture* **2003**, *18*, 35–46. [CrossRef]

20. Pasciuto, I.; Bergamini, E.; Iosa, M.; Vannozzi, G.; Cappozzo, A. Overcoming the limitations of the Harmonic Ratio for the reliable assessment of gait symmetry. *J. Biomech.* **2017**, *53*, 84–89. [CrossRef]

21. Iosa, M.; Fusco, A.; Morone, G.; Paolucci, S. Development and Decline of Upright Gait Stability. *Front. Aging Neurosci.* **2014**, *6*, 14. [CrossRef]

22. Lowry, K.A.; Smiley-Oyen, A.L.; Carrel, A.J.; Kerr, J.P. Walking stability using harmonic ratios in Parkinson's disease. *Mov. Disord.* **2009**, *24*, 261–267. [CrossRef] [PubMed]

23. Menz, H.B.; Latt, M.D.; Tiedemann, A.; San Kwan, M.M.; Lord, S.R. Reliability of the GAITRite walkway system for the quantification of temporo-spatial parameters of gait in young and older people. *Gait Posture* **2004**, *20*, 20–25. [CrossRef]

24. Pau, M.; Mandaresu, S.; Pilloni, G.; Porta, M.; Coghe, G.; Marrosu, M.G.; Cocco, E. Smoothness of gait detects early alterations of walking in persons with multiple sclerosis without disability. *Gait Posture* **2017**, *58*, 307–309. [CrossRef] [PubMed]

25. Leban, B.; Cimolin, V.; Porta, M.; Arippa, F.; Pilloni, G.; Galli, M.; Pau, M. Age-Related Changes in Smoothness of Gait of Healthy Children and Early Adolescents. *J. Mot. Behav.* **2019**, *52*, 694–702. [CrossRef]

26. Cimolin, V.; Cau, N.; Sartorio, A.; Capodaglio, P.; Galli, M.; Tringali, G.; Leban, B.; Porta, M.; Pau, M. Symmetry of Gait in Underweight, Normal and Overweight Children and Adolescents. *Sensors* **2019**, *19*, 2054. [CrossRef]

27. Cimolin, V.; Pau, M.; Cau, N.; Leban, B.; Porta, M.; Capodaglio, P.; Sartorio, A.; Grugni, G.; Galli, M. Changes in symmetry during gait in adults with Prader-Willi syndrome. *Comput. Methods Biomech. Biomed. Eng.* **2020**, *2020*, 1–8. [CrossRef]

28. Belluscio, V.; Bergamini, E.; Salatino, G.; Marro, T.; Gentili, P.; Iosa, M.; Morelli, D.; Vannozzi, G. Dynamic balance assessment during gait in children with Down and Prader-Willi syndromes using inertial sensors. *Hum. Mov. Sci.* **2019**, *63*, 53–61. [CrossRef]

29. Pau, M.; Mulas, I.; Putzu, V.; Asoni, G.; Viale, D.; Mameli, I.; Leban, B.; Allali, G. Smoothness of Gait in Healthy and Cognitively Impaired Individuals: A Study on Italian Elderly Using Wearable Inertial Sensor. *Sensors* **2020**, *20*, 3577. [CrossRef]

30. Brach, J.S.; McGurl, D.; Wert, D.; VanSwearingen, J.M.; Perera, S.; Cham, R.; Studenski, S. Validation of a Measure of Smoothness of Walking. *J. Gerontol. Ser. A Boil. Sci. Med. Sci.* **2010**, *66*, 136–141. [CrossRef]

31. Cimolin, V.; Galli, M. Summary measures for clinical gait analysis: A literature review. *Gait Posture* **2014**, *39*, 1005–1010. [CrossRef]

32. Mizuike, C.; Ohgi, S.; Morita, S. Analysis of stroke patient walking dynamics using a triaxial accelerometer. *Gait Posture* **2009**, *30*, 60–64. [CrossRef] [PubMed]

33. Iosa, M.; Fusco, A.; Morone, G.; Pratesi, L.; Coiro, P.; Venturiero, V.; De Angelis, D.; Bragoni, M.; Paolucci, S. Assessment of upper-body dynamic stability during walking in patients with subacute stroke. *J. Rehabil. Res. Dev.* **2012**, *49*, 439–450. [CrossRef] [PubMed]

34. Iosa, M.; Picerno, P.; Paolucci, S.; Morone, G. Wearable Inertial Sensors for Human Movement Analysis. *Expert Rev. Med. Devices* **2016**, *13*, 641–659. [CrossRef] [PubMed]

35. Isho, T.; Usuda, S. Association of trunk control with mobility performance and accelerometry-based gait characteristics in hemiparetic patients with subacute stroke. *Gait Posture* **2016**, *44*, 89–93. [CrossRef]

36. Yen, C.L.; Chang, K.C.; Wu, C.Y.; Hsieh, Y.W. The relationship between trunk acceleration parameters and kinematic characteristics during walking in patients with stroke. *J. Phys. Ther. Sci.* **2019**, *31*, 638–644. [CrossRef]

37. Buckley, T.A.; Oldham, J.R.; Watson, D.J.; Murray, N.G.; Munkasy, B.A.; Evans, K.M. Repetitive Head Impacts in Football Do Not Impair Dynamic Postural Control. *Med. Sci. Sports Exerc.* **2019**, *51*, 132–140. [CrossRef]

38. Kottink, A.I.; Tenniglo, M.J.; de Vries, W.H.; Hermens, H.J.; Buurke, J.H. Effects of an implantable two-channel peroneal nerve stimulator versus conventional walking device on spatiotemporal parameters and kinematics of hemiparetic gait. *J. Rehabil. Med.* **2012**, *44*, 51–57. [CrossRef]

39. Bae, Y.H.; Ko, Y.J.; Chang, W.H.; Lee, J.H.; Lee, K.B.; Park, Y.J.; Ha, H.G.; Kim, Y.H. Effects of Robot-assisted Gait Training Combined with Functional Electrical Stimulation on Recovery of Locomotor Mobility in Chronic Stroke Patients: A Randomized Controlled Trial. *J. Phys. Ther. Sci.* **2014**, *26*, 1949–1953. [CrossRef]

40. Hwang, D.-Y.; Lee, H.-J.; Lee, G.-C.; Lee, S.-M. Treadmill training with tilt sensor functional electrical stimulation for improving balance, gait, and muscle architecture of tibialis anterior of survivors with chronic stroke: A randomized controlled trial. *Technol. Health Care* **2015**, *23*, 443–452. [CrossRef]

41. Ring, H.; Treger, I.; Gruendlinger, L.; Hausdorff, J.M. Neuroprosthesis for Footdrop Compared with an Ankle-Foot Orthosis: Effects on Postural Control during Walking. *J. Stroke Cerebrovasc. Dis.* **2009**, *18*, 41–47. [CrossRef]

42. Sheffler, L.R.; Hennessey, M.T.; Naples, G.G.; Chae, J. Peroneal Nerve Stimulation versus an Ankle Foot Orthosis for Correction of Footdrop in Stroke: Impact on Functional Ambulation. *Neurorehabilit. Neural Repair* **2006**, *20*, 355–360. [CrossRef] [PubMed]

43. Kottink, A.I.; Hermens, H.J.; Nene, A.V.; Tenniglo, M.J.; van der Aa, H.E.; Buschman, H.P.; Ijzerman, M.J. A randomized controlled trial of an implantable 2-channel peroneal nerve stimulator on walking speed and activity in poststroke hemiplegia. *Arch. Phys. Med. Rehabil.* **2007**, *88*, 971–978. [CrossRef] [PubMed]

44. Fatone, S.; Gard, S.A.; Malas, B.S. Effect of ankle-foot orthosis alignment and foot-plate length on the gait of adults with poststroke hemiplegia. *Arch. Phys. Med. Rehabil.* **2009**, *90*, 810–818. [CrossRef] [PubMed]

45. von Elm, E.; Altman, D.G.; Egger, M.; Pocock, S.J.; Gotzsche, P.C.; Vandenbroucke, J.P.; Initiative, S. The Strengthening the Reporting of Observational Studies in Epidemiology (STROBE) Statement: Guidelines for reporting observational studies. *Int. J. Surg.* **2014**, *12*, 1495–1499. [CrossRef] [PubMed]

46. Daly, J.J.; Zimbelman, J.; Roenigk, K.L.; McCabe, J.P.; Rogers, J.M.; Butler, K.; Burdsall, R.; Holcomb, J.P.; Marsolais, E.B.; Ruff, R.L. Recovery of Coordinated Gait: Randomized Controlled Stroke Trial of Functional Electrical Stimulation (FES) Versus No FES, with Weight-Supported Treadmill and Over-Ground Training. *Neurorehabilit. Neural Repair* **2011**, *25*, 588–596. [CrossRef]

47. Almeida, O.P. Mini mental state examination and the diagnosis of dementia in Brazil. *Arq. Neuropsiquiatr.* **1998**, *56*, 605–612. [CrossRef]

48. Bohannon, R.W.; Andrews, A.W. Interrater reliability of hand-held dynamometry. *Phys. Ther.* **1987**, *67*, 931–933. [CrossRef]

49. Damiano, D.L.; Prosser, L.A.; Curatalo, L.A.; Alter, K.E. Muscle Plasticity and Ankle Control after Repetitive Use of a Functional Electrical Stimulation Device for Foot Drop in Cerebral Palsy. *Neurorehabilit. Neural Repair* **2012**, *27*, 200–207. [CrossRef]

50. Everaert, D.G.; Stein, R.B.; Abrams, G.M.; Dromerick, A.W.; Francisco, G.E.; Hafner, B.J.; Huskey, T.N.; Munin, M.C.; Nolan, K.J.; Kufta, C.V. Effect of a foot-drop stimulator and ankle-foot orthosis on walking performance after stroke: A multicenter randomized controlled trial. *Neurorehabil. Neural Repair* **2013**, *27*, 579–591. [CrossRef]

51. Kleiner, A.; Galli, M.; Gaglione, M.; Hildebrand, D.; Sale, P.; Albertini, G.; Stocchi, F.; De Pandis, M.F. The Parkinsonian Gait Spatiotemporal Parameters Quantified by a Single Inertial Sensor before and after Automated Mechanical Peripheral Stimulation Treatment. *Parkinsons Dis.* **2015**, *2015*, 390512. [CrossRef]

52. Pau, M.; Corona, F.; Pilloni, G.; Porta, M.; Coghe, G.; Cocco, E. Do gait patterns differ in men and women with multiple sclerosis? *Mult. Scler. Relat. Disord.* **2017**, *18*, 202–208. [CrossRef] [PubMed]

53. Pau, M.; Coghe, G.; Corona, F.; Marrosu, M.G.; Cocco, E. Effect of spasticity on kinematics of gait and muscular activation in people with Multiple Sclerosis. *J. Neurol. Sci.* **2015**, *358*, 339–344. [CrossRef] [PubMed]

54. Brandes, M.; Zijlstra, W.; Heikens, S.; Van Lummel, R.; Rosenbaum, D. Accelerometry based assessment of gait parameters in children. *Gait Posture* **2006**, *24*, 482–486. [CrossRef] [PubMed]

55. Buganè, F.; Benedetti, M.G.; Casadio, G.; Attala, S.; Biagi, F.; Manca, M.; Leardini, A. Estimation of spatial-temporal gait parameters in level walking based on a single accelerometer: Validation on normal subjects by standard gait analysis. *Comput. Methods Programs Biomed.* **2012**, *108*, 129–137. [CrossRef] [PubMed]

56. Bohannon, R.W.; Andrews, A.W.; Glenney, S.S. Minimal Clinically Important Difference for Comfortable Speed as a Measure of Gait Performance in Patients Undergoing Inpatient Rehabilitation after Stroke. *J. Phys. Ther. Sci.* **2013**, *25*, 1223–1225. [CrossRef] [PubMed]

57. Fulk, G.D.; Ludwig, M.; Dunning, K.; Golden, S.; Boyne, P.; West, T. Estimating Clinically Important Change in Gait Speed in People with Stroke Undergoing Outpatient Rehabilitation. *J. Neurol. Phys. Ther.* **2011**, *35*, 82–89. [CrossRef]

58. Halliday, S.E.; Winter, D.A.; Frank, J.S.; Patla, A.E.; Prince, F. The initiation of gait in young, elderly, and Parkinson's disease subjects. *Gait Posture* **1998**, *8*, 8–14. [CrossRef]

59. Rossi, S.A.; Doyle, W.; Skinner, H.B. Gait initiation of persons with below-knee amputation: The characterization and comparison of force profiles. *J. Rehabil. Res. Dev.* **1995**, *32*, 120–127.

60. Pérennou, D. Weight bearing asymmetry in standing hemiparetic patients. *J. Neurol. Neurosurg. Psychiatry* **2005**, *76*, 621. [CrossRef]

61. Roerdink, M.; Geurts, A.C.H.; De Haart, M.; Beek, P. On the Relative Contribution of the Paretic Leg to the Control of Posture after Stroke. *Neurorehabilit. Neural Repair* **2008**, *23*, 267–274. [CrossRef]

62. Roerdink, M.; De Haart, M.; Daffertshofer, A.; Donker, S.F.; Geurts, A.C.; Beek, P.J. Dynamical structure of center-of-pressure trajectories in patients recovering from stroke. *Exp. Brain Res.* **2006**, *174*, 256–269. [CrossRef] [PubMed]

63. Sackley, C.M. Falls, sway, and symmetry of weight-bearing after stroke. *Int. Disabil. Stud.* **1991**, *13*, 1–4. [CrossRef] [PubMed]

64. Genthon, N.; Rougier, P.; Gissot, A.-S.; Froger, J.; Pélissier, J.; Pérennou, D. Contribution of Each Lower Limb to Upright Standing in Stroke Patients. *Stroke* **2008**, *39*, 1793–1799. [CrossRef] [PubMed]

65. Chow, J.W.; Stokic, D.S. Relations between knee and ankle muscle coactivation and temporospatial gait measures in patients without hypertonia early after stroke. *Exp. Brain Res.* **2020**, *238*, 2909–2919. [CrossRef]

66. Tamaya, V.C.; Wim, S.; Herssens, N.; Van De Walle, P.; Willem, D.H.; Steven, T.; Ann, H. Trunk biomechanics during walking after sub-acute stroke and its relation to lower limb impairments. *Clin. Biomech.* **2020**, *75*, 105013. [CrossRef]

67. Calvo-Lobo, C.; Useros-Olmo, A.I.; Almazán-Polo, J.; Becerro-De-Bengoa-Vallejo, R.; Losa-Iglesias, M.E.; Palomo-López, P.; Rodríguez-Sanz, D.; López-López, D. Rehabilitative ultrasound imaging of the bilateral intrinsic plantar muscles and fascia in post-stroke survivors with hemiparesis: A case-control study. *Int. J. Med. Sci.* **2017**, *15*, 907–914. [CrossRef]
68. Nolan, K.J.; Yarossi, M.; McLaughlin, P. Changes in center of pressure displacement with the use of a foot drop stimulator in individuals with stroke. *Clin. Biomech.* **2015**, *30*, 755–761. [CrossRef]
69. Everaert, D.G.; Thompson, A.K.; Chong, S.L.; Stein, R.B. Does functional electrical stimulation for foot drop strengthen corticospinal connections? *Neurorehabil. Neural Repair* **2010**, *24*, 168–177. [CrossRef]
70. Brodie, M.; Menz, H.B.; Smith, S.T.; Delbaere, K.; Lord, S.R. Good Lateral Harmonic Stability Combined with Adequate Gait Speed Is Required for Low Fall Risk in Older People. *Gerontology* **2015**, *61*, 69–78. [CrossRef]
71. Asai, T.; Misu, S.; Sawa, R.; Doi, T.; Yamada, M. The association between fear of falling and smoothness of lower trunk oscillation in gait varies according to gait speed in community-dwelling older adults. *J. Neuroeng. Rehabil.* **2017**, *14*, 5. [CrossRef]
72. O'Connor, S.M.; Kuo, A.D. Direction-Dependent Control of Balance during Walking and Standing. *J. Neurophysiol.* **2009**, *102*, 1411–1419. [CrossRef] [PubMed]
73. Bellanca, J.L.; Lowry, K.A.; Vanswearingen, J.M.; Brach, J.S.; Redfern, M.S. Harmonic ratios: A quantification of step to step symmetry. *J. Biomech.* **2013**, *46*, 828–831. [CrossRef] [PubMed]

 *sensors*

*Article*

# Design and Validation of an E-Textile-Based Wearable Sock for Remote Gait and Postural Assessment

**Federica Amitrano** [1,2], **Armando Coccia** [1,2], **Carlo Ricciardi** [2,3], **Leandro Donisi** [2,3],
**Giuseppe Cesarelli** [2,4], **Edda Maria Capodaglio** [2] **and Giovanni D'Addio** [2,*]

[1]  Department of Information Technologies and Electrical Engineering, University of Naples 'Federico II',
    80125 Naples, Italy; federica.amitrano@unina.it (F.A.); armando.coccia@unina.it (A.C.)
[2]  Scientific Clinical Institutes ICS Maugeri SPA SB, 27100 Pavia (PV), Italy; carloricciardi.93@gmail.com (C.R.);
    leandro.donisi@unina.it (L.D.); giuseppe.cesarelli@unina.it (G.C.); edda.capodaglio@icsmaugeri.it (E.M.C.)
[3]  Department of Advanced Biomedical Sciences, University of Naples Federico II, 80131 Naples, Italy
[4]  Department of Chemical, Materials and Production Engineering, University of Naples Federico II,
    80125 Naples, Italy
*   Correspondence: gianni.daddio@icsmaugeri.it; Tel.: +39-082-490-9640 or +39-082-490-9001

Received: 22 October 2020; Accepted: 19 November 2020; Published: 23 November 2020

**Abstract:** This paper presents a new wearable e-textile based system, named SWEET Sock, for biomedical signals remote monitoring. The system includes a textile sensing sock, an electronic unit for data transmission, a custom-made Android application for real-time signal visualization, and a software desktop for advanced digital signal processing. The device allows the acquisition of angular velocities of the lower limbs and plantar pressure signals, which are postprocessed to have a complete and schematic overview of patient's clinical status, regarding gait and postural assessment. In this work, device performances are validated by evaluating the agreement between the prototype and an optoelectronic system for gait analysis on a set of free walk acquisitions. Results show good agreement between the systems in the assessment of gait cycle time and cadence, while the presence of systematic and proportional errors are pointed out for swing and stance time parameters. Worse results were obtained in the comparison of spatial metrics. The "wearability" of the system and its comfortable use make it suitable to be used in domestic environment for the continuous remote health monitoring of de-hospitalized patients but also in the ergonomic assessment of health workers, thanks to its low invasiveness.

**Keywords:** wearable devices; e-textile; gait analysis; m-health; plantar pressure; validation; Internet of Things

---

## 1. Introduction

The term Electronic-Textiles, or E-Textiles, refers to a wide range of studies and products that extend the usefulness and functionalities of common fabrics. The innovative feature taken by this novel application regards the embedding of digital components, such as batteries, LEDs, and, in general, electronic components, in common fabrics. Thus, through E-Textile technology, every kind of digital application can be potentially developed on a textile substrate. This attractive opportunity is bringing a revolution in the market of wearable devices, with the involvement of big companies which are trying to shift from the wearable electronic hardware to the more comfortable electronic textiles. The market of wearable technologies has a compound annual growth rate of 15.5%, with great opportunities of expansion, it is expected to reach 51.6 billion USD by 2022 (IDTechEx). E-textile is gradually covering this market, offering cheap and comfortable solution in different sectors, such as fashion, entertainment, military and defense, space exploration, health, and wellness.

Healthcare remains one of the most interesting and promising markets: e-textile features are very suitable for the development of innovative medical devices or applications that can potentially establish significant cost reductions for healthcare systems. Wearable devices for health monitoring can be easily used by patient in domestic environment and, when they are integrated in a complete communication chain, they allow smart remote monitoring with great benefits for caregivers and patient himself. E-textile sensitive fabrics can be developed to acquire and react to clinical signals detectable on body, with some interesting advantages: first, the nature of fabrics makes them the best solution to realize sensors in direct contact with the skin; second, fabrics are flexible and well adaptable to human body offering technological possibilities not available with the common electronics; and third, fabrics are cheap, comfortable, washable, and easily customizable [1]. Thus, smart biomedical clothes potentially represent an innovative tool for the continuous monitoring of vital signs, combining the function of sophisticated medical devices with the comfort and ease of use of clothing products.

Moreover, the opportunity to integrate these innovative devices in IoT networks makes possible to establish smart solution for remote health monitoring, exploring the growing field of m-health and supporting cost reduction in healthcare system by facilitating early hospital discharges. Many E-Textile solutions for health monitoring have been proposed in literature, but most of them are blocked in the research field and are not intended to flow to the pragmatic healthcare world. Regulatory issues regarding patient safety, privacy, data management [2,3], and the need of a safe degree of reliability for device performances represent the main obstacles to the large commercial diffusion of such types of devices.

This manuscript presents a prototypical system, based on an e-textile sensing sock, able to collect the angular velocities of lower limbs, using Inertial Measurement Units (IMUs), and the plantar pressures, by means of textile sensors. Our aim is to provide a wearable and portable system for the assessment of both postural and gait tasks, exploiting the recent advances in the field of e-textile, electronic and signal processing. In particular the system is intended to provide the assessment of spatio-temporal gait parameters by processing the angular velocities signals while the pressure signals will be used to assess Center of Pressure (COP) displacements during static postural tests.

Static posturography in clinical environment is usually achieved by means of commercial platform systems. These systems include a big number of sensors arranged in a matrix resulting in high spatial resolution and high accuracy [4]. However, platform systems are expensive, not portable, and require a trained technician to be used. In-shoe systems can overcome the usability limitations of platforms, enabling measurements of plantar pressure distribution within a shoe, in indoor and outdoor environments. In [5–7], three insoles with, respectively, 10, 4, and 3 sensors are used to measure the COP for the assessment of balance. All these applications are based on force sensing resistors (FSRs), whose hard structure can reduce the comfort for the user. Moreover, insoles create an additional layer inside the shoe, which can essentially change the distribution of plantar load of the foot compared to the natural in-shoe condition [8]. Textile pressure sensors represent an attractive solution because they improve comfort for users and their thickness ensure no distortion of plantar pressure. Several experimental custom-made smart socks, with textile pressure sensors embedded, are described in literature. Most of them are developed for the assessment of spatio-temporal gait parameters [9–11], while other solutions [8,12] provide for postural assessment in dynamic tasks. Nevertheless, the latter offers only a qualitative representation of pressures distribution during walking tasks. Unlike these, the proposed system uses the textile pressure sensors not for gait analysis nor for dynamic postural assessment, but pressure signals are considered and processed to provide quantitative estimation and analysis of COP displacement during static tasks.

Regarding gait assessment, we decided to exclusively exploit kinematic data collected from IMUs because plantar pressure signals would not provide significant support for the estimation of spatio-temporal parameters. IMUs are nowadays broadly used in biomedical field. These devices are light, small, and can be easily integrated in electronic circuits, so they are very suitable for wearable application. Different kinds of IMU-based medical applications are available in literature, from the activity classification [13–18] to the balance assessment [19,20], but gait analysis is the most explored [21,22]. IMUs overcome the limitations of laboratory measurements enabling the assessment of spatio-temporal gait parameters in indoor and outdoor environments. Moreover, IMUs are cheaper and more practical than full gait analysis systems, thus broadening the range of its potential users. As reported in [23], gait analysis is typically gained using the accelerometer, while the gyroscope is arguably the next most commonly used sensor. The different gait phases can be detected from angular velocities, measured by gyroscopes attached to lower limbs [24]. Accelerometers by themselves can measure angular rotation but they cannot give a good result as gyroscopes. Thus, gyroscopes are often used in fusion with accelerometer readings [25–27], when deployed together such as in an IMU, or alone [28–31] in the assessment of gait parameters.

There is a variety of commercially available IMU-based systems for gait analysis that are currently used in clinical environment, such as Opal by APDM or G-Sensor by BTS. They are wearable and portable systems, but they are expensive and require the presence of a technician to place sensors and carry out the acquisition using the computer software. Therefore, they cannot be used in domestic environment nor without the supervision of an expert. In contrast, our system is intended to be used in real-life conditions without any aid, as it only requires to wear socks and follow the easy steps guided by a mobile application that can be installed on the patient's smartphone.

In this manuscript, we describe the details of prototype design and development. We also provide a validation analysis of the system concerning the assessment of spatio-temporal gait parameters deriving from IMU signals digital processing. This analysis has been obtained by performing comparative assessments with a stereophotogrammetry system for gait analysis, used in clinical environment and considered to be the gold standard in this kind of assessment.

## 2. Materials and Methods

The aim of this work is to present the novel wearable device SWEET Sock for remote health monitoring and to validate its performances in the acquisition and analysis of angular velocity signals of the lower limbs for the assessment of spatio-temporal gait parameters. The first version of this device, presented in [32], has been improved with new more efficient textile and electronic components and through the addition of a set of signal processing algorithms. In this chapter, we will present in detail the units making up the update version of the system and the materials and methods used to perform the validation analysis.

### 2.1. Wearable Device: SWEET Sock

SWEET Sock is a wearable sensing device which allows the acquisition of accelerometric and pressure signals. It can be integrated in a complete system for remote health monitoring, presented in the schematic diagram in Figure 1.

**Figure 1.** System Architecture: (**1**) SWEET Sock—Textile Unit; (**2**) SWEET Sock—Control Unit; (**3**) SWEET App; (**4**) Web Server; (**5**) SWEET Lab.

The wearable sensor unit allows the acquisition of bio-signals when connected to the analogue front-end located in the electronic unit. This unit also contains a microcontroller and allows data transmission through an integrated Bluetooth Low Energy (BLE) module. A custom-made Android mobile application has been developed to receive and visualize real-time signals on a smartphone, and to upload data on a dedicated web server afterwards. This is a restricted area that is accessible after prior authentication, exclusively by authorized and appointed health professionals, who can download, analyze, and process data using the custom-made MATLAB desktop software.

In the following sections, the functional modules of the system are individually presented.

### 2.1.1. Wearable Sensing Unit

The wearable sensing unit consists of a commercial sports sock in which three pressure sensors, in e-textile technology, have been integrated as sensing elements in three strategic points of the foot arch. The number and placement of sensors were based on anatomical considerations: in standing position, the main force transmitted onto the foot originates at the bones of the lower leg. At the ankle, this force is divided into three smaller forces in the style of a tripod. Within the foot, one of these three forces is directly transmitted onto the calcaneus, the second one onto the first metatarsal, and the third one is distributed across the second to fifth metatarsal [33]. We therefore decided to use three pressure sensors per foot: one under the heel (HEEL), one under the first metatarsal bone (MTB1), and one under the fifth metatarsal bone (MTB5) (Figure 2c). Besides the experimental device presented in [33], also the commercial smart socks Sensoria are designed with the same number and placement of the pressure sensors. The performances of the latter in static postural assessment have been also investigated, with good results, in comparison with a stabilometric platform [34]. The use of the minimum number of sensors needed for the analysis reduces the complexity of textile design and can improve the comfort and wearability for users.

Sensors have been realized by using 2-by-4 cm sheets of EeonTex fabric, a conductive and nonwoven microfiber with piezo-resistive functionality (surface resistivity 2000 ohm/sq), offering a reduction of the electrical resistance to the application of force. Their characterization was carried out with load tests using a controlled mechanical clamp with decreasing/increasing loads [32]. The three conductive sensors have been covered by non-conductive fabric to prevent degradation by contact with the skin and are thin enough to provide postural monitoring at natural in-shoe conditions, without distortion of plantar pressure. A conductive ribbon (5 mm tick), with a resistance of less than 0.1 ohm per cm, has been used to connect sensors to the output connectors of the wearable unit. Compared to the conductive wires available on the market, the ribbon has a lower resistance (0.1 vs. 0.9 ohm per cm) and is more robust as it does not break due to stretch. The design of conductive pathways provides a placement of all connectors of the data acquisition system, represented by snap buttons, on the lateral part of the sock, which essentially improves the system usability. The textile connections have been sewn on the side of the sock avoiding, when possible, the passage under the sole of the feet, where they could be deteriorated. Connection lengths have also been minimized by studying the shortest path in order to reduce noise and interference. Figure 2 shows the complete device with its sartorial design.

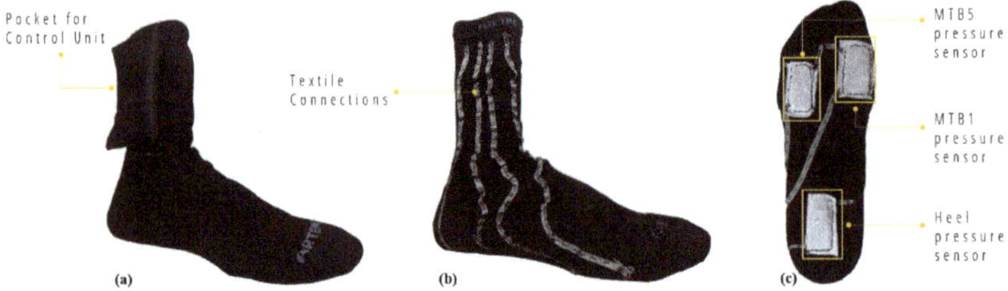


**Figure 2.** SWEET Sock sensing unit: (**a**) external view; (**b**) internal view of textile connections; (**c**) textile pressure sensors.

### 2.1.2. Electronic Unit

The electronic unit is a compact module containing all the electric and electronic elements to allow acquisition, digitalization, storage, and wireless transmission of the signals.

A conditioning circuit, for each conductive sensor, has been realized in order to read a voltage signal proportional to the applied force. This circuit is realized by means of a voltage divider consisting of two resistors: one of which is of known value and the other represented by the e-textile sensor. The known resistance value is fixed to 18 kohm, around which the conductive sensor resistance ranges, to reach the condition of maximal sensitivity. The IMU FLORA 9-DOF (Adafruit Inc.: New York, NY, USA) has been integrated in the electronic unit to acquire gyroscopic signal. It consists of a small electronic board mounting LSM9DS1 module, a system-in-package featuring a 3D digital linear acceleration sensor, a 3D digital angular rate sensor, and a 3D digital magnetic sensor.

A LilyPad Simblee™ BLE Board (Sparkfun Inc.: Niwot, CO, USA) has been used as the microcontroller. It provides the digitalization of pressure signals, and it is connected to Flora IMU through the I2C serial bus interface. LilyPad Simblee also allows to send data via Bluetooth Low-Energy protocol (BLE, or Bluetooth 4.0), using Simblee™ Bluetooth® Smart Module integrated on the shield. BLE technology represents a perfect trade-off between energy consumption, latency, piconet size, and throughput. Its control features are implemented exploiting the ARM® Cortex M0 microcontroller that can be programmed using the Arduino IDE. The control unit is programmed to sample pressure analogue signals with a sample period of 15 ms (66.7 Hz), and to receive digital data from the gyroscope with the same rate. Data are collected in 16-bytes-sized packets (2 bytes for each information: Packet, Time, x-y-z axes of the gyroscope, MTB1, MTB5, and HEEL pressure data) and real-time sent, via BLE, to the smartphone using SWEET App. Other signals deriving from IMUs (signals from accelerometer and magnetometer) are not recorded by the device because they do not provide any essential information for the planned assessments. We actually choose to implement a gyroscope-based algorithm to evaluate all spatio-temporal metrics because accelerometer signals are affected by gravity and are sensitive to sensor location [35]. When using accelerometers, it is important that they are placed in the same location each time as the signal is affected by how far from the center of rotation they are. The advantage of using a shank mounted gyroscope compared to accelerometers is that, as long as the gyroscope is recording data in the correct plane, it does not matter where on the shank the sensor is placed [36,37]. This reduction in the amount of acquired and sent data allows to improve signals sampling and sending rate.

All modules making up the electronic unit are powered by a 190 mAh/3.7 V lithium battery, placed on the back of the same unit. The electronic unit is housed in a 3D-printed plastic case (73 mm × 52 mm × 21 mm). On the top part of the case, 4 snap buttons allow the connection to the wearable sensing unit, in order to provide the input signals for the analogue front ends. In Figure 3 the electronic unit, with its main details, is shown.

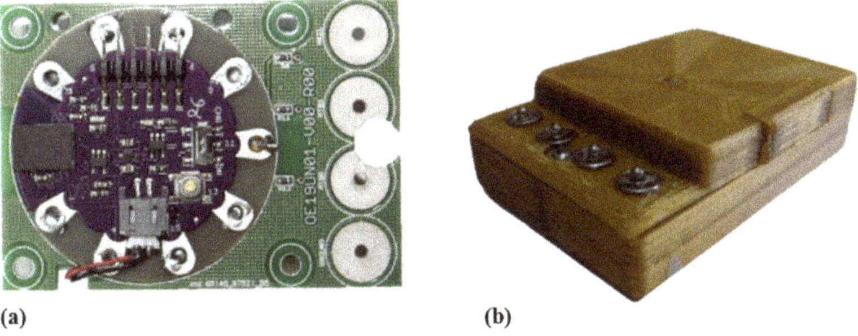

(a)       (b)

**Figure 3.** SWEET Sock ElectronicUnit: (**a**) internal electronic unit; (**b**) complete unit external view.

### 2.1.3. SWEET App

SWEET App is a custom-made Java language application for mobile devices requiring Android 6.0 or higher operating system and BLE technology. The application allows the smartphone to communicate and receive data coming from the electronic unit, via BLE protocol. When the application is started it is possible to associate and connect the wearable device, using its MAC address. Then, the measurement session can start, data are transferred from the electronic unit to the mobile device, which allows signals real time plotting. At the end of the session data are automatically saved in a ".csv" file, which is stored locally and can be uploaded at any time to a dedicated web server. In Figure 4 the main frames of the app are shown.

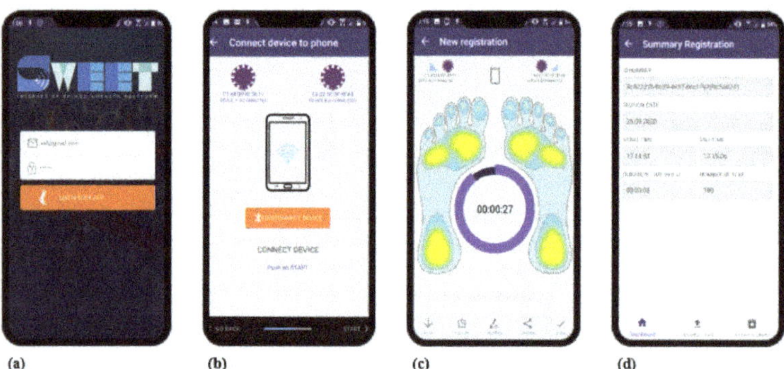

(a)  (b)  (c)  (d)

**Figure 4.** SWEET App main frames: (**a**) login; (**b**) unit connection; (**c**) signal recording; (**d**) results summary.

2.1.4. Signal Processing Algorithms

A custom-made Matlab GUI software, named SWEET Lab, has been developed to allow signal visualization and digital processing. Health professionals have the possibility to download data from the server and analyze them using the tools offered by this software. Pressure and gyroscope signals gathered by the hardware are individually processed to respectively perform posturographic assessment and spatio-temporal gait analysis. The two types of signal were not integrated because they are used in the analysis of two separate phases: pressure signals for static postural assessment while angular velocities in dynamic walking tasks analysis.

A gyroscope-based algorithm for gait analysis has been developed. The angular velocity signals on the sagittal plane are selected and low-pass filtered with 5th order Butterworth filter (cut-off frequency 5 Hz) to reduce noise. Mid-swing, heel-strike, and toe-off events are then identified on the filtered signals for both feet, using a threshold-based algorithm [38]. The starting point of the algorithm is the identification of the time events corresponding to the mid-swing, identified as the local maximum peaks of the signal. In the next step, local minimum peaks prior and after the mid-swing point are selected as, respectively, toe-off and heel-strike time events. Starting from these gait events times, all temporal parameters of gait analysis are calculated. In Table 1, the list of temporal parameters is provided with a description clearly outlining the methods used to calculate them. Spatial parameters are assessed using a single pendulum model described in [36], where the distance from the foot to the top vertex of the rotation is modeled as equal to the height of the subject multiplied by a scaling factor. Equation (1) shows how the stride length is calculated:

$$StrideLength(m) = S \times H \times 2(1 - \cos\theta) \tag{1}$$

$S$ represents the scaling factor chosen equal to 0.52 [36], H represents subject height [$m$] and $\theta$ is the angular displacement in the sagittal plane during the stride [$rad$], assessed by integration of the gyroscope signal.

Plantar pressure signals collected by the sensorized socks are used to perform sway analysis, as a systematic assessment of the readiness and stability of the human body to achieve and maintain equilibrium. This analysis starts with the estimation of the center of pressure (COP), whose displacement during stand task is a meaningful parameter for a quantitative evaluation of the ability to maintain equilibrium. At each instant, COP coordinates in the medio-lateral ($X_{COP}$) and antero-posterior ($Y_{COP}$) directions have been calculated by processing raw pressure data according to the following Equation (2),

$$X_{COP} = \frac{\sum_{i=1}^{N} X_i P_i}{\sum_{i=1}^{N} P_i}$$

$$Y_{COP} = \frac{\sum_{i=1}^{N} Y_i P_i}{\sum_{i=1}^{N} P_i}$$

$$(2)$$

where $N$ denotes the total number of sensors, and $X$ and $Y$ are the sensor coordinate inside the whole foot shape area and P the pressure value. The resulting signals express COP displacement along time in the medio-lateral (ML) and antero-posterior (AP) directions, with respect to a reference point located in the middle between the feet. The mono-dimensional representations of these signals constitute the ML and AP stabilograms, while the combined bidimensional plot is referred to as statokinesigram, representing the ground projection of the COP during the stand task.

**Table 1.** Spatio-temporal gait parameters.

| Temporal Measures | |
| --- | --- |
| **Variable** | **Description** |
| Gait Cycle Time (GCT) [s] | Defined as the time between two successive heel strikes of the same foot. |
| Stance Time [s] | The amount of time a foot is in contact with the ground within a single gait cycle. It is the time between the heel-strike and the successive toe-off of the same foot. |
| Stance Phase [%] | Stance time expressed in percentage of the GCT. |
| Swing Time [s] | Duration of the swing phase, in which the foot is not in contact with the ground. It is calculated as the time between the toe-off and the successive heel strike of the same foot. |
| Swing Phase [%] | Swing time expressed in percentage of the GCT. |
| Single Support [%] | Part of the GCT in which a single foot is in contact with the ground. It is the time between the toe-off of the opposite foot and the successive heel-strike of the opposite foot, expressed in percentage of the GCT. |
| Double Support [%] | Part of the GCT in which both feet are in contact with the ground. It is the time between the heel-strike of a foot and the successive toe-off of the opposite foot, expressed in percentage of the GCT. |
| Cadence [steps/min] | Number of steps per minute. |
| **Spatial Measures** | |
| **Variable** | **Description** |
| Stride Length [m] | Distance covered during GCT. |
| Stride Velocity [m/s] | Defined as the ratio between Stride Length and GCT. |

Signals are filtered with a low-pass 4th-order Butterworth digital filter with a cut-off frequency of 5 Hz [39], and then analyzed in time domain to calculate a set of parameters describing the stability of the subject during the task (Table 2) [34,40,41].

Stabilometric signals are also analyzed in frequency domain. The Matlab periodogram algorithm is used to estimate power spectral density (PSD), modified using the Hamming window. Frequency assessment is provided by means of a set of measures describing the distribution of PSD, such as peak and centroidal frequencies, band powers, and others. All the parameters assessed are listed in Table 2. The description clarifies the methods used to evaluate both spatial and frequency domain metrics starting from stabilometric signals and ground projection of the COP trajectory.

**Table 2.** Static postural assessment parameters.

| Time Domain Measures | |
| --- | --- |
| **Variable** | **Description** |
| Mean COP coordinates [cm] | ML and AP mean COP displacements during time. |
| Mean Distance [cm] | Mean distance of COP trajectory from the center of the trajectory itself. |
| COP Trajectory Range [cm] | Maximum distance between 2 points of COP trajectory in ML and AP directions. |
| Root Mean Square (RMS) [cm] | RMS of COP trajectory. It is provided also for single ML and AP directions. |
| Angle form AP axis [deg] | Mean angle formed by the segments composing COP trajectory and AP direction. |
| Sway Path [cm] | Total length of COP trajectory, computed as the sum of distances between successive points of the trajectory. |
| Mean Velocity [cm/s] | Mean velocity of COP trajectory, computed as the ratio between sway path length and duration of the test. |
| 95% Ellipse Area [cm$^2$] | Area of 95% confidence ellipse encompassing the COP trajectory in transverse plane. |
| 95% Ellipse Angle [deg] | 95% confidence ellipse inclination with respect to the ML direction. |
| **Frequency Domain Measures** | |
| **Variable** | **Description** |
| Peak Frequency [Hz] | Peak frequency for ML and AP power spectrum. |
| Median Frequency [Hz] | Frequency below which the 50th percentile of total power is present. |
| 80% Frequency [Hz] | Frequency below which the 80th percentile of total power is present. |
| Centroidal Frequency [Hz] | Spectral centroid of power spectrum. It indicates where the center of mass of the spectrum is located. |
| Band Power [cm$^2$] | Power comprised in low [0.1–0.2 Hz], mid [0.2–0.3 Hz], and high [0.3–1 Hz] frequency bands, expressed as absolute and percentage values. |

## 2.2. Validation Analysis

This manuscript presents a validation analysis concerning SWEET Sock gait assessment.

In [42], a first validation analysis was performed by comparing the raw accelerometric and plantar pressure signals acquired by the prototype with those recorded by reference systems. Following the results obtained, in this work we want to proceed the process of validation of device performances exploring the results of gait assessment, in order to carry out any possible unconformity in measurement and/or processing phases managed by the new prototype. We compared spatio-temporal gait parameters calculated by SWEET Sock with those found by an optoelectronic stereophotogrammetric system. The comparison has been carried out by means of statistical methods. This section describes the methods used for data acquisition and analysis.

### 2.2.1. Stereophotogrammetric System for Gait Analysis

The reference system chosen for the validation analysis is SMART-DX 700 by BTS Bioengineering, an optoelectronic stereophotogrammetric system used for movement analysis. Stereophotogrammetry is usually considered a "gold standard" in gait analysis when used appropriately. The system is made of 6 infrared digital cameras, with a sensor resolution of 1.5 megapixel, an acquisition frequency from 250 fps (at maximum resolution) to 1000 fps and an accuracy lower than 0.1 mm. The recognition of body segments during movement is achieved through the use of twenty-two retro-reflective passive markers (diameter 14 mm), which are attached to subject's skin at specific landmarks. Video data are processed on a PC workstation running SMART Clinic software, able to store and compute a set of parameters concerning kinematic (spatiotemporal parameters, joint angles) and dynamic (forces exchanged).

### 2.2.2. Experimental Setup

One-hundred-and-eight records were acquired on three healthy subjects: two males (aged 27 and 26) and one female (aged 25). Participants were free of neurological, muscular, and skeletal comorbidities affecting mobility and gait. The subject wore the sensorized socks connected to the electronic unit and was equipped with the markers of the stereophotogrammetric system, in order to perform simultaneous recording of the walking tasks with the two systems under test (Figure 5). The markers were attached to subject's skin according to the protocol described by Davis et al. [43].

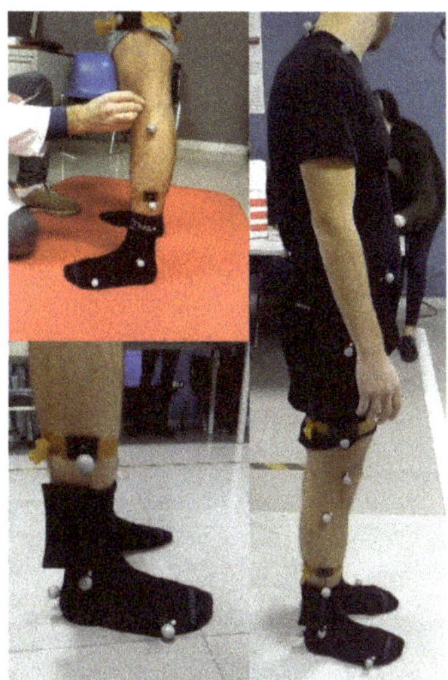

**Figure 5.** Subject equipped with both systems: SWEET Sock and reflective markers.

The trials involved free walking tests on a 11 m walkway in the movement analysis laboratory of University Hospital "Ruggi D'Aragona" of Salerno (Italy). Each subject was instructed to perform eight independent trials respectively at preferred, slow and fast self-selected walking speed. After that, the use of a metronome was introduced to force subjects walking at fixed normal, slow and high speed. Metronome rate was set at 100%, 67%, and 133% of the average cadence previously assessed for each subject over 5 free walking tests using the accelerometers-based gait analysis system Opal by APDM. Subjects performed four walking trials at each speed imposed by metronome. The trials were performed at different walking speed in order to obtain a dataset covering a wider range of values. Doing so, we expect a more specific characterization of the relationship existing between the two methods over all the range of measurement.

In order to validate the proposed e-textile wearable system, the gait analysis parameters obtained from this device have been compared with those obtained by the reference system. Starting from gyroscope signals measured by SWEET Sock, spatio-temporal gait parameters were computed by the custom-made MATLAB algorithms shown in the previous paragraph. The corresponding parameters assessed by the reference system were retrieved from the reports generated by SMART CLINIC software.

The following spatiotemporal parameters were considered for the benchmarking analysis; Gait Cycle Time (s), Cadence (step/min), Stance Time (s), Swing Time (s), and Step Length (m).

### 2.2.3. Statistical Analysis

The agreement between measurements computed by the two systems—SWEET Sock and SMART-DX 700—was investigated by means of two-tailed paired *t*-test, Passing–Bablok regression, and Bland–Altman analysis. The paired *t*-test has been performed for all the parameters selected for the analysis, in its parametric or nonparametric form (Wilcoxon matched pairs signed-rank test) in according to D'Agostino–Pearson omnibus normality test result. With the paired *t*-test, the null hypothesis of no difference between the two systems in mean values of each spatio-temporal parameter was tested.

A two-tail test was used and the nominal alpha level was set to 0.05 [44]. In combination with the *t*-test, the linear correlation between each pair of measurements has been assessed, using Pearson's correlation coefficient (r). The agreement was further investigated using PB regression and BA plots, with the aim to find out any proportional or constant systematic error between the two methods of measurement.

Passing–Bablok regression is a method proposed in 1983 for testing the agreement of two sets of measurement achieved by different systems [45]. The novelties taken by this method, with respect to the standard linear regression are that it is based on nonparametric model, it is not sensitive towards outliers, and it assumes imprecision in both measurement methods and that errors in both methods have the same distribution, not necessarily normal. As quantitative outcomes, this method returns slope (proportional systematic error) and intercept (constant systematic error) of the fitting linear model. The quantitative-based rules to accept the agreement between systems are whether the confidence intervals (CI) of slope and intercept contain respectively 1 and 0 [45].

Bland–Altman analysis is a graphical method based on the plots of the differences between two measurements against their averages, and it is the most popular method used to measure agreement between two measurement systems [46]. If the differences are randomly distributed around the zero-value axis, no proportional nor systematic error is underlined by the analysis. Quantitative assessment is given through the bias, as the mean of the differences, and the limits of agreement (LoA) assessed as the bias $\pm1.96$ times standard deviation of the differences [47,48]. If the differences between methods do not have a normal and/or symmetric distribution, LoA are considered to be between the 2.5% and 97.5% percentiles. Significant statistical errors are said to be present if the confidence interval does not contain zero value. Bland and Altman propose to accept the agreement between the methods under test if this interval contains zero value [47].

Statistical analyses were performed using R software (ver. 4.0.3).

## 3. Results

We approached the analysis of agreement between the two methods of measurement performing a paired *t*-test on all the parameters considered for the analysis. For each parameter, the values deriving from all the trials performed were considered, with no separation between subjects or walking speeds adopted. Table 3 shows mean and standard deviation values of each analyzed parameter dataset for each system of measure. The results of the two tailed paired *t*-test, with a confidence interval of 95%, are reported using a symbol in accordance with the following convention: ns *p*-value > 0.05, * *p*-value < 0.05, ** *p*-value < 0.01, *** *p*-value < 0.001, **** *p*-value < 0.0001. The hypothesis of no difference between systems was tested, so lower *p*-values suggest rejecting the accordance of systems. In the same table Pearson's r values are reported.

The Bland–Altman analysis produces the plots shown in Figures 6a–10a. They provide a qualitative assessment of the distribution of the differences between methods. The descriptive numeric values deriving from the analysis are reported in Table 4. The bias represents the mean of the differences between the measures computed by the systems, it is provided with the limits of its 95% CI. In the plots, biases are reported as continue red lines, while the red dashed lines represent the corresponding confidence intervals. The LoA reported in table are also shown in the graphical representations as black dashed lines. They are assessed as the 2.5 and 97.5 percentiles of differences, as they do not have a symmetric gaussian distribution.

The last analysis on data was performed using Passing–Bablok regression. In addition to the previous analyses, this analysis can reveal the presence of a trend between the measures of the two systems, thus indicating a proportional error in the tested method according to the slope of the fitting regression line. Figures 6b–10b show the scatter plot of the dataset for each parameter, with the Passing–Bablok regression line in black. The shaded area around the regression line represents its CI, while the red dashed line corresponds to the reference identity line, to which the regression line should be tend in a scenario of perfect agreement. In the Passing–Bablok plots, Pearson's correlation coefficient (r) is also shown because high values of r justify the choice to perform a linear regression analysis. The quantitative outcomes of Passing–Bablok analysis are reported in Table 5: slope and intercept of the regression line are listed for each parameter, along with the corresponding 95% CI limits.

**Table 3.** Paired-*T* test.

| Variable | SWEET (mean ± std) | BTS (mean ± std) | *p*-Value Summary [1] | Pearson's r |
|---|---|---|---|---|
| Gait Cycle Time [s] | 1.15 ± 0.25 | 1.15 ± 0.26 | ns | 0.992 |
| Cadence [step/min] | 109.30 ± 21.85 | 109.60 ± 22.25 | * | 0.996 |
| Stance Time [s] | 0.63 ± 0.19 | 0.70 ± 0.18 | **** | 0.994 |
| Swing Time [s] | 0.52 ± 0.07 | 0.45 ± 0.08 | **** | 0.969 |
| Step Length [m] | 0.73 ± 0.08 | 0.68 ± 0.10 | **** | 0.283 |

[1] ns $p > 0.05$, * $p < 0.05$, ** $p < 0.01$, *** $p < 0.001$, **** $p < 0.0001$.

**Table 4.** Bland–Altman analysis.

| Variable | Bias | Lower Bound Bias CI | Upper Bound Bias CI | Lower Bound LoA | Upper Bound LoA |
|---|---|---|---|---|---|
| Gait Cycle Time [s] | 0.00 | −0.01 | 0.01 | −0.06 | 0.05 |
| Cadence [step/min] | −0.35 | −0.74 | 0.03 | −3.83 | 3.26 |
| Stance Time [s] | −0.07 | −0.07 | −0.06 | −0.11 | −0.01 |
| Swing Time [s] | 0.07 | 0.07 | 0.08 | 0.04 | 0.10 |
| Step Length [m] | 0.06 | 0.03 | 0.08 | −0.13 | 0.25 |

**Table 5.** Passing–Bablok regression analysis.

| Variable | Slope | Lower Bound Slope CI | Upper Bound Slope CI | Intercept | Lower Bound Intercept CI | Upper Bound Intercept CI |
|---|---|---|---|---|---|---|
| Gait Cycle Time [s] | 1.00 | 0.99 | 1.02 | 0.00 | −0.02 | 0.02 |
| Cadence [step/min] | 0.99 | 0.97 | 1.00 | 0.74 | −0.95 | 2.38 |
| Stance Time [s] | 1.06 | 1.03 | 1.08 | −0.11 | −0.13 | −0.09 |
| Swing Time [s] | 0.90 | 0.87 | 0.94 | 0.12 | 0.10 | 0.13 |
| Step Length [m] | 0.70 | 0.52 | 0.95 | 0.25 | 0.08 | 0.36 |

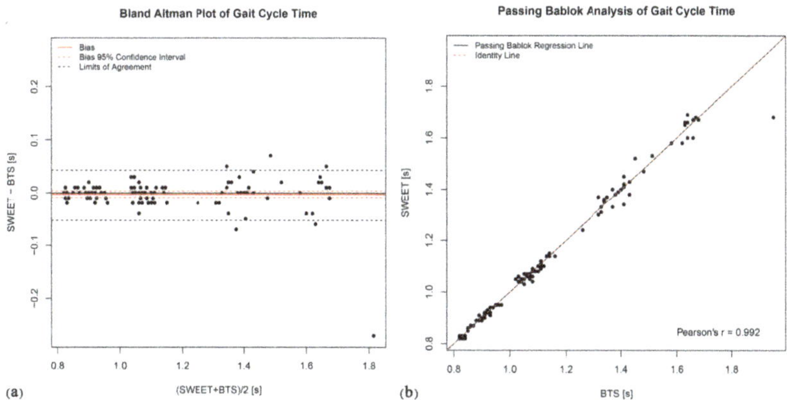


**Figure 6.** Gait cycle time: (**a**) Bland–Altman plot; (**b**) Passing–Bablok regression analysis.

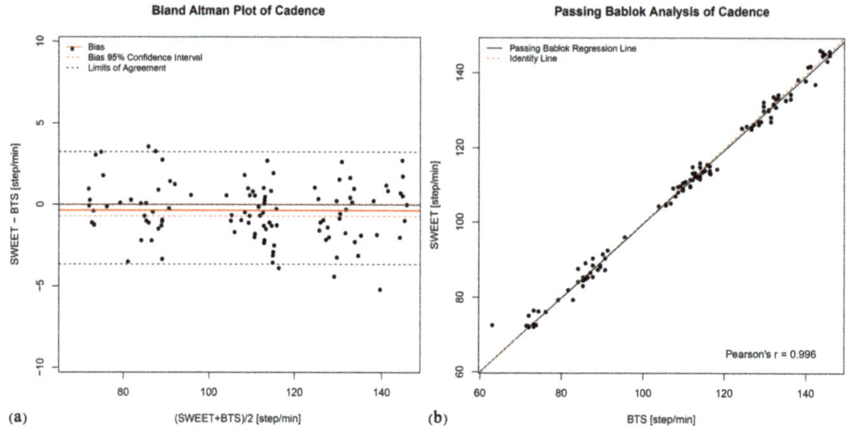

**Figure 7.** Cadence: (**a**) Bland–Altman plot; (**b**) Passing–Bablok regression analysis.

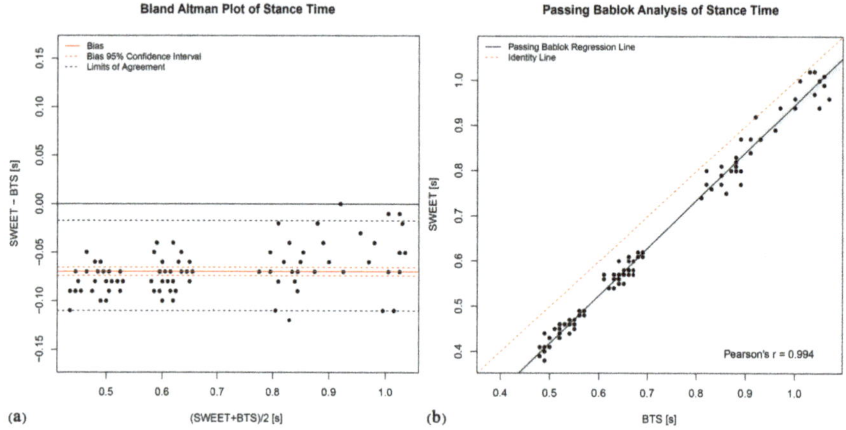

**Figure 8.** Stance Time: (**a**) Bland–Altman plot; (**b**) Passing–Bablok regression analysis.

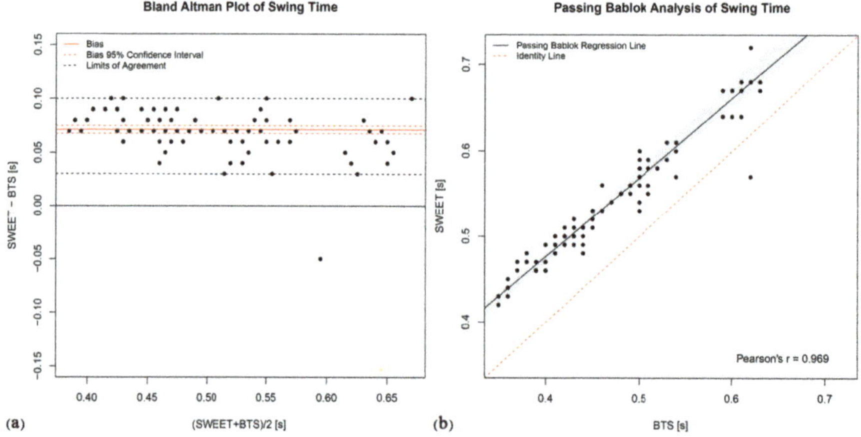

**Figure 9.** Swing Time: (**a**) Bland–Altman plot; (**b**) Passing–Bablok regression analysis.

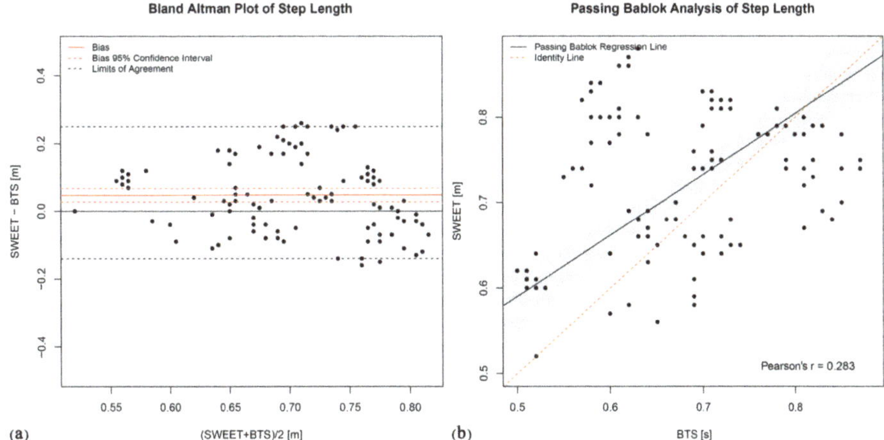

**Figure 10.** Step length: (**a**) Bland–Altman plot; (**b**) Passing–Bablok regression analysis.

## 4. Discussion

This work aims to evaluate the agreement between a novel wearable and portable device for gait analysis and the gold standard of stereo-photogrammetry system. The comparative analysis has been performed on a selected group of the principal temporal and spatial parameters assessed in gait analysis by both systems. Three different statistical methods were used to properly characterize the relationship between the measurement systems under test: paired *t*-test, Bland–Altman plots, and Passing–Bablok regression analysis.

In the assessment of the mean gait cycle time, significant agreement has been pointed out by the statistical analysis. The paired *t*-test leads to a non-significant *p*-value ($p > 0.05$), suggesting to accept the hypothesis of no difference between systems. The bias value in the Bland–Altman analysis is null (0.00 from Table 4) and the LoA are very low (in the order of few hundredths of a second). The Pearson's correlation coefficient is very high (0.992), supporting the concept of a linear dependence between the measures, explored by means of Passing–Bablok analysis. The regression line obtained with this method coincides with the identity line (slope = 1.00, intercept = 0.00), confirming the significant agreement between the two methods in assessing gait cycle time.

Concerning the measure of cadence, a deeper discussion is required. The *T*-test result suggests to refuse the hypothesis of absence of difference between the methods, but with low significance ($0.05 < p$-value $< 0.01$). The bias pointed out by Bland–Altman analysis is very low ($-0.35$, about 0.3% of the average value of cadence), with its 95% CI containing the zero value and limited to few units of steps per minute ($-0.74$ to 0.03). Passing–Bablok regression is legitimated by a high value of Pearson's r (0.996): its slope is very close to 1 (0.99 with CI of 0.97–1.00), the intercept is different from 0 (0.74) but its CI contains this value ($-0.95$ to 2.38). Starting from these results and analyzing the Bland–Altman Plot in Figure 7a, we can observe that the SWEET system slightly underestimates the value of cadence compared to BTS system. Further exploring data, we identified the cause of the non-perfect agreement in the different range of steps analyzed by the two systems. The reference system SMART-DX 700 by BTS performs gait analysis on a limited range of steps, contained in the central 3 or 4 strides of the walking trial, as they are completely included in the field of view of the cameras. The detected volume cannot be extended because it is limited by the configuration of the system which considers the limited volume of the laboratory. Instead, SWEET Sock system elaborates the entire signal coming from the IMUs, removing only the first and the last steps performed to start and stop walking. The analysis of the punctual values of cadence assessed in each single step of the walking trial by SWEET Sock system clarify that in the first and last part of walking a lower step cadence is adopted. Figure 11 shows, for each step of the walking trial, the average of the differences between the punctual cadence assessed by SWEET and the mean step cadence suggested by BTS system. We can observe that in the first and last part of walking the difference is higher in absolute value, while in the middle steps it is reduced. Therefore, we can affirm that probably a better agreement would have been obtained if the same range of steps were analyzed by the two systems. We have chosen not to do so for two reasons: the first is that in SMART-DX 700 the steps to be considered in the analysis have to be chosen manually, while the signal processing of SWEET Sock is entirely automatic, and second because we have chosen not to modify the methods of analysis of SWEET system, which can provide more accurate results by taking into account the entire walking trial.

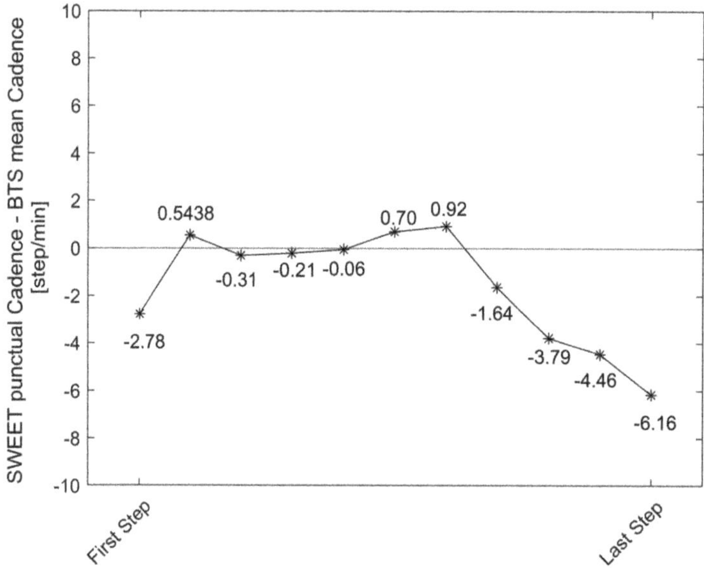


**Figure 11.** Mean difference between the punctual cadence assessed by SWEET and the mean step cadence suggested by BTS system for each step of the walking trial.

Stance and swing phase durations are complementary parameters, because they are the two parts composing the gait cycle time. Gait cycle time is defined as the time between two successive initial contacts of the same foot. Stance phase duration is the time between the initial contact and the successive terminal contact of the same foot, while swing time goes from the terminal contact to the subsequent initial contact. The complementarity of these parameters is perfectly reflected in the results of the statistical analyses. The *T*-test identified a significative statistical difference between the systems ($p$-values < 0.0001), even if a linear correlation exists in both stance and swing phase durationsas shown by Pearson's r values, respectively 0.994 and 0.969. The Bland–Altman plots clearly show that SWEET system underestimates Stance time compared to BTS system (bias = −0.07), and therefore overestimates of the same quantity the Swing time (bias = 0.07). Passing–Bablok results confirm the presence of a systematic error in the measures: intercepts' CIs are symmetric for the two variables and do not contain zero value (stance Cis = −0.13 to −0.09, swing Cis = 0.10 to 0.13). It also points out a proportional error proven by the fact that the slopes of the two regression lines are different from 1 (the CIs are respectively from 1.03 to 1.08 and, symmetrically, from 0.87 to 0.94). Therefore, the difference between the methods of measures is made of a constant part and a proportional part which grows when the value of the parameter is increased. The error is to be probably addressed to the wrong detection of the initial and terminal contact of the foot with the ground, made by SWEET system through the analysis of the filtered gyroscope signal in accordance to the rules illustrated by

Doheny et al. in [36]. Although the gait cycle time shows very good agreement, it does not mean that the initial contacts are well identified in the signal, because they could be all translated in time of the same quantity, still resulting in good output values. To understand the error a further analysis is required on the mutual position of initial and terminal contacts identified on gyroscope signals.

The last parameter is the step length, which has been selected to investigate the performances of SWEET system in the assessment of spatial measures. Results of the statistical analysis are not very encouraging. *T*-test points out a significative statistical difference between the measures of the systems ($p < 0.0001$), that is confirmed by Bland–Altman analysis. Actually, even if the CI of bias includes the zero, it is quite wide ($-0.13$ to $0.25$ m) for the precision required in this spatial metric. Moreover, the reduced value of Pearson's coefficient shows that no linear correlation exists between the measures (r = 0.283), so it does not make sense to perform the Passing–Bablok regression analysis. Actually Passing–Bablok regression line in Figure 10b does not fit accurately the points, which are distributed with no detectable trend. These results allow to affirm that there is not agreement between the systems in the assessment of the step length. Moreover, in this case the cause of the error could be probably found in the processing of the gyroscope signal that lead to the assessment of the spatial parameters. The algorithm proposed in [36] is based on modeling the movement of the shank as a single pendulum, thus deriving the spatial parameters from the calculation of the angle covered by the foot during the swing phase and using geometrical consideration. A further analysis is required to understand if this model is too simplistic to represent leg swing during gait or if other aspects (device positioning, signal filtering, etc.) cause errors in the measure of spatial parameters in SWEET Sock system. Our first purpose is to try maintaining a gyroscope-based algorithm for gait assessment, by considering other more specific models proposed in literature regarding the movement of the shank during the swing phase. An example is the double segment gait model involving both shank and thigh proposed by Aminian et al. in [24]. Doing so we can avoid the use of other sensors data, such as linear accelerations, keeping the gyroscope advantages explored in the description of the electronic unit, and avoiding the reconfiguration of the entire system.

We explored scientific literature to find out and analyze other results from gait analysis systems based on similar measuring principles. Some works exist regarding validation analysis of wearable systems for gait analysis based on processing of kinematic signals. These studies address comparative analyses with clinical instruments, such as instrumented treadmill [49], force platform [50] or pressure sensitive walkway (GAITRite) [35,51,52]. No works presenting a comparative analysis with the gold standard (stereophotogrammetry system) has been found. Results from the analyzed works show a common trend: temporal parameters present a better agreement than spatial metrics. Among temporal parameters, step time and GCT show the best agreement, while stance and swing phases measurements are moderately correlated with reference measures. Results presented in this article are in accordance with this trend, confirming the poor performances of IMU-based systems in assessing gait spatial metrics. Only in [35] spatial metrics show a good agreement level, that could be caused by the different placement of IMUs, placed on both feet rather than on shanks. Results from the works in [35,49] demonstrated that foot placement allow a better measurement of spatial gait parameters. However, we did not choose this placement because it can worsen the comfort and wearability of the system for users and preclude its in-shoes use.

*Comfort Assessment*

In addition to the validation of technical performance, the wearability and comfort assessment was carried out in order to evaluate the acceptance of the system by final users and to identify possible areas of improvement in terms of design. To carry out this conformity assessment, an already validated methodology was used, specifically the Comfort Rating Scales (CRSs).

The wearability evaluation of a device is a multidimensional analysis: wearable devices affect the wearer in different ways. Among the effects to be taken into consideration, there are those related to comfort. When wearing something, the level of comfort can be affected by several aspects, such as device size and weight, how it affects movement, and pain.

The design of the sock has been implemented in order to achieve the greatest comfort for the user. The integrated pressure sensors are made of textile material, therefore are flexible and imperceptible on the skin. The electronic unit has also been designed to be as comfortable as possible for the user: it is light and it can be connected to the textile sock without the need to use bands. In fact, the use of the latter could cause discomfort to the user due to the presence of a narrow element tied to the limb.

In addition to physical factors, comfort may be affected by psychological responses such as embarrassment or anxiety. Consequently, Knight and Baber proposed that comfort should be measured across a number of dimensions and for such task they developed the Comfort Rating Scales (CRSs) [53].

The CRSs provide a quick and easy-to-use tool to assess the comfort of wearable devices, which attempt to gain a comprehensive assessment of the comfort status of the wearer of any item of technology by measuring comfort across the six dimensions described in Table 6. In rating perceptions of comfort, the scorer simply marks on the scale his or her level of agreement, from low (0) to high (20), with the statements made in the "description" column of Table 6. According to Knight and Baber, this range was considered large enough to elicit a range of responses that could be used for detailed analysis [53].

The three participants involved in the study were invited to fill in the CRSs to provide a judgment on their comfort. Table 6 shows the scores assigned, for each field, by the subjects involved in the study.

Although the evaluation was carried out on only three people, it provides a preliminary measure of the comfort of the prototype device. Knight et al. [54] have proposed five Wearability Levels (WLs), determined by proportioning the scales into equal parts (Table 7). The mean score of Emotion dimension is in the WL2 suggesting that users show little embarrassment in wearing the system. All the other dimensions were rated in the WL1 proving a high wearability and comfort of the device. However, to better identify the wearability level of the device and how to improve it, future analysis will aim to make a significant assessment of comfort, testing the device on a wider cohort of subjects.

**Table 6.** Comfort rating scales.

| Title | Description | Subject 1 | Subject 2 | Subject 3 | Mean |
|---|---|---|---|---|---|
| Emotion | I am worried about how I look when I wear this device. I feel tense or on edge because I am wearing the device. | 7 | 4 | 7 | 6.0 |
| Attachment | I can feel the device on my body. I can feel the device moving. | 3 | 3 | 5 | 3.7 |
| Harm | The device is causing me some harm. The device is painful to wear. | 0 | 0 | 0 | 0.0 |
| Perceived change | Wearing the device makes me feel physically different. I feel strange wearing the device. | 5 | 0 | 0 | 1.7 |
| Movement | The device affects the way I move. The device inhibits or restricts my movement. | 5 | 2 | 1 | 2.7 |
| Anxiety | I do not feel secure wearing the device. | 0 | 0 | 0 | 0.0 |

**Table 7.** Wearability Levels.

| Wearability Level | CRS Score | Outcome |
|---|---|---|
| WL1 | 0–4 | System is wearable |
| WL2 | 5–8 | System is wearable, but changes may be necessary, further investigation is needed |
| WL3 | 9–12 | System is wearable, but changes are advised, uncomfortable |
| WL4 | 13–16 | System is not wearable, fatiguing, very uncomfortable |
| WL5 | 17–20 | System is not wearable, extremely stressful, and potentially harmful |

## 5. Conclusions

SWEET Sock is a new wearable and portable device for the measurement and analysis of biosignals, based on textile sensors, able to perform posturographic assessment and gait analysis. In this manuscript, we presented the development of the system and we illustrated the validation analysis of the performances of the novel system in gait assessment.

The sensing unit is a textile sock in which textile sensors and bus structures are integrated, making it possible to use the system during normal daily activities, without any discomfort. The system includes a mobile app for real time visualization of the acquired signals and a software desktop for off-line plotting and digital signal processing.

The analysis of the performances of the system in gait assessment was performed by comparing the results given by the novel system with the corresponding values computed by an optoelectronic stereophotogrammetric system (SMART-DX 700 by BTS Bioengineering) in the analysis of 108 walking trials at different walking speeds. Study results show that the agreement is not confirmed for all the spatio-temporal gait parameters analyzed. In particular, gait cycle time and cadence are the two parameters presenting the best agreement, even if the latter presents a small systematic difference between the values computed by the two systems. Stance and swing phase durations present both systematic and proportional errors in the comparison between the methods. Although both errors could be removed by taking into account this misalignment, a further analysis will be performed to understand and correct the problems directly in the processing phase. Worse results are achieved in the analysis of spatial parameters' agreement. The measures of step length provided by the two systems are not correlated. For this parameter, a further analysis is required to correct the issues in the computational process. Based on these findings, we can affirm that the novel system can be safely used in the evaluation of gait cycle time while some issues were found in the validation of the other temporal and spatial parameters. Future developments will concern the resolution of the problems encountered in this work and the execution of a similar validation analysis regarding the posturographic assessment provided by the system.

The innovative features of the system rely in the multiparametric approach in health monitoring and in its ease of use. The "wearability" of the system and its comfortable use make it very suitable to be used in domestic environment for the continuous remote health monitoring of de-hospitalized patients. The CRSs were used to assess the comfort of the wearable system. The scores provided by the subjects involved in the study, allow to assume a good level of comfort when the socks are used.

Another valid field of interest regards occupational ergonomics, related to the prevention of work-related musculoskeletal disorders (WRMSDs) in healthcare workers.

The use of SWEET Sock during working hours by nurses and therapists could help monitor postural and dynamic variables in activities most associated with exposure to biomechanical overload (i.e., frequent patient handling, pushing and pulling, awkward postures, prolonged standing, and significant sideways twisting).

The biomechanical advantage of using patient handling devices and technological aids, including exoskeletons, could be verified through the analysis of postural parameters. Gait analysis could help rethink preventive strategies aimed at work organization (for example by providing for the alternation of dynamic and static phases, and adequate recovery breaks). Last, but not least, balance analysis and COP coordinates could provide insights into the prevention of slips, trips, and falls, which are the second most common cause of injuries leading to lost working days in hospitals. The advantages combined in a minimally invasive device, together with the accuracy and reliability of the measurement, and the future opportunity of integration into IoT networks open new perspectives to increase the effectiveness of prevention and safety strategies in healthcare workers.

**Author Contributions:** Conceptualization, F.A. and A.C.; methodology, F.A. and L.D.; software, A.C.; validation, G.C.; formal analysis, F.A. and E.M.C.; investigation, C.R.; data curation, F.A., A.C. and C.R.; writing—original draft preparation, F.A. and E.M.C.; writing—review and editing, G.D.; visualization, F.A. and A.C.; supervision, G.D.; project administration, G.D. All authors have read and agreed to the published version of the manuscript.

**Funding:** This research was partially funded by SWEET—Smart WEarable E-Textile based m-health system an Horizon H2020 project financed by Ministry of Economic Development of Italy.

**Acknowledgments:** The authors thank M. Ramaglia of Adiramef company (CE, Italy) and A. Biancardi for their strong support in the development of SWEET Sock device.

## References


1. DeRossi, D.; Lymberis, A. Guest editorial new generation of smart wearable health systems and applications. *IEEE Trans. Inf. Technol. Biomed.* **2005**, *9*, 293–294.

2. Erdmier, C.; Hatcher, J.; Lee, M. Wearable device implications in the healthcare industry. *J. Med. Eng. Technol.* **2016**, *40*, 141–148. [PubMed]

3. Lewy, H. Wearable technologies—Future challenges for implementation in healthcare services. *Healthc. Technol. Lett.* **2015**, *2*, 2–5. [PubMed]

4. Orlin, M.N.; McPoil, T.G. Plantar Pressure Assessment. *Phys. Ther.* **2000**, *80*, 399–409. [PubMed]

5. Nagamune, K.; Yamada, M. A Wearable Measurement System for Sole Pressure to Calculate Center of Pressure in Sports Activity. In Proceedings of the 2018 IEEE International Conference on Systems, Man, and Cybernetics (SMC), Miyazaki, Japan, 7–10 October 2018; pp. 1333–1336.

6. Deng, X.; Dian, S.; WENG, T. Wearable plantar pressure detecting system based on FSR. *Transducer Microsyst. Technol.* **2013**, *32*, 81–86.

7. Abou Ghaida, H.; Mottet, S.; Goujon, J.M. A real time study of the human equilibrium using an instrumented insole with 3 pressure sensors. In Proceedings of the 2014 36th Annual International Conference of the IEEE Engineering in Medicine and Biology Society, Chicago, IL, USA, 26–30 August 2014; pp. 4968–4971.

8. Oks, A.; Katashev, A.; Eizentals, P.; Rozenstoka, S.; Suna, D. Smart socks: New effective method of gait monitoring for systems with limited number of plantar sensors. *Health Technol.* **2020**, 853–860.

9. Preece, S.J.; Kenney, L.P.; Major, M.J.; Dias, T.; Lay, E.; Fernandes, B.T. Automatic identification of gait events using an instrumented sock. *J. Neuroeng. Rehabil.* **2011**, *8*, 32.

10. Eizentals, P.; Katashev, A.; Oks, A. Gait analysis by using Smart Socks system. *IOP Conf. Ser. Mater. Sci. Eng.* **2018**, *459*, 012037.

11. Yang, C.; Chou, C.; Hu, J.; Hung, S.; Yang, C.; Wu, C.; Hsu, M.; Yang, T. A wireless gait analysis system by digital textile sensors. In Proceedings of the 2009 Annual International Conference of the IEEE Engineering in Medicine and Biology Society, Minneapolis, MN, USA, 2–6 September 2009; pp. 7256–7260.

12. Lin, X.; Seet, B. Battery-Free Smart Sock for Abnormal Relative Plantar Pressure Monitoring. *IEEE Trans. Biomed. Circuits Syst.* **2017**, *11*, 464–473.

13. Derawi, M.; Bours, P. Gait and activity recognition using commercial phones. *Comput. Secur.* **2013**, *39*, 137–144.

14. De, D.; Bharti, P.; Das, S.K.; Chellappan, S. Multimodal wearable sensing for fine-grained activity recognition in healthcare. *IEEE Internet Comput.* **2015**, *19*, 26–35.

15. Panahandeh, G.; Mohammadiha, N.; Leijon, A.; Händel, P. Continuous hidden Markov model for pedestrian activity classification and gait analysis. *IEEE Trans. Instrum. Meas.* **2013**, *62*, 1073–1083.

16. Alshurafa, N.; Xu, W.; Liu, J.J.; Huang, M.C.; Mortazavi, B.; Roberts, C.K.; Sarrafzadeh, M. Designing a robust activity recognition framework for health and exergaming using wearable sensors. *IEEE J. Biomed. Health Inform.* **2013**, *18*, 1636–1646. [PubMed]

17. Ghasemzadeh, H.; Amini, N.; Saeedi, R.; Sarrafzadeh, M. Power-aware computing in wearable sensor networks: An optimal feature selection. *IEEE Trans. Mob. Comput.* **2014**, *14*, 800–812.

18. Chen, B.; Zheng, E.; Wang, Q.; Wang, L. A new strategy for parameter optimization to improve phase-dependent locomotion mode recognition. *Neurocomputing* **2015**, *149*, 585–593.

19. Bertolotti, G.M.; Cristiani, A.M.; Colagiorgio, P.; Romano, F.; Bassani, E.; Caramia, N.; Ramat, S. A wearable and modular inertial unit for measuring limb movements and balance control abilities. *IEEE Sens. J.* **2015**, *16*, 790–797.

20. Tang, W.; Sazonov, E.S. Highly accurate recognition of human postures and activities through classification with rejection. *IEEE J. Biomed. Health Inform.* **2014**, *18*, 309–315.

21. Coccia, A.; Lanzillo, B.; Donisi, L.; Amitrano, F.; Cesarelli, G.; D'Addio, G. Repeatability of Spatio-Temporal Gait Measurements in Parkinson's Disease. In Proceedings of the 2020 IEEE International Symposium on Medical Measurements and Applications (MeMeA), Bari, Italy, 1 June–1 July 2020; pp. 1–6.

22. Donisi, L.; Coccia, A.; Amitrano, F.; Mercogliano, L.; Cesarelli, G.; D'Addio, G. Backpack Influence on Kinematic Parameters related to Timed Up and Go (TUG) Test in School Children. In Proceedings of the 2020 IEEE International Symposium on Medical Measurements and Applications (MeMeA), Bari, Italy, 1 June–1 July 2020; pp. 1–5.

23. Jarchi, D.; Pope, J.; Lee, T.K.; Tamjidi, L.; Mirzaei, A.; Sanei, S. A review on accelerometry-based gait analysis and emerging clinical applications. *IEEE Rev. Biomed. Eng.* **2018**, *11*, 177–194.

24. Tong, K.; Granat, M.H. A practical gait analysis system using gyroscopes. *Med Eng. Phys.* **1999**, *21*, 87–94.

25. Ngo, T.T.; Makihara, Y.; Nagahara, H.; Mukaigawa, Y.; Yagi, Y. Similar gait action recognition using an inertial sensor. *Pattern Recognit.* **2015**, *48*, 1289–1301.

26. Hsu, Y.L.; Chung, P.C.; Wang, W.H.; Pai, M.C.; Wang, C.Y.; Lin, C.W.; Wu, H.L.; Wang, J.S. Gait and balance analysis for patients with Alzheimer's disease using an inertial-sensor-based wearable instrument. *IEEE J. Biomed. Health Inform.* **2014**, *18*, 1822–1830. [PubMed]

27. Panero, E.; Digo, E.; Agostini, V.; Gastaldi, L. Comparison of different motion capture setups for gait analysis: Validation of spatio-temporal parameters estimation. In Proceedings of the 2018 IEEE International Symposium on Medical Measurements and Applications (MeMeA), Rome, Italy, 11–13 June 2018; pp. 1–6.

28. Bejarano, N.C.; Ambrosini, E.; Pedrocchi, A.; Ferrigno, G.; Monticone, M.; Ferrante, S. A novel adaptive, real-time algorithm to detect gait events from wearable sensors. *IEEE Trans. Neural Syst. Rehabil. Eng.* **2014**, *23*, 413–422.

29. Zhao, H.; Wang, Z.; Qiu, S.; Shen, Y.; Wang, J. IMU-based gait analysis for rehabilitation assessment of patients with gait disorders. In Proceedings of the 2017 4th International Conference on Systems and Informatics (ICSAI), Hangzhou, China, 11–13 November 2017; pp. 622–626.

30. Gujarathi, T.; Bhole, K. GAIT ANALYSIS USING IMU SENSOR. In Proceedings of the 2019 10th International Conference on Computing, Communication and Networking Technologies (ICCCNT), Kanpur, India, 6–8 July 2019; pp. 1–5.

31. Han, Y.C.; Wong, K.I.; Murray, I. Stride Length Estimation Based on a Single Shank's Gyroscope. *IEEE Sens. Lett.* **2019**, *3*, 7001804.

32. D'Addio, G.; Evangelista, S.; Donisi, L.; Biancardi, A.; Andreozzi, E.; Pagano, G.; Arpaia, P.; Cesarelli, M. Development of a Prototype E-Textile Sock. In Proceedings of the 2019 41st Annual International Conference of the IEEE Engineering in Medicine and Biology Society (EMBC), Berlin, Germany, 23–27 July 2019; pp. 1749–1752.

33. Holleczek, T.; Rüegg, A.; Harms, H.; Tröster, G. Textile pressure sensors for sports applications. In Proceedings of the 2010 IEEE SENSORS, Kona, HI, USA, 1–4 November 2010; pp. 732–737.

34. D'Addio, G.; Iuppariello, L.; Pagano, G.; Biancardi, A.; Lanzillo, B.; Pappone, N.; Cesarelli, M. New posturographic assessment by means of novel e-textile and wireless socks device. In Proceedings of the 2016 IEEE International Symposium on Medical Measurements and Applications (MeMeA), Benevento, Italy, 15–18 May 2016; pp. 1–5.

35. Morris, R.; Stuart, S.; McBarron, G.; Fino, P.C.; Mancini, M.; Curtze, C. Validity of Mobility Lab (version 2) for gait assessment in young adults, older adults and Parkinson's disease. *Physiol. Meas.* **2019**, *40*, 095003.

36. Doheny, E.P.; Foran, T.G.; Greene, B.R. A single gyroscope method for spatial gait analysis. In Proceedings of the 2010 Annual International Conference of the IEEE Engineering in Medicine and Biology, Buenos Aires, Argentina, 31 August–4 September 2010; pp. 1300–1303.

37. Rampp, A.; Barth, J.; Schülein, S.; Gaßmann, K.G.; Klucken, J.; Eskofier, B.M. Inertial sensor-based stride parameter calculation from gait sequences in geriatric patients. *IEEE Trans. Biomed. Eng.* **2014**, *62*, 1089–1097.

38. Salarian, A.; Russmann, H.; Vingerhoets, F.J.; Dehollain, C.; Blanc, Y.; Burkhard, P.R.; Aminian, K. Gait assessment in Parkinson's disease: toward an ambulatory system for long-term monitoring. *IEEE Trans. Biomed. Eng.* **2004**, *51*, 1434–1443.

39. Prieto, T.E.; Myklebust, J.B.; Hoffmann, R.G.; Lovett, E.G.; Myklebust, B.M. Measures of postural steadiness: differences between healthy young and elderly adults. *IEEE Trans. Biomed. Eng.* **1996**, *43*, 956–966.

40. Reinfelder, S.; Durlak, F.; Barth, J.; Klucken, J.; Eskofier, B.M. Wearable static posturography solution using a novel pressure sensor sole. In Proceedings of the 2014 36th Annual International Conference of the IEEE Engineering in Medicine and Biology Society, Chicago, IL, USA, 26–30 August 2014; pp. 2973–2976.

41. Agostini, V.; Aiello, E.; Fortunato, D.; Knaflitz, M.; Gastaldi, L. A wearable device to assess postural sway. In Proceedings of the 2019 IEEE 23rd International Symposium on Consumer Technologies (ISCT), Ancona, Italy, 19–21 June 2019; pp. 197–200.

42. Amitrano, F.; Donisi, L.; Coccia, A.; Biancardi, A.; Pagano, G.; D'Addio, G. Experimental Development and Validation of an E-Textile Sock Prototype. In Proceedings of the 2020 IEEE International Symposium on Medical Measurements and Applications (MeMeA), Bari, Italy, 1 June–1 July 2020; pp. 1–5.

43. Davis, R.B., III; Ounpuu, S.; Tyburski, D.; Gage, J.R. A gait analysis data collection and reduction technique. *Hum. Mov. Sci.* **1991**, *10*, 575–587.

44. Blair, R.C.; Cole, S.R. Two-sided equivalence testing of the difference between two means. *J. Mod. Appl. Stat. Methods* **2002**, *1*, 18. [CrossRef]

45. Passing, H.; Bablok, W. A New Biometrical Procedure for Testing the Equality of Measurements from Two Different Analytical Methods. Application of Linear Regression Procedures for Method Comparison Studies in Clinical Chemistry, Part I. *J. Clin. Chem. Clin. Biochem.* **1983**, *21*, 709–720.

46. Zaki, R.; Bulgiba, A.; Ismail, R.; Ismail, N.A. Statistical methods used to test for agreement of medical instruments measuring continuous variables in method comparison studies: a systematic review. *PLoS ONE* **2012**, *7*, e37908.

47. Altman, D.G.; Bland, J.M. Measurement in medicine: the analysis of method comparison studies. *J. R. Stat. Soc. Ser. D (The Statistician)* **1983**, *32*, 307–317.

48. Bland, J.M.; Altman, D.G. Measuring agreement in method comparison studies. *Stat. Methods Med Res.* **1999**, *8*, 135–160. [PubMed]

49. Washabaugh, E.P.; Kalyanaraman, T.; Adamczyk, P.G.; Claflin, E.S.; Krishnan, C. Validity and repeatability of inertial measurement units for measuring gait parameters. *Gait Posture* **2017**, *55*, 87–93. [PubMed]

50. Patterson, M.R.; Johnston, W.; O'Mahony, N.; O'Mahony, S.; Nolan, E.; Caulfield, B. Validation of temporal gait metrics from three IMU locations to the gold standard force plate. In Proceedings of the 2016 38th Annual International Conference of the IEEE Engineering in Medicine and Biology Society (EMBC), Orlando, FL, USA, 16–20 August 2016; pp. 667–671.

51. Moore, S.A.; Hickey, A.; Lord, S.; Del Din, S.; Godfrey, A.; Rochester, L. Comprehensive measurement of stroke gait characteristics with a single accelerometer in the laboratory and community: A feasibility, validity and reliability study. *J. Neuroeng. Rehabil.* **2017**, *14*, 130. [CrossRef] [PubMed]

52. De Ridder, R.; Lebleu, J.; Willems, T.; De Blaiser, C.; Detrembleur, C.; Roosen, P. Concurrent validity of a commercial wireless trunk triaxial accelerometer system for gait analysis. *J. Sport Rehabil.* **2019**, *28*. [CrossRef]

53. Knight, J.F.; Baber, C. A Tool to Assess the Comfort of Wearable Computers. *Hum. Factors* **2005**, *47*, 77–91.

54. Knight, J.F.; Deen-Williams, D.; Arvanitis, T.N.; Baber, C.; Sotiriou, S.; Anastopoulou, S.; Gargalakos, M. Assessing the wearability of wearable computers. In Proceedings of the 2006 10th IEEE International Symposium on Wearable Computers, Montreux, Switzerland, 11–14 October 2006; pp. 75–82.

**Publisher's Note:** MDPI stays neutral with regard to jurisdictional claims in published maps and institutional affiliations.

*Article*

# Assessment of Upper Limb Movement Impairments after Stroke Using Wearable Inertial Sensing

Anne Schwarz [1,2,*], Miguel M. C. Bhagubai [1], Gerjan Wolterink [1,3], Jeremia P. O. Held [2], Andreas R. Luft [2,4] and Peter H. Veltink [1]

[1]    Biomedical Signals and Systems (BSS), University of Twente, 7500 AE Enschede, The Netherlands; miguel.bhagubai12@gmail.com (M.M.C.B.); gerjan.wolterink@utwente.nl (G.W.); p.h.veltink@utwente.nl (P.H.V.)
[2]    Vascular Neurology and Neurorehabilitation, Department of Neurology, University Hospital Zurich, University of Zurich, 8091 Zurich, Switzerland; jeremia.held@usz.ch (J.P.O.H.); andreas.Luft@usz.ch (A.R.L.)
[3]    Robotics and Mechatronics group, University of Twente, 7500 AE Enschede, The Netherlands
[4]    Cereneo, Center for Neurology and Rehabilitation, 6354 Vitznau, Switzerland
*    Correspondence: anne.schwarz@usz.ch

Received: 7 July 2020; Accepted: 20 August 2020; Published: 24 August 2020

**Abstract:** Precise and objective assessments of upper limb movement quality after strokes in functional task conditions are an important prerequisite to improve understanding of the pathophysiology of movement deficits and to prove the effectiveness of interventions. Herein, a wearable inertial sensing system was used to capture movements from the fingers to the trunk in 10 chronic stroke subjects when performing reach-to-grasp activities with the affected and non-affected upper limb. It was investigated whether the factors, tested arm, object weight, and target height, affect the expressions of range of motion in trunk compensation and flexion-extension of the elbow, wrist, and finger during object displacement. The relationship between these metrics and clinically measured impairment was explored. Nine subjects were included in the analysis, as one had to be excluded due to defective data. The tested arm and target height showed strong effects on all metrics, while an increased object weight showed effects on trunk compensation. High inter- and intrasubject variability was found in all metrics without clear relationships to clinical measures. Relating all metrics to each other resulted in significant negative correlations between trunk compensation and elbow flexion-extension in the affected arm. The findings support the clinical usability of sensor-based motion analysis.

**Keywords:** upper extremity; stroke; biomechanical phenomena; kinematics; inertial measurement systems; motion analysis

## 1. Introduction

Human hand and arm function contribute to a wide range of activities in daily life, ranging from sensory functions to interacting with the environment and to functions that have a strong motor component like the manipulation of objects in grasping [1]. Hand and arm functionalities including object manipulation and physical interactions with the environment rely on the ability to control prehensile finger forces to perform specific grasp types [2,3] and ability to control both the distal and proximal joints of the upper limb in a goal-directed manner [4], for example when transporting the hand to reach the location of a desired object and forming the fingers for grasping [5].

In subjects, experiencing upper limb impairments due to a stroke, these complex hand- and arm-grasping functionalities are defective [6]. Stroke is known as the leading cause of disability in the world [7], defined as a disruption in brain cell perfusion that leads to cell death and losses in network connectivity and multimodal impairments [8]. In particular, infarctions of the middle cerebral artery affecting the primary motor cortex and the integrity of the corticospinal tract have been associated with upper limb movement deficits [9,10], such as weakness, decreased interjoint coordination and in particular diminished finger dexterity [11,12]. Of these motor performance aspects, weakness caused by stroke indicates the inability to activate certain upper limb muscles or segments, whereas interjoint coordination is defined as the ability to control all upper limb joints or segments in a spatially and temporally efficient manner. The differentiation between weakness and interjoint coordination during upper limb activities is rather precisely definable, as both show strong associations with each other [13] and other stroke-related impairments, such as spasticity [14]. Phenotypes of stroke-related interjoint coordination deficits include the appearance of the pathological flexor synergy in reaching, increased trunk movements to compensate for the upper limb limitations and a decreased finger dexterity for prehensile grasp application. The pathological flexor synergy was defined as a stereotypical co-activation of elbow flexion and shoulder abduction [15] that becomes visible in reaching [14], in arm-load related reductions in upper limb workspace [16], and in a diminished ability to extend the fingers [17,18]. Engaging trunk movements in reaching has been considered as movement strategies to compensate for upper limb motor impairments with associations to the level of impairment [19]. These stroke-related movement abnormalities might become present in isolation or combination and chronically manifested depending on the severity of deficit and cerebral region affected [11], thereby presenting a continuous challenge for treatment approaches.

Being able to capture these upper limb movement characteristics is important to improve the understanding of stroke-related movement deficits, including their possible underlying dysfunctions, and to further investigate effectiveness of approaches to influence these deficits [20]. In this regard, it needs to be considered that upper limb movements can be assessed on different levels. Approaches to evaluate and assess upper limb movement deficits after strokes range from more or less extensive qualitative descriptions of body functions and activities in therapeutic records of clinical practice, over clinical scales that mostly rely on observer-based scoring and time-efficiency measures, to instrumentations and technologies for kinematic motion analysis. Although clinical assessments, such as the Fugl–Meyer Assessment of the Upper Extremity (FMA-UE) or the Action Research Arm Test (ARAT), have demonstrated excellent reliability and validity as assessment tools [21], their level of information detail, mostly due to the gross ordinal scoring nature of relatively complex defined movement items, does not allow to sufficiently discriminate physiological and above-mentioned pathological movement behavior [20].

Kinematic assessments on the other hand are supposed to offer fine-grained and objective outcomes on movement quality and have shown to detect stroke-related movement impairments in terms of longer movement times, greater trunk displacement and less elbow extension in reaching movements [22,23]. However, the widespread application of kinematic measurements in clinical practice faces several barriers. First, the high variety of measurement systems with different considerations on interaction forces, movement tasks and different metric derivations hampers the comparability and conclusion drawing [24]. Secondly, investigations of complete motion kinematics including trunk and finger motions were sparse. Thirdly, most of the measurement systems being used were optoelectronic and robotic systems that are based on fixed laboratory environments and expensive equipment [25]. Being able to perform comprehensive upper-limb kinematic analysis outside of the laboratory, in flexible environments with the least possible influence on movement behavior would facilitate implementation of kinematic measurements of qualitative aspects movement behavior in clinical practice. In setting up this pilot study, it was aimed to address the outlined limitations by extensively measuring and quantifying reach-to-grasp movements after stroke, with respect to interjoint coordination determined by trunk compensation and flexion-extension of the elbow, wrist and fingers, specifically quantified during active grasp and object displacement. A portable inertial system was used to measure complete upper limb kinematics, from the trunk to the fingertip, including fingertip force sensing in flexible experimental tasks and set-up environments. Different task characteristics, such as the target locations and the object to be grasped were considered in the experimental design to investigate influences of additional arm load and workspace relations including increased mechanical work demands in movements against gravity on metrics for determining inter-joint coordination. It is assumed that the reaching movement might result in different joint executions with respect to different target locations in the workspace, e.g., features of the pathological flexor pathology might become more pronounced in target positions with higher anti-gravitational mechanical work and with more distance from the body center. Likewise, grasping different object weights results in different additional armloads, that could affect the ability to perform unaffected reaching.

The primary study goal was to evaluate spatiotemporal kinematic metrics for the assessment of upper limb movements after stroke. It was first questioned whether changes in the kinematic range of motion (ROM) in terms of joint angle ranges can be attributed to the factor tested arm, object weight and target height during object displacement. The second question was how far the kinematic metrics relate to clinically measured upper limb impairment. The third question related to the correlation between each of the joint range metrics to evaluate potential joint coupling, such as the pathological flexor synergy between shoulder flexion-extension, elbow flexion-extension and trunk compensation.

## 2. Materials and Methods

### 2.1. Study Design and Participants

This pilot study was set up to investigate upper limb motion primitives from proximal to distal function in stroke subjects by use of a wearable inertial sensing system. The study was approved by cantonal ethics in Zurich (BASEC-No: Req-2019-00417) and carried out in accordance with the declaration of Helsinki. Subjects after stroke were recruited from a University Hospital Zurich Stroke Registry and invited for a single-session measurement of two hours at the Clinic of Neurology of the University Hospital Zurich, Switzerland.

Subjects were included if they were at least 18 years old, able to give informed consent and had been diagnosed with unilateral stroke at least six months before the study onset with associated upper limb impairments. Subjects had to have at least partial ability to move the arm against gravity and to perform finger movements for basic grasp function. Exclusion criteria were pre-existing deficits of the upper limb, such as orthopedic impairments, severely increased muscle tone with limitation in range of motion in the upper limb (Modified Ashworth Scale of >2 in one of the upper limb muscle groups), severe sensory deficits in the upper limb (absence of light touch in the hand and fingers), and severe communication or cognitive deficits that cause inability to follow the procedures. Participant characteristics of interest included the gender, age, stroke location side, time since stroke, stroke affected cerebral perfusion territory and the severity of upper limb motor impairment, as measured with FMA-UE. The FMA-UE is a cumulative numerical scoring system to evaluate motor function after stroke, which consists of an arm, wrist, hand and coordination subsection to account for independent severity and recovery patterns, presented in a full score range from 0 to 66 points [26].

### 2.2. Measurement System

The wearable inertial sensing measurement system was a modified version of the inertial measurement unit (IMU)-based hand and finger sensing system, reported and evaluated by Kortier et al. [27]. It was composed of eight IMUs, with triaxial accelerometers and gyroscopes, based on a micro-controller-based sensing system principles of the PowerGlove [27,28] that were covered by 3D-printed housings, and combined with force sensors. The IMUs were placed and fixated at the sternum, shoulder, upper arm, lower arm, hand, thumb, index and fingers with medical tape or 3D-printed flexible straps (Figure 1). Additionally, the finger IMUs were combined with force-sensitive resistors (FSR) to detect interaction forces between the object to be grasped or manipulated and the finger pad. The upper arm IMU was placed on the lateral side of the arm, close to the elbow and the lower arm IMU was placed on the dorsal side of the forearm, close to the wrist. The hand sensor was placed on the back of the hand, and the thumb, index and middle finger IMUs were attached on the fingertips of the respective fingers, with the force-sensitive resistors fixed on the finger pad by the IMU housing's strap. Each pair of triaxial accelerometers and gyroscopes (ST LSM330DLC manufactured by STMicroelectronics, Geneva, Switzerland) were contained within small printed circuit boards (PCBs). The separately encased IMUs were connected via flexible cabling strips, forming two separate strings (arm string—containing the sternum, shoulder, upper and lower arm sensors; and hand string—containing the hand and fingers sensors). Signals from both sensor strings were collected and connected through a bus master/microcontroller (Atmel XMEGA manufactured by Atmel, California, USA) and streamed real-time via the USB channel onto a PC for control in Matlab software (MATLAB version 2016b, The Mathwork, Natick, MA, USA). Acceleration data was collected with a sampling frequency of 100Hz and gyroscope data with a frequency of 200 Hz. Both were low-pass filtered by using a Butterworth filter with a cut-off frequency of 10 Hz

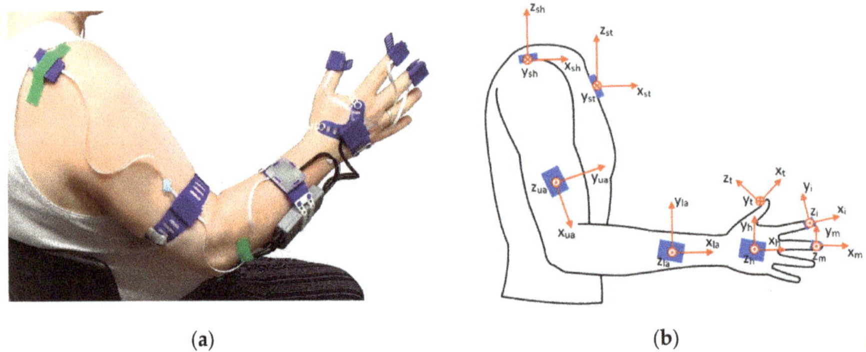

(a)       (b)

**Figure 1.** Wearable inertial sensing system: (**a**) system set-up; (**b**) anatomical frame definition per segment.

*2.3. Kinematic Reconstruction*

All sensors were calibrated each day prior to the measurements by placing them inside a box with orthogonal sides, which was turned over 90 degrees in all three orthogonal directions. The accelerometer bias in the different axes and the gyroscope static bias was measured before the whole experiment per subject and compensated during the measurements [29]. The kinematic reconstruction was based on the estimation of the sensors' orientation, which is taken from the acceleration and angular velocity measures of the IMUs. In order to estimate the orientation of the limb segments, a sensor-to-segment calibration was performed, as well as a definition of a common global frame for all sensors. The sensor-to-segment calibration was carried out to determine the upper body anatomical axes of the limb segments (joints) relative to the corresponding sensors by performing ten different postures and movements that were based on Luinge et al. [30] and Ricci et al. [31]. The equipped test person was assisted by a trained research clinician to perform the calibration protocol, which consisted in eight static positions and two dynamic movements, as shown in Table 1.

**Table 1.** Sensor-to-segment calibration protocol.

| No | Calibration Position/Movement | Anatomical Axes | |
|:---:|:---:|:---:|:---:|
| | | Left Arm | Right Arm |
| 1 | Hand held in pronation flat on the metal box | $z_h, z_i, z_m$ | $z_h, z_i, z_m$ |
| 2 | Hand held in sagittal plane with 90° elbow flexion | $-y_h, -y_i, -y_m$ | $y_h, y_i, y_m$ |
| 3 | Thumb held flat on the metal box | $z_t$ | $z_t$ |
| 4 | Thumb held in sagittal plane with the hand in pronation | $y_t$ | $-y_t$ |
| 5 | Forearm held in pronation along the transversal plane | $z_{la}$ | $z_{la}$ |
| 6 | Forearm motion from supination to pronation in elbow flexion | $x_{la}$ | $-x_{la}$ |
| 7 | Upper arm held parallel to the sagittal plane with 90° elbow flexion | $-x_{ua}$ | $-x_{ua}$ |
| 8 | Shoulder horizontal abduction with 90° elbow flexion | $z_{ua}$ | $z_{ua}$ |
| 9 | Standing straight | $z_{sh}, z_{st}$ | $z_{sh}, z_{st}$ |
| 10 | Bending forward by hip flexion until around 60° | $y_{sh}, y_{st}$ | $-y_{sh}, -y_{st}$ |

In the static positions, the gravity vector measured by the accelerometers represents one of the axes. In the dynamic movements, the angular velocity, depending on the rotation direction, also represents the rotation around a specific anatomical axis of a body segment. For each anatomical frame, two different axes were measured using either the accelerometer or gyroscope, depending on the segment. The third axis is calculated using the cross-product of the previous two axes. Subsequently, an orthonormal coordinate axis was based on these three axes. The last two movements, standing straight and bending by hip flexion, were used to determine the global frame and initial sensor orientation estimation [32]. The static neutral pose, with the arm stretched along the body and the fingers extended, gives the common vertical axis by measuring the gravity vector in all sensors. The hip flexion movement is performed with the arms extended along the body for the definition of the horizontal axis of the global frame. With the sensor-to-segment alignment and the common global frame for every IMU, it is possible to reconstruct the movement of the trunk, arm, hand, and fingers. Integration drift of the angular velocity over time was corrected by applying a Madgwick filter to correct for the inclination error of the sensor with respect to the gravitational component of the accelerometers [33]. Drifts in the gyroscope orientation were reduced by zero-velocity updates, following the methods of Kirking et al. [34] where if the norm of the angular velocity is below 3°/s is defined to be static in terms of actual sensor movements.

The joint angles are defined as the angle between two anatomical axes of adjacent limb segments of the respective joint as indicated in Figure 1b. Positive angles indicate flexion, abduction or supination of a joint and a negative angle indicates extension, adduction or pronation.

Three trunk compensation angles were calculated by comparing the projected trunk axes onto the global frames corresponding to the static neutral pose, consisting of trunk flexion (rotation in the sagittal plane around the $y$-axis of the sternum), lateral rotation (rotation around the $x$-axis of the sternum), and torsion (rotation around the $z$-axis of the sternum). *Shoulder flexion/extension* was defined as the angular variation of the upper arm's $x$-axis ($x_{ua}$) in the frontal plane (defined by the $x$-$z$ plane of the sternum's frame). Shoulder abduction/adduction is determined by relating the upper arm ($x_{ua}$) to the sternum's frame in the frontal plane (defined by the $y$-$z$ plane of the sternum, see Figure 1b). Elbow flexion/extension was determined by the angle between the upper ($x_{ua}$) and the lower arm's ($x_{la}$) $x$-axis. Forearm supination/pronation was defined by the mean orientation variation around the $x$-axis of the lower arm ($x_{la}$). Wrist flexion/extension was defined by the angle between the $x$-axis of the lower arm ($x_{la}$) and the hand's $x$-axis ($x_h$). The finger flexion/extension (thumb, index finger and middle finger) was defined as the angle between the $x$-axis of the hand ($x_h$) and the fingertip frames ($x_m$, $x_i$ and $x_t$).

*2.4. Experimental Protocol*

At the beginning, each participant was interviewed about demographic information and assessed for upper limb impairments by use of the FMA-UE [26]. The experimental protocol was performed on both limbs separately, starting with the non-affected limb (NAF) followed by the affected limb (AF), to study differences between pathological and physiological movement behavior. The protocol consisted in performing reach-to-grasp and displace different types of cubic blocks to different target positions. The participant was positioned sitting in front of a table with the tested arm held in 90° elbow flexion and the palm facing down on the table. Three markers were defined on the table for placing the hand and fingers in the starting position. The target positions were determined by each participants' maximal arm length in four pre-defined target locations as shown in Figure 2 and mirrored between both upper limbs. This task set-up was adapted from the ARAT [35], which evaluates the ability to grasp and displace, for example wooden blocks, onto a 37 cm high to-shelf.

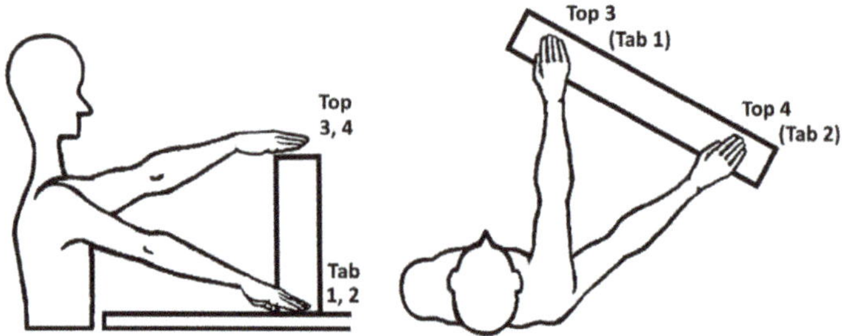


**Figure 2.** Experimental set up in sagittal and top view including the target locations: Tab 1; in ipsilateral arm length, Tab 2; in abducted arm length, Top 3; ipsilateral arm length, Top 4; in abducted arm length. Block objects: BL (big light block, 108 g), BW (big wooden block, 490 g) and BH (big heavy block, 1008 g).

The target locations at table height (1, ipsilateral arm length and 2, abducted arm length) and at top-shelf height (3, ipsilateral arm length and 4, abducted arm length) were selected to explore kinematic expressions in a relevant arm workspace and observe effects of arm loading in movements against gravity. The 10 cm block objects to be grasped varied in three different weights: 108 g (BL, big light block), 490 g (BW, big wooden block) and 1008 g (BH, big heavy block) to investigate influence of additional load during object grasp and displacement. The weight range of the object was based on the weight of the standardized weight of the wooden block (490 g) that is used in the Action Research Arm Test. The 1kg weight was selected as it corresponds to objects, relevant for daily-life functioning daily life, e.g., when manipulating a 1 l bottle of water. The lighter block was included to enable the movement analysis with only little additional weight load. The order of blocks was randomized in advance to avoid task-related physical fatigue during the experiments. This resulted in a combination of 12 task conditions per tested arm, that were each repeated three times.

After donning the system, the sensor-to-segment calibration protocol was performed with manual guidance of a therapist to assure proper execution of the static positions and the dynamic movements. Each position was measured for at least five seconds and checked online by an experienced engineer. For accurate global frame definition and sensor orientation estimations, the last two calibration movements were performed before each measurement trial. This procedure allowed to reduce drift in the sensor data during measurements. In between the three repetitions, each subject was asked to avoid extra movements of the tested arm and go back to the starting position as soon as the movement task was finished. This procedure allowed the subjects to rest for about 10 s between the trials. The whole experiment was expected to be performed within a maximum of 2 h. After the system was donned, each participant rated the wearing comfort of the system, possible limitations of gross movements due to the cables and limitations of grasping due to the fingertip sensors on a 5-point Likert-scale.

### 2.5. Feature Extraction

To enable task-specific spatiotemporal analysis of the reach-to-grasp movements, each movement trial was segmented into three phases: (1) reach, (2) displacement and (3) return by determining the time points of movement onset, grasp, release and movement end. Movement start and movement end were detected using a threshold detection algorithm for the upper arm's IMU angular velocity norm, where a threshold of 0.1 rad/s was used [36] to account for relevant limb motion. The force data of the fingertips on the table was also used as an onset and offset indicator, where applicable. The moment of grasping was defined by the detection of finger reaction and interaction forces, whereas the release is defined by the decrease in force signal to the lowest value of force, as displayed in Figure 3. In cases where no force profile was detected due to low interaction forces or because the finger contact points deviating from the force sensor placements, the grasp and release time points were identified via the joint angle profiles. The release time point was defined by the changes from finger flexion to extension including the maximum elbow extension and shoulder flexion, that represent the moment of maximal reach to target position. The duration of each movement phase was calculated as the time between the delimiting time points of each phase.

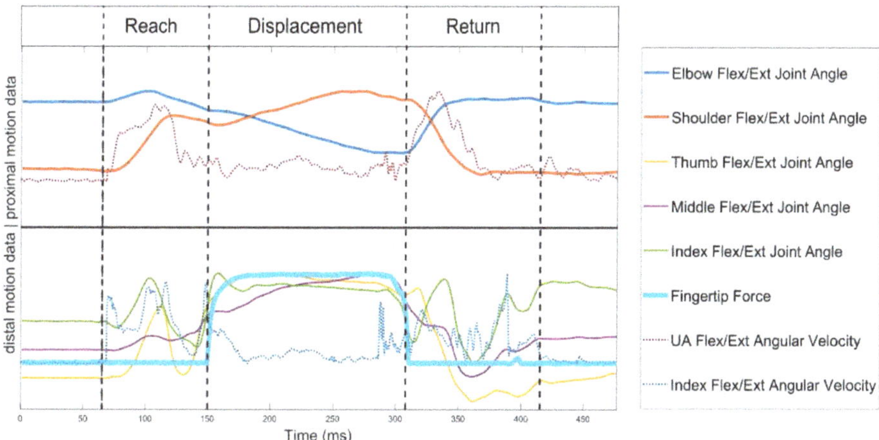


**Figure 3.** Proximal (shoulder, elbow), distal (finger) motion data and force signal for phase segmentation. The data is scaled to fit the plot, not the actual measured values on the $y$-axis.

For validation of the relevant expected differences between physiological and pathological movement behavior in the study sample, movement time and active range of motion of the main degrees of freedom (DOF) were compared between the affected and non-affected side. Movement time was defined as the time between movement start and end, detected by the 0.1 rad/s threshold. The DOF included trunk compensation, shoulder flexion-extension, shoulder abduction-adduction, elbow flexion-extension, forearm supination-pronation, wrist flexion-extension, thumb, middle finger and index finger flexion-extension for the entire task analysis per target location of the reach-to-grasp movement.

The primary outcome parameters, range of motion in trunk displacement, elbow, wrist, and finger flexion-extension were defined as the difference between the maximum and minimum angle of the joint during the period of object displacement, because they were expected to show expressions of the pathological flexor synergy and compensatory trunk movements. Kinematic parameters of interest to determine interjoint-coordination during the reach-to-grasp movement were defined as the joint ranges in trunk displacement, elbow, wrist, and finger flexion-extension within the displacement phase of the task. Trunk compensation in degrees was used as a metric to quantify the amount of compensatory trunk inclination during the upper limb movement and was defined by the square root of the sum of squares of the ranges in all three trunk compensation angles. Range of motion in elbow and wrist flexion-extension were calculated by taking the difference between the maximum and minimum joint angle measured in the displacement phase, as a metric for quantifying the pathological flexor synergy. The range of motion in finger flexion-extension was calculated as the mean between the range of the index and the middle fingers for each movement execution to consider distal characteristics of the pathological flexor synergy.

Statistical Analysis

All outcome parameters were visually inspected in histograms and presented descriptively by means and standard deviations.

Differences in range of motion in trunk displacement, elbow, wrist, and finger flexion-extension during object displacement were analyzed with respect to tested arm, object weight and target height, by considering the average of the three repetitions per subject and task condition. A linear mixed model analysis was applied to investigate significant differences and interactions between the independent factors, arm (AF, NAF), object (BL, BW, BH) and target height (Tab, Top), on the dependent variable of the metrics on joint range of motion, as presented in the model: Metric$-$a1 $\times$ Arm + a2 $\times$ Weight + a3 $\times$ Height + a4 $\times$ Subject. The linear mixed model analysis was selected as it takes into account the repeated measures experimental design and inner subject effects in a nested structure of the dependent variables.

The analysis of the relationship between the displacement phase kinematics of trunk displacement, elbow, wrist, and finger flexion-extension and the individual impairment level, as determined with the clinical FMA-UE test, was explored by plotting the median joint ranges including the upper and lower boundaries of the interquartile range of the affected arm against the measured impairment with the FMA-UE. Statistical testing for answering the second and third research question was performed by cross correlations based on Spearman rank correlations to investigate the relationships between FMA-UE, trunk displacement, elbow flexion-extension, wrist flexion-extension, and finger flexion-extension. All statistical tests were performed using Matlab (MATLAB version 2016b, The Mathwork, Natick, MA, USA) and SPSS (SPSS version 26.0, IBM Corp., Armonk, NY, USA) with a significance level of $p = 0.05$, indicating significances of $p = 0.01$ and $p = 0.001$ specifically.

## 3. Results

Kinematic measurements were gathered in 10 chronic stroke subjects within a recruitment period of 8 days in July 2019. One subject performed only two of the three block conditions due to time constraints. The data of the remaining blocks were discarded due to incomplete and incorrect sensor-to-segment calibration data. This resulted in a total of nine out of 10 subjects, who were included in the data analysis, adding up to 324 affected and non-affected side motion data sets. All participants rated the measurement system to be comfortable to wear. One subject rated some influence on the gross movements due to the cable wires of the sensing system. Three of the participants reported impedance of grasp due to the finger sensors.

### 3.1. Demographics

The demographics of the study participants are shown in Table 2, consisting of four right-side dominant and five left-side affected subjects. Upper limb impairments were measured with the FMA-UE score, ranging from 28 to 46 out of 66 points. Subjects with strokes in the perfusion territory of the middle and posterior cerebral artery showed slight increased upper limb impairments (FMA-UE mean 32.6) when compared to those with strokes in the anterior cerebral artery area (FMA-UE mean 43). According to a group analysis of the upper limb capacity-levels in relation to FMA-UE score [37], this sample included one subject with poor capacity (FMA-UE 23–31), eight subjects showing limited capacity (FMA-UE 32–47) and no subject with notable capacity (FMA-UE 48–52) or full function (FMA-UE 53–66).

**Table 2.** Demographic and clinical characteristics of the participants ($n = 10$). Subject S02 was discarded from the data analysis.

| Part ID | M/F | Age in Years | Month Since Stroke | Type | Cerebral Artery Territory Affected | Affected Body Side | FMA-UE Total/66 | FMA-UE Arm/36 | FMA-UE Wrist/10 | FMA-UE Hand/14 |
|---|---|---|---|---|---|---|---|---|---|---|
| S01 | M | 72 | 73 | Isch | PCA | L | 35 | 22 | 6 | 10 |
| S02 | M | 73 | 30 | Isch | MCA | R | 33 | 22 | 7 | 9 |
| S03 | M | 62 | 181 | Hem | (BG) | R | 33 | 26 | 7 | 5 |
| S04 | F | 61 | 70 | Isch | ACA | L | 40 | 22 | 6 | 13 |
| S05 | M | 57 | 45 | Isch | ACA | L | 46 | 31 | 8 | 11 |
| S06 | F | 70 | 21 | Isch | PCA | R | 34 | 22 | 5 | 9 |
| S07 | F | 70 | 33 | Isch | MCA | L | 34 | 22 | 6 | 11 |
| S08 | M | 59 | 9 | Isch | MCA | R | 32 | 16 | 4 | 11 |
| S09 | F | 42 | 20 | Isch | MCA/ICA | R | 32 | 24 | 3 | 6 |
| S10 | M | 42 | 13 | Isch | MCA/PCA | L | 28 | 25 | 6 | 2 |
| Median (IQR) | | 61.5 (57.5–70) | 31.5 (20.3–63.8) | | | | 33.5 (32.3–34.8) | 22 (22–24) | 6 (5.3–6.8) | 9.5 (6.8–11) |
| Mean ± SD | | 60.8 ± 11.4 | 49.5 ± 51.2 | | | | 34.7 ± 4.9 | 23.2 ± 3.8 | 5.8 ± 1.5 | 8.7 ± 3.4 |

Legend: ACA, anterior cerebral artery; (BG), basal ganglia; FMA-UE Total, Fugl–Meyer Assessment of the Upper Extremity; FMA-UE arm, FMA-UE arm subsection; FMA-UE wrist, FMA-UE wrist subsection, FMA-UE hand, FMA-UE hand subsection; Hem, hemorrhage; ICA, internal carotid artery; Isch, ischemic; L, left body side; MCA, middle cerebral artery; M/F, male/female; PCA, posterior cerebral artery; R, right body side.

### 3.2. Upper Limb Kinematic Measures

The automated detection algorithms were successfully applied in 56.1% of the data. Corrections had to be made in 47.8% of the NAF data and 52.2% of the AF data. Failures in automated detection were 76.5% related to inconsistent or low force profiles and in 23.5% related to jerky and noisy angular velocities or joint angle profiles and manually corrected.

Statistically significant higher movement times were found in the AF (mean 4.9 ± 1.6 s) when compared to the NAF (mean 2.8 ± 5.4 s) for the whole task execution ($p < 0.000$) and accordingly for all subphases ($p < 0.000$) of the reach-to-grasp movement. The mean difference in ROM across the main DOF between the AF and the NAF was 10.0 ± 6.9 degrees across the investigated joints, ranging from 0.2 to 28.7 degrees. The differences in range of motion between the AF and NAF were statistically significant across target locations for shoulder flexion-extension, elbow flexion-extension, wrist supination-pronation, thumb, and index finger flexion-extension, as shown in the Appendix A, Table A1. Range of motion in flexion-extension of the shoulder and the elbow were consistently lower in the AF when compared to the NAF, indicating a limited ability to elevate the arm and extend the elbow in reaching. Trunk compensation was significantly different between AF and NAF for the two abducted target locations. A higher mean flexion-extension range was detected for both the index and the middle finger of the AF compared to the NAF, besides lower flexion-extension ranges in the thumb of the affected side for all target positions.

### 3.3. Influences of the Factors, Arm, Object Weight and Target Height on Joint Range of Motion

For each of the primary kinematic features (trunk compensation, elbow, wrist, and finger flexion/extension), significant differences in range of motion of the displacement phase can be attributed to the factors tested (arm, object weight, target location). The results of estimates for the independent fixed factors arm (affected side vs. non-affected side), object (BL, BW, BH) and target height (table location vs. top location) on the selected DOF are shown in Table 3.

**Table 3.** Statistical significance of the effects of the independent fixed factors arm, object, and height on the dependent variables of the selected joint range metrics. The factor object including post-hoc pairwise testing between the three levels (BL, BW, BH).

| Factor | Trunk Compensation | Elbow Flexion Extension | Wrist Flexion Extension | Finger Flexion Extension |
|---|---|---|---|---|
| Arm (AF vs. NAF) | 0.006 ** | 0.000 *** | 0.000 *** | 0.000 *** |
| Object (BL, BW, BH) | 0.022 * | 0.146 | 0.401 | 0.588 |
| −BL vs. BW | 1.000 | 0.680 | 1.000 | 1.000 |
| −BL vs. BH | 0.026 * | 1.000 | 0.543 | 1.000 |
| −BW vs. BH | 0.067 | 0.156 | 1.000 | 1.000 |
| Height (Tab vs. Top) | 0.006 ** | 0.000 *** | 0.040 * | 0.006 ** |

Legend: *, **, *** indicate statistical significance of $p < 0.05$, $p < 0.01$ and $p < 0.001$, respectively. AF, affected side; BH, heavy block; BL, light block; BW, wooden block; NAF, non-affected side; Tab, table target position; Top, top location.

The factor of the tested arm showed significant effects on trunk compensation with larger range of motion in the AF (mean 9.4 ± 1.2 degrees) when compared to the NAF (mean 8.2 ± 1.1 degrees) with $F = 8.327$, $p = 0.006$. Elbow flexion-extension was significantly lower in the AF (mean 44.3 ± 3.9 degrees) than in the NAF (mean 54.2 ± 4.6 degrees) resulting in significant effects of the arm tested with $F = 23.385$, $p = 0.000$. Higher ranges in wrist flexion-extension were found in the AF (mean 29.4 ± 4.2 degrees) than in the NAF (21.2 ± 2.7 degrees) with $F = 30.798$, $p = 0.000$ and in finger flexion-extensions of the AF (mean 99.6 ± 11.4 degrees) when compared to the NAF (mean 77.1 ± 9.0 degrees) with $F = 29.553$, $p = 0.000$.

Significant effects for the fixed factor of object weight were found on the metric of trunk compensation ($F = 4.238$, $p = 0.022$). Considering post-hoc pairwise testing, trunk displacement was significantly larger when displacing the big heavy block (mean 10.2 ± 1.9 degrees) when compared to the displacement of the big light block (mean 7.9 ± 1.4, $p = 0.026$) and non-significantly larger in comparison to the big wooden block (mean 8.4 ± 1.3 degrees, $p = 0.067$).

The factor height showed significant effects on all DOF. Displacement to the top height location resulted in significantly higher trunk compensation (mean 9.5 ± 1.3 degrees) when compared to table locations (mean 8.1 ± 1.1 degrees, $p = 0.006$). The highest statistically significant effect was found in elbow flexion-extension with pronouncedly increased range of motion in the top height location (mean 61.3 ± 4.4 degrees) when compared to the table locations (mean 37.2 ± 3.9 degrees, $p = 0.000$) with $F = 147.742$, $p = 0.000$. Likewise, ranges in wrist flexion-extension and finger flexion were increased in the top locations with a wrist flexion-extension mean of 23.7 ± 3.4 degrees in the table locations when compared to a mean of 26.9 ± 3.8 degrees in the top locations ($F = 4.354$, $p = 0.040$) and a finger flexion-extension mean of 82.1 ± 9.9 degrees in the table locations and a mean of 94.4 ± 11.1 degrees in the table locations ($F = 7.920$, $p = 0.006$).

*3.4. Relationship between Kinematic Parameters and Clinical Measures of Impairment*

For investigating the relationship between the individual participants' impairment level, as indicated by the FMA-UE score, and the joint ranges of the affected side during displacement, the FMA-UE score was plotted against the subjects median range of motion in trunk compensation and flexion-extension of the elbow, wrist, and fingers as visualized in Figure 4a–d. Three repetitions, three block weights and four target positions were considered for each subject resulting in 36 trials per subject and tested arm, represented by median and interquartile range. There was no significant correlations found between the FMA-UE and the individuals mean trunk compensation (r = 0.11, $p = 0.78$), elbow flexion/extension (r = 0.00, $p = 1.00$), wrist flexion/extension (r = −0.12, $p = 0.77$) and finger flexion/extension (r = −0.28, $p = 0.46$).

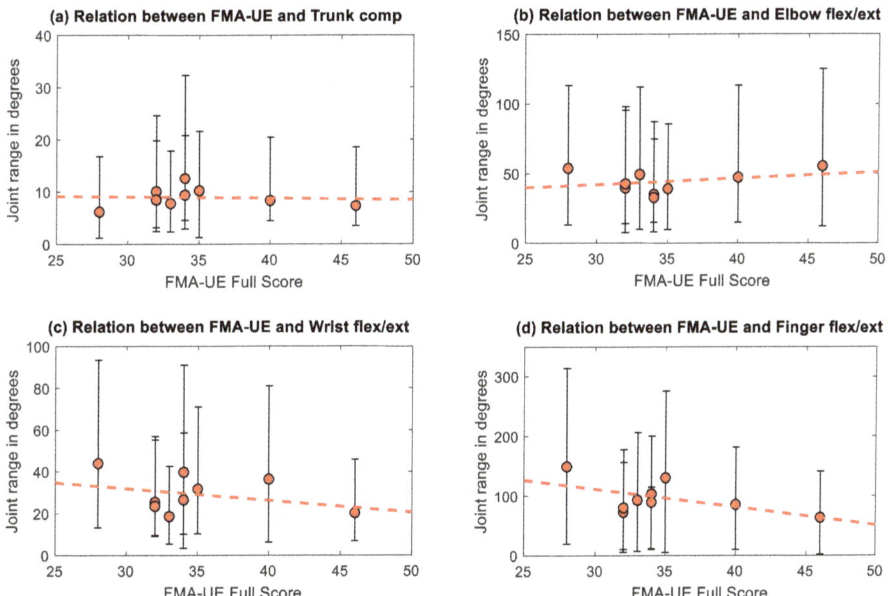


**Figure 4.** Subjects median joint range of (**a**) trunk compensation, (**b**) elbow, (**c**) wrist, and (**d**) finger flexion/extension of the affected side in relation to impairment level (FMA-UE score ranging from 0–66 points). Error bars represent the interquartile range over all trials performed by each of the nine subjects and the regression lines over the subjects are included for each metric.

In a sub analysis, the relationship between the kinematic metric and the related FMA-UE subsection was explored. The correlation between the FMA-UE arm section and trunk compensation resulted in $r = -0.57$ ($p = 0.11$). Elbow flexion/extension correlated statistically significantly with the FMA-UE arm subsection with $r = 0.68$ ($p = 0.04$). The relationship between the FMA-UE wrist subsection and wrist flexion/extension ($r = 0.00$, $p = 0.99$) as well as between the FMA-UE hand subsection and finger flexion/extension ($r = -0.56$, $p = 0.11$) was not conclusive.

*3.5. Relationship between the Selected Joint Range Metrics*

Similarly, the relationship between the selected joint range metrics did not result in significant correlations, except for trunk compensation and elbow flexion/extension. A statistically significant correlation was found between the mean trunk compensation and the elbow flexion/extension in the AF with a negative relationship ($r = -0.88$, $p = 0.0031$) as shown in Table 4. In the NAF statistically significant correlations were found between wrist and finger flexion/extension with strong positive correlations ($r = 0.72$, $p = 0.0369$).

**Table 4.** Confusion matrix of the Spearman rank correlation coefficients between the selected joint range metrics of the AF and the NAF side.

| AF | Trunk Comp | Elbow Flex/Ext | Wrist Flex/Ext | Finger Flex/Ext | NAF | Trunk Comp | Elbow Flex/Ext | Wrist Flex/Ext | Finger Flex/Ext |
|---|---|---|---|---|---|---|---|---|---|
| Trunk Comp | 1.00 | −0.88 ** | 0.05 | 0.10 | Trunk Comp | 1.00 | −0.35 | 0.08 | 0.17 |
| Elbow Flex/Ext | . | 1.00 | −0.32 | −0.20 | Elbow Flex/Ext | . | 1.00 | 0.35 | −0.03 |
| Wrist Flex/Ext | . | . | 1.00 | 0.53 | Wrist Flex/Ext | . | . | 1.00 | 0.72 * |
| Finger Flex/Ext | . | . | . | 1.00 | Finger Flex/Ext | . | . | . | 1.00 |

* indicates the statistical significance of the correlation with $p < 0.05$ and ** indicating statistical significance of the correlation with $p < 0.01$.

The relationship between the statistically significant correlations between the DOF, elbow flexion/extension joint ranges against trunk compensation and wrist against finger flexion/extension joint ranges were further evaluated by visualizing, as presented in Figure 5. The linear regression line between the trunk and the elbow joint ranges of the AF was defined by $y = -3.6x + 76$. Linear regression between the wrist and finger flexion/extension joint ranges was described by $y = 2.1x + 34$.

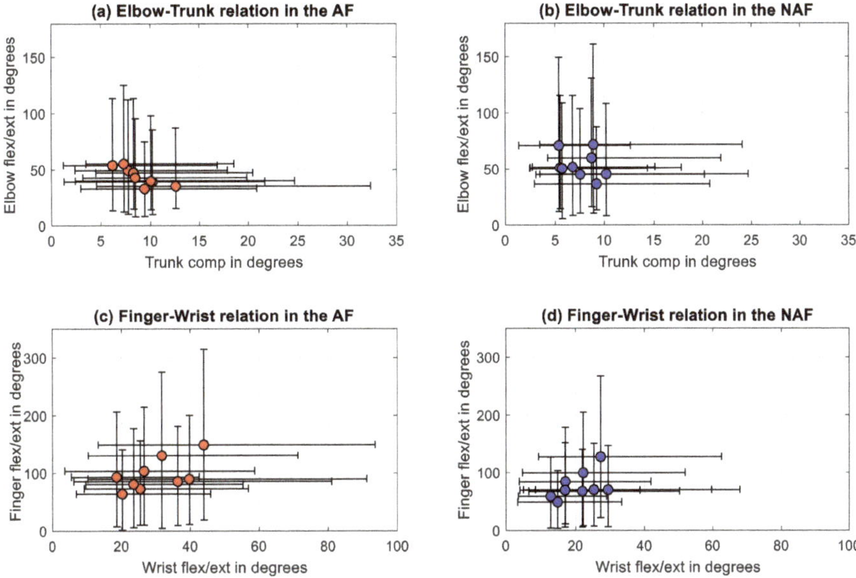

**Figure 5.** Correlations between (**a**) trunk compensation and elbow flexion/extension in the affected side, (**b**) trunk compensation and elbow flexion/extension in the non-affected side, (**c**) wrist and finger flexion/extension of the affected side, and (**d**) wrist and finger flexion/extension of the non-affected side.

## 4. Discussion

In this pilot study, sensor-based upper limb kinematic measurements of reach-to-grasp and displacement activities executed by chronic stroke subjects were used to examine and relate characteristics of movement impairments and to explore the influences of additional weight loads and mechanical work requirements on the upper limb kinematics. Movement impairments, such as longer movement times and decreased range of motion across the upper limb DOF were found in the affected when compared to the non-affected side, supported by existing literature [23,38]. Besides the evidence for weakness and impaired interjoint coordination in the affected upper limb, illustrated in the consistently decreased shoulder flexion and elbow extension for the whole task execution, this study focused on investigating the expression of pathological coupling between the trunk, elbow, wrist and fingers during object displacement within maximal arm length, as most clearly represented in Table 4. In order to include distally pronounced aspects of movement behavior in the kinematic analysis of object grasping and displacement, the range of motion of the wrist and the fingers' flexion-extension has been included in the analysis. The increased finger flexion in the AF when compared to the NAF expands on the characterization of the pathological flexor synergy and confirms previous research by Miller et al. [39] and Lan et al. [17] that described and increased difficulty to release the finger flexion with increased arm load. The significant positive correlation between finger and wrist flexion/extension in the non-affected upper limb, as shown in Table 4, could be interpreted as a physiological movement synergy allowing the subject to perform efficient grasp function. Herein, factors impacting the force and mechanical work demands were examined to prove the load-dependent appearance of pathological joint coupling in the upper extremity.

On the level of trunk compensatory movements, increased trunk movements were found in tasks with the affected arm when displacing the heavy block, that could be related to a compensation of weakness in the proximal shoulder muscles or weakness of the trunk muscles themselves. If trunk weakness itself was present in the investigated population, this could account as one explanation for why trunk compensatory movements were also detected in the non-affected side of the data set. It can be assumed that trunk weakness itself would diminish the ability to counterbalance an additional arm weight with either the affected or the non-affected limb. Another explanation for increased trunk movements in the non-affected side could be based on the fact that the NAF arm might deviate from complete healthy movement behavior due to indirect deficits in the non-crossing pathways from the ipsilesional cortex [40]. Nevertheless, these findings are in line with Repnik et al., 2018, who investigated the parameters movement time, smoothness, hand trajectory similarity and trunk stability in stroke subjects and healthy subjects when performing the ARAT and found similarly differences in trunk movements, especially early at movement onset, besides also noting occasional trunk motions above 10° in healthy subjects [41]. These findings suggest that the trunk compensation feature should be further studied with respect to diagnostic sensitivity and specificity to quantify stroke-related upper limb impairments. Apart from trunk compensation, an increase of the object weight showed no significant effects on the features of flexion-extension range of motion of the elbow, wrist, and fingers.

Differences related to the target height factor were detected in all tested features and can be partially explained by the different movement trajectory and positioning of the block object with respect to the hand posture between the top shelf and the table locations. The differences in wrist and finger flexion/extension can in part be explained by differences in hand positioning with respect to the target location, e.g., the hand might be more flexed in the wrist when displacing the block to the top shelf. Nevertheless, the strongest effect of target height was found in the elbow flexion/extension ROM, with a mean increase of elbow range of motion 24.0 ± 5.9 degrees in the top shelf locations when compared to the table target locations. This study finding was surprising, since all four target locations were defined by the maximum arm length to assure the requirement of complete elbow extension at the end of the displacement phase. Furthermore, the increased elbow flexion/extension motion in movements with increased gravity impact stand in contrast to previous research and the hierarchical structure of the synergistic movement patterns [15,26], that presume an increased difficulty of uncoupling elbow flexion from shoulder flexion with increased load and motion. The present study's findings, contrarily, could suggest that range of motion in elbow extension is increased in target positions that have a larger distance to the subjects' body center and require increased mechanical work against gravity. Despite the tentativeness of these results and the small study population, these outcomes could open new intervention strategies and should be addressed in future research with larger study samples to investigate possible underlying mechanisms. If the identification of factors that influence the increase or decrease in pathological joint coupling is possible, new intervention approaches would be opened to sustain stroke-related movement impairments. Including gradual decrease or increase of the armload has shown benefits for determining the severity of pathological joint coordination and providing patient-centered interventions, as indicated by Ellis and colleagues [42]. The examination of the influence of task conditions on the selected DOFs support the definition of the task-dependent and dynamic appearance of the pathological flexor synergy [13,17,18]. In the current study, the body of research on task-dependent changes based on planar movement task evaluations were extended to evaluations of reach-to-grasp activities in non-laboratory environments with a close linkage to functional activities of daily life. The fact, that we did not find significant effects of the object weight on the upper limb features, elbow, wrist, and finger flexion-extension except for trunk compensation, might be due to the range of object weight selected, from 100 g to 1 kg. Considering previous research on arm loading during reaching reported a maximum additional load of 50% of the arm weight [18] that would result in about 2 kg for an average person of 80 kg and an arm weight of around 5% of the body weight. Nevertheless, the subjects included in this study showed considerable difficulty in grasping and displacing the 1 kg heavy block.

These findings on movement condition effects on the relevant kinematic features stress the importance of considering task-dependent influences, such as gravitational forces and biomechanical constraints, when assessing and treating stroke-related upper limb impairments. Cortes et al. 2017 studied arm motor control in a planar robotic device and found a non-linear relationship between two-dimensional pointing parameters and scores from clinical scales incorporating antigravity strength demands. The authors suggested that arm motor control plateaus at 5 weeks post-stroke, whereas strength improvement, as measured by clinical scales, continues to improve up to 54 weeks post-stroke [43]. In this regard, it would be interesting to extend these objectives to three-dimensional movement tasks and investigate whether arm motor control, when measured in more complex reach-to-grasp movements by use of less motion impeding measurement systems, follows a similar recovery scheme when compared to 2D arm motor control and clinical scales. The usage of wearable sensing allows movement quality to be tracked in terms of kinematics in a less obstructive and more flexible way.

Another question addressed in this study was the relationship between the kinematic features of the displacement phase and the clinically measured individual impairment level. No clear correlations were found between the kinematic metrics and the FMA-UE, whereas trunk compensation and elbow flexion/extension showed strong correlation with the FMA-UE arm subsection, as well as the correlation between the FMA-UE hand subsection and finger flexion/extension. These findings are in line with existing research [24,44] and support the fact that kinematic parameters are, rather, complementary than redundant to standard clinical scales and potentially add clinically relevant information. The large interquartile ranges in all measured DOF in all study subjects illustrates the large variability in movement execution especially in non-cyclical discrete motions. The negative correlation between trunk compensation and elbow flexion/extension in the movements of the affected limb can be interpreted as an expression of the pathological joint coupling in stroke, where trunk compensation is increased relatively to the lack of active range of motion in the elbow during reaching. The significant positive correlation between the wrist and finger flexion/extension in the non-affected side could account for the appearance of physiological movement synergies during grasping and displacement that is less strong including larger interquartile ranges in both joints and non-significant in the affected side. These results support the use of the selected spatiotemporal features by use of non-laboratory kinematic movement analysis to assess aspects upper limb movement quality and impairments after stroke. Capturing and analyzing the relevant joint ranges during functional activities provides additional complementary information concerning how functional movements are performed and thereby help to overcome limitations of most existing clinical scales. Being able to detect the main aspects of movement quality and impairments allows selecting and monitoring changes in functional outcome and planning interventions that target these aspects. Future research should consider and re-evaluate the outcome features and task considerations presented herein on larger sample sizes to further underpin existing evidence of sufficient validity and reliability for metrics of joint range of motion and trunk displacement [24,44]. Furthermore, analysis of the assessments' clinimetric properties should be extended to domains sensitivity and specificity for differentiation physiological and pathological movement behavior.

### 4.1. Implementation of the Device and Analysis Methods in Clinical Practice

Wearable devices for assessments of motor function have been an ongoing research direction over the last decade. Portable devices facilitate the setup time and do not require patients to be directed to specific labs for measurements. The presented system potentiates the objective monitoring of the patients' impairments and provides the therapists an additional and more precise information about the movements' profile. The collective use of visual observations by the clinician and objectively measured patient movements using a sensing system as proposed in the current study system is intended to be used as means to provide better diagnostic and, thus, better therapy outcomes by providing a more thorough evaluation. Further research should focus on a clinician's point of view in the usability of the system in the clinic. By instructing therapists on how to use and analyze the distributed measuring system and its output, it is possible to obtain feedback, both from the patient and therapist, on its usability and relevance. In future, and after iterating the development steps of the device and methodology based on the feedback received, objective measurements with these types of system can become the standard for motor function evaluation.

*4.2. Strength and Limitations*

As a main limitation of this pilot study, the small sample size of the study needs to be considered as a factor that suppresses the robustness and degree of reliance of the findings and results presented. However, the investigated sample was homogenic with respect to a limited upper limb capacity, as determined by Hoonhorst et al., 2014, and allowed exploration of the applicability of the multisensory wearable system in the target population at an early device development stage [37]. The study sample included was intended to be able to perform reach-to-grasp and displace movements, which excludes more severely affected subjects. Nevertheless, besides the similar overall upper limb impairments, the included subjects showed reasonable variation in terms of the deficit distribution in the corresponding limb segments, as depicted in Table 2.

The principal idea of combining multiple sensing modalities, such as inertial sensing and other signal quantities, in a wearable system for upper limb kinematic motion analysis was considered as a strength of the device used, as this allows both for simplification or extension of the measurement modalities and enables the conduction of neurophysiological and biomechanical experiments on post-stroke upper limb movements in relatively unrestricted measurement surroundings. The wearable measurement system presented here combined complete kinematic motion analysis of the main DOFs of the upper limb kinematic chain and interaction force measurements at the fingertip, that have shown to be a powerful tool in reach-to-grasp detection and could further inform through measurements of grasp control. Although, we could confirm the application for assessing upper limb movements in chronic stroke subjects in this pilot study, the usability in clinical practice, including set-up, running and analyzing and the selected outcomes, would need to be addressed in future research.

Unfortunately, the force-sensitive resistor sensors used in this study showed limitations in capturing low forces per area and diminished flexibility to adapt to the shape of the finger pad and the grasped object. Therefore, grasp force could not be quantified as an outcome measure apart from the phase segmentation detection. An advanced version of flexible fingerprint sensors, as described in Wolterink et al. [45] is intended to be incorporated in the next generation of this multisensory measurement device. Detecting normal and shear force during grasp can provide further insights into movement control and effectiveness [46]. The combination of kinetic and kinematic measurements would allow to further study grasp control and stroke-related deficits, such as force limitation due to weakness or findings on force overshoot [2]. Effective grasping is undertaken by placing single fingers perpendicularly to the object surface [47]. This could be further explored in subjects after stroke with more adequate kinetic measurements.

Another considerable limitation relates to the systems' measurement accuracy. Similar to other IMU sensors, the systems' measurement accuracy depends on a successful sensor and sensor-to-segment calibration, appropriate filtering and fusion algorithms and reliable segment and joint angle definitions [25]. The accuracy of measurements was assured by updates of the global frame orientation definition and the avoidance of unnecessary extra movements prior to each task execution, which lasted not longer than nine seconds.

The detection of phases related to the movement primitives of reaching, object transport and return was feasible by a set of automatic detection algorithms in 47.8% of the affected upper limb data and 52.2% of the non-affected movement data. The observer-based validation of the points for phase discrimination and manual correction of defective time points to differentiate movement phases remain limited to subjective decision-making and time-consuming in processing. The grasping and release point, defined by an increase and decrease of the force profile and/or angular velocity in flexion-extension of the index finger, could show deviations due to inconsistent finger motion and force signals. In particular, the point of object release was difficult to detect when no distal signal peaks were detectable and could be affected by a systematic error if, for example, only maximum elbow extension is used to determine object release, which has to be considered rather as an indirect assumption than a proof of object release. Additionally, periods of transition or "dead time" between the phases need to be considered, as for example at movement start and end, where indifferent minor motions could affect the threshold detection. The application of improved flexible fingertip force sensors would reasonably improve the accuracy and reliability of time-points for phase detection of reach, displacement and return that are in alignment with studies on comparable movement analysis [2,41]. The accurate and time-efficient detection of motion primitive phases of reach-to-grasp activities is a relevant requirement for comparable and repeatable motion analysis of upper limb function.

Finally, we acknowledge that beside movement time and joint range of motion, several other kinematic parameters could have been investigated, such as hand trajectories or smoothness measures to complement the picture of movement quality and impairments. Based on the fact that signal information for the parameter calculation is provided by the system, this could be addressed in future studies using this multisensory measurement device. The data acquired in this study was publically made available for transparent reporting and re-evaluation and extension of the results [48]. To realize the long-term goal of upper limb kinematic assessments in clinical practice, this pilot study investigated metrics that were appropriate to detect and quantify impaired movement behavior after stroke by use of a wearable inertial measurement system. Even though the suggested metrics were derived from well-defined movement tasks, it is reasonable to include these metrics in existing analysis, that have been proven to be useful in the evaluation of non-structured daily-life activities [49,50]. Additionally, considering movement task characteristics and factors influencing the movement behavior were $p$ to enable the evaluation of subject-specific motion aspects and assessing the dynamics of the impairments.

## 5. Conclusions

This pilot study demonstrates the applicability of sensor-based kinematic motion analysis of functional reach-to-grasp and displacement movements in chronic stroke subjects with limited upper limb capacity by use of a wearable inertial sensing system. Relevant features to determine upper limb upper limb movement quality were suggested and examined for influences caused by the tested arm, object weight, target height factors and with respect to clinically measured impairment level. Range of motion in trunk displacement, elbow, wrist, and finger flexion-extension showed considerable differences between the AF and the NAF. Effects on metrics for interjoint coordination, as defined by the features, trunk compensation, elbow, wrist, and finger flexion-extension during displacement were found for the factors of an increase in object weight and target height. Hence, the factor's object weight and target height were suggested to study expressions of the pathological flexor synergy in functional reach-to-grasp movements with different task conditions. The significant correlations between elbow flexion/extension and trunk compensation detected in the affected upper limb support the appearance of pathological joint coupling during object displacement. Range of motion in elbow flexion-extension tended to be lower in the affected side when compared to the non-affected. The finger flexion-extension ROM showed significant differences between the AF and NAF and between the target heights, supporting further evaluation of this feature to quantify distally pronounced aspects of the pathological flexor synergy. These findings support the assessment of kinematic features of reach-to-grasp and displacement movements by use of IMUs and, therefore, help in paving the path towards clinically meaningful and feasible upper limb kinematic assessments in stroke research and clinical practice. The additional investigations on the effect of additional arm load and target height revealed relevant findings in the field of neurophysiology with respect to pathological joint coupling after stroke and highlight important considerations for upper limb kinematic assessments and possible treatment strategies to restore quality of movement in order to regain functionality in activities of daily life.

**Author Contributions:** Conceptualization, A.S. and M.M.C.B., G.W., J.P.O.H. and P.H.V.; methodology, A.S., M.M.C.B., G.W., J.P.O.H. and P.H.V.; software, M.M.C.B., G.W. and P.H.V.; validation, M.M.C.B., G.W. and P.H.V.; formal analysis, A.S., M.M.C.B.; investigation, A.S., M.M.C.B.; resources, P.H.V.; data curation, A.S., M.M.C.B.; writing—original draft preparation, A.S.; writing—review and editing, A.S., M.M.C.B., G.W., J.P.O.H., P.H.V.; visualization, A.S. and M.M.C.B.; supervision, P.H.V.; project administration, A.S., J.P.O.H., P.H.V.; funding acquisition, A.R.L., P.H.V. All authors have read and agreed to the published version of the manuscript.

**Funding:** This research was funded by the European Union's Horizon 2020 research and innovation program under grant agreement No 688857, SoftPro.

**Acknowledgments:** We would like to thank all study participants for their participation. Furthermore, thanks go to Jacques S.D. Getrouw, who worked on the initial set up steps of the inertial measurement system and Ed Droog who continuously assisted in technical support of the system.

**Conflicts of Interest:** The authors declare no conflict of interest. The funders had no role in the design of the study; in the collection, analyses, or interpretation of data; in the writing of the manuscript; or in the decision to publish the results.

## Appendix A

As a proof of concept validation, differences between the affected and non-affected sides in movement time and range of motion of the main DOFs were tested for statistically significant differences between the dependent groups by use of the Wilcoxon signed rank test, as illustrated in Figure A1 and Table A1. For movement time, all trial of the affected and non-affected sides were considered, whereas range of motion was compared between affected and non-affected sides, evaluation all trials separately for each target location.

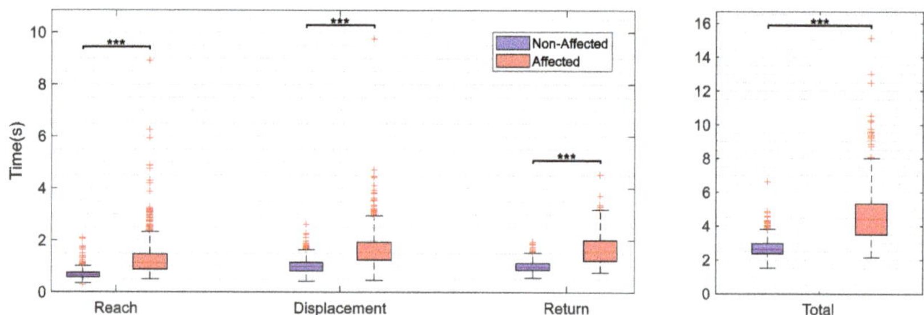


**Figure A1.** Movement time in seconds represented for the phases of reach, displacement and return and the total movement time. Data include all trials, object weights and target positions for the affected side (AF) and non-affected side (NAF).

Table A1. Differences in range of motion (AF vs. NAF) for the main upper limb degrees of freedom per target location.

| Target Side | Trunk Comp | Shoulder Flex/Ext | Shoulder Abd/Add | Elbow Flex/Ext | Forearm Sup/Pro | Wrist Flex/Ext | Thumb Flex/Ext | Index Flex/Ext | Middle Flex/Ext |
|---|---|---|---|---|---|---|---|---|---|
| Tab 1 AF | 17.37 ± 5.81 | 35.21 ± 10.65 | 24.02 ± 9.60 | 48.47 ± 18.11 | 43.98 ± 22.90 | 42.68 ± 15.83 | 82.48 ± 33.57 | 98.64 ± 37.73 | 101.19 ± 31.54 |
| Tab 1 NAF | 16.47 ± 6.10 | 49.03 ± 9.39 | 25.38 ± 10.69 | 63.62 ± 14.18 | 38.00 ± 10.53 | 37.19 ± 10.44 | 98.46 ± 39.26 | 76.13 ± 31.54 | 93.13 ± 27.47 |
| *p*-value | 0.149 | *** | 0.446 | *** | *** | * | *** | *** | *** |
| Tab 2 AF | 13.18 ± 7.63 | 25.93 ± 11.42 | 35.66 ± 9.05 | 51.92 ± 12.64 | 49.12 ± 23.24 | 45.00 ± 14.47 | 81.11 ± 32.19 | 95.65 ± 40.08 | 97.38 ± 32.33 |
| Tab 2 NAF | 8.17 ± 4.80 | 35.83 ± 10.00 | 33.71 ± 7.87 | 65.91 ± 14.62 | 38.00 ± 10.53 | 40.09 ± 13.22 | 97.48 ± 40.74 | 74.25 ± 32.03 | 90.64 ± 28.58 |
| *p*-value | *** | *** | * | *** | *** | * | ** | *** | *** |
| Top 3 AF | 15.41 ± 5.06 | 47.36 ± 11.53 | 31.41 ± 10.41 | 58.20 ± 17.34 | 60.63 ± 30.96 | 48.47 ± 18.09 | 85.89 ± 38.89 | 107.74 ± 41.14 | 103.75 ± 28.04 |
| Top 3 NAF | 15.63 ± 6.08 | 60.21 ± 12.12 | 29.71 ± 11.05 | 71.87 ± 13.12 | 45.78 ± 18.03 | 44.38 ± 14.00 | 100.45 ± 40.75 | 85.54 ± 30.15 | 98.94 ± 28.34 |
| *p*-value | 0.287 | *** | 0.116 | *** | *** | 0.264 | *** | *** | * |
| Top 4 AF | 16.16 ± 9.03 | 33.41 ± 11.08 | 45.44 ± 10.98 | 65.36 ± 12.89 | 72.55 ± 27.12 | 51.06 ± 19.96 | 86.28 ± 38.53 | 110.81 ± 39.35 | 103.24 ± 32.21 |
| Top 4 NAF | 9.69 ± 3.80 | 41.23 ± 12.80 | 39.63 ± 10.44 | 77.44 ± 12.55 | 63.66 ± 16.27 | 46.88 ± 14.23 | 100.56 ± 44.34 | 82.13 ± 31.31 | 99.67 ± 29.52 |
| *p*-value | *** | *** | *** | *** | * | 0.060 | * | *** | 0.105 |

Means and standard deviations (±) are presented for each DOF and target location (Tab 1, Tab 2, Top 3, Top 4) *n* the affected (AF) and non-affected side (NAF). Statistical significance of *p* < 0.05 is indicated by *, of *p* < 0.01 by ** and of *p* < 0.001 by ***. Legend: Abd/Add, abduction/adduction; Flex/Ext, flexion/extension; Sup/Pro, supination/pronation; Trunk Comp, trunk compensation.

## References


1. Jones, L.A.; Lederman, S.J. *Human Hand Function*; Oxford University Press: New York, NY, USA, 2006.
2. Parry, R.; Soria, S.M.; Pradat-Diehl, P.; Marchand-Pauvert, V.; Jarrassé, N.; Roby-Brami, A. Effects of Hand Configuration on the Grasping, Holding, and Placement of an Instrumented Object in Patients With Hemiparesis. *Front. Neurol.* **2019**, *10*, 240. [CrossRef] [PubMed]
3. Feix, T.; Romero, J.; Schmiedmayer, H.-B.; Dollar, A.M.; Kragic, D. The GRASP Taxonomy of Human Grasp Types. *IEEE Trans. Hum. Mach. Syst.* **2016**, *46*, 66–77. [CrossRef]
4. Bernstein, N.A. *The Coordination and Regulation of Movements*; Pergamon Press Ltd.: London, UK, 1967.
5. Jeannerod, M. The representing brain: Neural correlates of motor intention and imagery. *Behav. Brain Sci.* **1994**, *17*, 187–202. [CrossRef]
6. Ekstrand, E.; Rylander, L.; Lexell, J.; Brogårdh, C. Perceived ability to perform daily hand activities after stroke and associated factors: A cross-sectional study. *BMC Neurol.* **2016**, *16*, 208. [CrossRef]
7. Vos, T.; Allen, C.; Arora, M.; Barber, R.M.; Bhutta, Z.A.; Brown, A.; Carter, A.; Casey, D.C.; Charlson, F.J.; Chen, A.Z.; et al. Global, regional, and national incidence, prevalence, and years lived with disability for 310 diseases and injuries, 1990–2015: A systematic analysis for the Global Burden of Disease Study 2015. *Lancet* **2016**, *388*, 1545–1602. [CrossRef]
8. Sacco, R.L.; Kasner, S.E.; Broderick, J.P.; Caplan, L.R.; Connors, J.; Culebras, A.; Elkind, M.S.; George, M.G.; Hamdan, A.D.; Higashida, R.T.; et al. An Updated Definition of Stroke for the 21st Century: A statement for healthcare professionals from the American Heart Association/ American Stroke Association. *Stroke* **2013**, *44*, 2064–2089. [CrossRef]
9. Byblow, W.D.; Stinear, C.M.; Barber, P.A.; Petoe, M.A.; Ackerley, S.J. Proportional recovery after stroke depends on corticomotor integrity. *Ann. Neurol.* **2015**, *78*, 848–859. [CrossRef]
10. Stinear, C.M.; Barber, P.A.; Smale, P.R.; Coxon, J.P.; Fleming, M.K.; Byblow, W.D. Functional potential in chronic stroke patients depends on corticospinal tract integrity. *Brain* **2007**, *130*, 170–180. [CrossRef]
11. Raghavan, P. Upper Limb Motor Impairment After Stroke. *Phys. Med. Rehabil. Clin. N. Am.* **2015**, *26*, 599–610. [CrossRef]
12. Santello, M.; Lang, C.E. Are Movement Disorders and Sensorimotor Injuries Pathologic Synergies? When Normal Multi-Joint Movement Synergies Become Pathologic. *Front. Hum. Neurosci.* **2015**, *8*, 1050. [CrossRef]
13. Sukal-Moulton, T.; Ellis, M.D.; Dewald, J.P. Shoulder abduction-induced reductions in reaching work area following hemiparetic stroke: Neuroscientific implications. *Exp. Brain Res.* **2007**, *183*, 215–223. [CrossRef] [PubMed]
14. Levin, M.F. Interjoint coordination during pointing movements is disrupted in spastic hemiparesis. *Brain* **1996**, *119*, 281–293. [CrossRef] [PubMed]
15. Twitchell, T.E. The Restoration of Motor Function Following Hemiplegia in Man. *Brain* **1951**, *74*, 443–480. [CrossRef] [PubMed]
16. Ellis, M.D.; Sukal-Moulton, T.; Dewald, J.P.A. Progressive shoulder abduction loading is a crucial element of arm rehabilitation in chronic stroke. *Neurorehabil. Neural Repair* **2009**, *23*, 862–869. [CrossRef]
17. Lan, Y.; Yao, J.; Dewald, J.P. The Impact of Shoulder Abduction Loading on Volitional Hand Opening and Grasping in Chronic Hemiparetic Stroke. *Neurorehabil. Neural Repair* **2017**, *31*, 521–529. [CrossRef]
18. Ellis, M.D.; Lan, Y.; Yao, J.; Dewald, J.P. Robotic quantification of upper extremity loss of independent joint control or flexion synergy in individuals with hemiparetic stroke: A review of paradigms addressing the effects of shoulder abduction loading. *J. Neuroeng. Rehabil.* **2016**, *13*, 95. [CrossRef]
19. Cirstea, C.M.; Levin, M.F. Compensatory strategies for reaching in stroke. *Brain* **2000**, *123*, 940–953. [CrossRef]
20. Kwakkel, G.; Lannin, N.A.; Borschmann, K.; English, C.; Ali, M.; Churilov, L.; Saposnik, G.; Winstein, C.; van Wegen, E.E.H.; Wolf, S.L.; et al. Standardized measurement of sensorimotor recovery in stroke trials: Consensus-based core recommendations from the Stroke Recovery and Rehabilitation Roundtable. *Int. J. Stroke* **2017**, *12*, 451–461. [CrossRef]
21. Platz, T.; Pinkowski, C.; van Wijck, F.; Kim, I.-H.; di Bella, P.; Johnson, G. Reliability and validity of arm function assessment with standardized guidelines for the Fugl-Meyer Test, Action Research Arm Test and Box and Block Test: A multicentre study. *Clin. Rehabil.* **2005**, *19*, 404–411. [CrossRef]
22. Collins, K.C.; Kennedy, N.C.; Clark, A.; Pomeroy, V.M. Getting a kinematic handle on reach-to-grasp: A meta-analysis. *Physiotherapy* **2018**, *104*, 153–166. [CrossRef]

23. Murphy, M.A.; Häger, C.K. Kinematic analysis of the upper extremity after stroke—How far have we reached and what have we grasped? *Phys. Ther. Rev.* **2015**, *20*, 137–155. [CrossRef]

24. Schwarz, A.; Kanzler, C.M.; Lambercy, O.; Luft, A.R.; Veerbeek, J.M. Systematic Review on Kinematic Assessments of Upper Limb Movements After Stroke. *Stroke* **2019**, *50*, 718–727. [CrossRef] [PubMed]

25. Walmsley, C.P.; Williams, S.A.; Grisbrook, T.L.; Elliott, C.; Imms, C.; Campbell, A. Measurement of Upper Limb Range of Motion Using Wearable Sensors: A Systematic Review. *Sports Med. Open* **2018**, *4*, 53. [CrossRef] [PubMed]

26. Fugl-Meyer, A.R.; Jääskö, L.; Leyman, I.; Olsson, S.; Steglind, S. The post-stroke hemiplegic patient. 1. a method for evaluation of physical performance. *Scand. J. Rehabil. Med.* **1975**, *7*, 13–31. [PubMed]

27. Kortier, H.G.; Sluiter, V.I.; Roetenberg, D.; Veltink, P.H. Assessment of hand kinematics using inertial and magnetic sensors. *J. Neuroeng. Rehabil.* **2014**, *11*, 70. [CrossRef] [PubMed]

28. van den Noort, J.C.; Kortier, H.G.; van Beek, N.; Veeger, D.H.E.J.; Veltink, P.H. Measuring 3D Hand and Finger Kinematics—A Comparison between Inertial Sensing and an Opto-Electronic Marker System. *PLoS ONE* **2016**, *11*, e0164889. [CrossRef]

29. Brodie, M.A.; Walmsley, A.; Page, W.; Brodie, M.A. The static accuracy and calibration of inertial measurement units for 3D orientation. *Comput. Methods Biomech. Biomed. Eng.* **2008**, *11*, 641–648. [CrossRef]

30. Luinge, H.; Veltink, P.H.; Baten, C.T.M. Ambulatory measurement of arm orientation. *J. Biomech.* **2007**, *40*, 78–85. [CrossRef]

31. Ricci, L.; Formica, D.; Sparaci, L.; Lasorsa, F.R.; Taffoni, F.; Tamilia, E.; Guglielmelli, E. A New Calibration Methodology for Thorax and Upper Limbs Motion Capture in Children Using Magneto and Inertial Sensors. *Sensors* **2014**, *14*, 1057–1072. [CrossRef]

32. Kong, W.; Sessa, S.; Zecca, M.; Takanishi, A. Anatomical Calibration through Post-Processing of Standard Motion Tests Data. *Sensors* **2016**, *16*, 2011. [CrossRef]

33. Madgwick, S.O.H.; Harrison, A.J.L.; Vaidyanathan, R. Estimation of IMU and MARG orientation using a gradient descent algorithm. In Proceedings of the 2011 IEEE International Conference on Rehabilitation Robotics, Zurich, Switzerland, 29 June–1 July 2011; pp. 1–7.

34. Kirking, B.; El-Gohary, M.; Kwon, Y. The feasibility of shoulder motion tracking during activities of daily living using inertial measurement units. *Gait Posture* **2016**, *49*, 47–53. [CrossRef] [PubMed]

35. Lyle, R.C. A performance test for assessment of upper limb function in physical rehabilitation treatment and research. *Int. J. Rehabil. Res.* **1981**, *4*, 483–492. [CrossRef] [PubMed]

36. de Vries, J.; van Ommeren, A.; Prange-Lasonder, G.; Rietman, J.; Veltink, P.H. Detection of the intention to grasp during reach movements. *J. Rehabil. Assist. Technol. Eng.* **2018**, *5*, 1–9. [CrossRef]

37. Hoonhorst, M.H.; Nijland, R.H.; van der Berg, J.S.; Emmelot, C.H.; Kollen, B.J.; Kwakkel, G. How Do Fugl-Meyer Arm Motor Scores Relate to Dexterity According to the Action Research Arm Test at 6 Months Poststroke? *Arch. Phys. Med. Rehabil.* **2015**, *96*, 1845–1849. [CrossRef] [PubMed]

38. van Kordelaar, J.; van Wegen, E.E.H.; Kwakkel, G. Unraveling the interaction between pathological upper limb synergies and compensatory trunk movements during reach-to-grasp after stroke: A cross-sectional study. *Exp. Brain Res.* **2012**, *221*, 251–262. [CrossRef] [PubMed]

39. Miller, J.C.; Dewald, J.P.A. Involuntary paretic wrist/finger flexion forces and EMG increase with shoulder abduction load in individuals with chronic stroke. *Clin. Neurophysiol.* **2012**, *123*, 1216–1225. [CrossRef] [PubMed]

40. Nowak, D.A.; Grefkes, C.; Dafotakis, M.; Eickhoff, S.; Küst, J.; Karbe, H.; Fink, G.R. Effects of Low-Frequency Repetitive Transcranial Magnetic Stimulation of the Contralesional Primary Motor Cortex on Movement Kinematics and Neural Activity in Subcortical Stroke. *Arch. Neurol.* **2008**, *65*, 741–747. [CrossRef]

41. Repnik, E.; Puh, U.; Goljar, N.; Munih, M.; Mihelj, M. Using Inertial Measurement Units and Electromyography to Quantify Movement during Action Research Arm Test Execution. *Sensors* **2018**, *18*, 2767. [CrossRef]

42. Ellis, M.D.; Carmona, C.; Drogos, J.M.; Traxel, S.; Dewald, J.P. Progressive abduction loading therapy targeting flexion synergy to regain reaching function in chronic stroke: Preliminary results from an RCT. In Proceedings of the 2016 38th Annual International Conference of the IEEE Engineering in Medicine and Biology Society (EMBC), Orlando, FL, USA, 6–20 August 2016; pp. 5837–5840.

43. Cortes, J.C.; Goldsmith, J.; Harran, M.D.; Xu, J.; Kim, N.; Schambra, H.M.; Luft, A.R.; Celnik, P.; Krakauer, J.W.; Kitago, T. A Short and Distinct Time Window for Recovery of Arm Motor Control Early After Stroke Revealed With a Global Measure of Trajectory Kinematics. *Neurorehabil. Neural Repair* **2017**, *31*, 552–560. [CrossRef]

44. Kanzler, C.M.; Rinderknecht, M.D.; Schwarz, A.; Lamers, I.; Gagnon, C.; Held, J.; Feys, P.; Luft, A.R.; Gassert, R.; Lambercy, O. A data-driven framework for selecting and validating digital health metrics: Use-case in neurological sensorimotor impairments. *NPJ Digit. Med.* **2020**, *3*, 80. [CrossRef]

45. Wolterink, G.J.W.; Sanders, R.G.P.; Krijnen, G. Thin, Flexible, Capacitive Force Sensors Based on Anisotropy in 3D-Printed Structures. In Proceedings of the 2018 IEEE Sensors Applications Symposium (SAS), New Delhi, India, 28–31 October 2018; pp. 1–4. [CrossRef]

46. Nowak, D.A.; Rosenkranz, K.; Topka, H.; Rothwell, J. Disturbances of grip force behaviour in focal hand dystonia: Evidence for a generalised impairment of sensory-motor integration? *J. Neurol. Neurosurg. Psychiatry* **2005**, *76*, 953–959. [CrossRef] [PubMed]

47. Cuijpers, R.H.; Smeets, J.B.; Brenner, E. On the Relation Between Object Shape and Grasping Kinematics. *J. Neurophysiol.* **2004**, *91*, 2598–2606. [CrossRef] [PubMed]

48. Schwarz, A.; Bhagubai, M.M.C.; Wolterink, G.; Held, J.P.O.; Luft, A.R.; Veltink, P.H. Kinematics of reach-to-grasp and displacement after stroke [Data set]. Zenodo. 2020. Available online: https://zenodo.org/record/3930752#.X0EGBjVCRhE (accessed on 22 August 2020). [CrossRef]

49. van Meulen, F.B.; Klaassen, B.; Held, J.; Reenalda, J.; Buurke, J.H.; van Beijnum, B.F.; Luft, A.; Veltink, P. Objective Evaluation of the Quality of Movement in Daily Life after Stroke. *Front. Bioeng. Biotechnol.* **2016**, *3*, 210. [CrossRef] [PubMed]

50. Held, J.P.O.; Klaassen, B.; Eenhoorn, A.; van Beijnum, B.-J.F.; Buurke, J.H.; Veltink, P.; Luft, A.R. Inertial Sensor Measurements of Upper-Limb Kinematics in Stroke Patients in Clinic and Home Environment. *Front. Bioeng. Biotechnol.* **2018**, *6*, 27. [CrossRef] [PubMed]

*Article*

# Smoothness of Gait in Healthy and Cognitively Impaired Individuals: A Study on Italian Elderly Using Wearable Inertial Sensor

Massimiliano Pau [1,*], Ilaria Mulas [1], Valeria Putzu [2], Gesuina Asoni [2], Daniela Viale [2], Irene Mameli [2], Bruno Leban [1] and Gilles Allali [3,4]

[1] Department of Mechanical, Chemical and Materials Engineering, Piazza d'Armi, 09123 Cagliari, Italy; imulas89@gmail.com (I.M.); bruno.leban@dimcm.unica.it (B.L.)

[2] Center for Cognitive Disorders and Dementia, Geriatric Unit SS. Trinità Hospital, Via Romagna 16, 09127 Cagliari, Italy; valeria.putzu@atssardegna.it (V.P.); gesuina.asoni@atssardegna.it (G.A.); danielaviale@tiscali.it (D.V.); dott.irenemameli@gmail.com (I.M.)

[3] Department of Clinical Neurosciences, Division of Neurology, Geneva University Hospitals and Faculty of Medicine, University of Geneva, 1205 Geneva, Switzerland; Gilles.Allali@hcuge.ch

[4] Department of Neurology, Division of Cognitive & Motor Aging, Albert Einstein College of Medicine, Yeshiva University, Bronx, NY 10461, USA

* Correspondence: massimiliano.pau@dimcm.unica.it; Tel.: +39-070-6753264

Received: 4 June 2020; Accepted: 23 June 2020; Published: 24 June 2020

**Abstract:** The main purpose of the present study was to compare the smoothness of gait in older adults with and without cognitive impairments, using the harmonic ratio (HR), a metric derived from trunk accelerations. Ninety older adults aged over 65 (age: 78.9 ± 4.8 years; 62% female) underwent instrumental gait analysis, performed using a wearable inertial sensor and cognitive assessment with the Mini Mental State Examination (MMSE) and Addenbrooke's Cognitive Examination Revised (ACE-R). They were stratified into three groups based on their MMSE performance: healthy controls (HC), early and advanced cognitive decline (ECD, ACD). The spatio-temporal and smoothness of gait parameters, the latter expressed through HR in anteroposterior (AP), vertical (V) and mediolateral (ML) directions, were derived from trunk acceleration data. The existence of a relationship between gait parameters and degree of cognitive impairment was also explored. The results show that individuals with ECD and ACD exhibited significantly slower speed and shorter stride length, as well as reduced values of HR in the AP and V directions compared to HC, while no significant differences were found between ECD and ACD in any of the investigated parameters. Gait speed, stride length and HR in all directions were found to be moderately correlated with both MMSE and ACE-R scores. Such findings suggest that, in addition to the known changes in gait speed and stride length, important reductions in smoothness of gait are likely to occur in older adults, owing to early/prodromal stages of cognitive impairment. Given the peculiar nature of these metrics, which refers to overall body stability during gait, the calculation of HR may result in being useful in improving the characterization of gait patterns in older adults with cognitive impairments.

**Keywords:** gait; smoothness; older adults; accelerometer; inertial measurement unit (IMU)

---

## 1. Introduction

Optimal locomotion capabilities represent a critical element in ensuring successful aging. Mobility is not only an important co-factor that influences life expectancy [1,2], but also plays a relevant role in the self-perception of aging [3], social participation [4], independence and overall quality of life [5].

The physiologic decline in quality and the effectiveness of sensory, vestibular and proprioceptive inputs, associated with the loss of muscle strength [6–8], alter several main features of gait pattern.

Elderly individuals present reduced gait speed, stride length and cadence, as well as increased stance and double support phase duration [9]. Taken together, these features indicate the adoption of cautious gait, a strategy necessary to counteract the loss of stability and, thus, reduce the risk of falls [10].

Although gait has long been considered mostly an automatic task, in the last decades, it has been postulated that cognitive performances (mainly executive functions) provide an essential contribution, especially through the regulation of postural control (strongly implicated during walking), owing to their role in the management of axial musculature and in the integration of visual, vestibular, proprioceptive and sensory feedback. The sum of required cognitive resources becomes even more relevant when environmental conditions tend to reduce the automaticity of the task, as occurs in the case of uneven terrain and in the presence of concurrent motor/cognitive tasks (i.e., dual-task). Instability thus increases and overall gait performance may be compromised as a result [11]. It has also been observed that early disturbances in cognitive processes, such as attention, executive functions and working memory, often coexist with slower gait speed, increased stride time variability and greater instability [12–14].

Owing to its essentiality for most activities of daily living and considering that a walking test can easily be performed even by an individual with severe cognitive impairment, gait is probably the most thoroughly investigated motor task in describing the impact of cognitive performance on overall mobility. While basic information on speed can be obtained from simple timed tests carried out using a stopwatch (like the 10-m walking test), fine details on the kinematics and kinetics of gait require more complex equipment, such as motion capture systems, force platforms and surface electromyography. In this scenario, for more than a decade, interest in the possibility of employing accelerometers and inertial measurement units (IMUs i.e., devices composed of tri-axial accelerometer, gyroscope and magnetometer) in human movement analysis has been increasing [15]. To date, low-cost, wearable and miniaturized IMUs featuring high reliability and easiness of use are available. Their performance is increasingly close to those of more expensive and complex equipment. Such devices have been successfully employed to perform several tests on balance, gait and functional mobility under ecological conditions in older adults, with and without cognitive impairments [16–21]. Particularly attractive for daily clinical routines is the use of a simple setup consisting of a single unit [22], since the analysis can be performed by a non-specialized person (e.g., nurse, physical therapist, physician) in a clinical/ambulatory setting, and under very ecologic conditions in a relatively short time.

A gait analysis assisted by IMUs can provide a large set of parameters, which includes the main spatio-temporal parameters (i.e., speed, cadence, step/stride length and duration of stance, swing and double support phases), as well as indicators of variability, regularity and symmetry (see the review by Jarchi et al. [23] for details). In particular, the specific processing of trunk accelerations allows the extraction of less conventional metrics which, in some cases, are able to reveal subtle changes in gait that might occur, well before they become detectable in terms of conventional spatio-temporal parameters. Among them, great interest has been raised by the so-called "smoothness" of gait [24] (also defined as "step-to-step" symmetry [25]). Such a feature, quantitatively identified by a parameter called harmonic ratio (HR), provides information about overall body movement during gait, in particular with regards to its stability [26], which is different from the typical spatio-temporal parameters, which are rather focused on lower-limb movement at the distal level. The study of HR has aroused significant interest among researchers of human movement, as it allows the detecting of gait alterations in individuals with neurologic and orthopedic conditions and characterizes the changes associated with aging [27–29].

*Use of Accelerometers and IMU to Analyze Gait in Elderly with and without Cognitive Impairment*

Accelerometers (alone or as part of IMU) have been used for almost three decades to investigate a wide range of aspects correlated with mobility and posture in older adults. In particular, early applications were focused on the analysis of trunk accelerations during gait, to obtain information on stability and smoothness [30,31], but with the advancement of the hardware's technology, as well as with the refinement of the signal processing techniques, even other movement features were explored. To date, gait analysis represents the most widespread example of application for such class of devices. In the elderly population, gait analysis is typically employed to assess spatio-temporal parameters which are useful to estimate, for example, the risk of falls, or to assess the extent of functional limitations associated with orthopedic and neurologic conditions [20]. Given the simplicity of use and the fact that no special preparation of the individual to test is needed, IMU are gaining increasing popularity in the clinical testing of elderly with mild cognitive impairments, Alzheimer's disease, or other types of dementia [17,21]. As a result, these studies allowed one to detect the existence of peculiar gait alterations (i.e., reduction of walking speed and stride length, increased variability and asymmetry, etc.), which reflect the modifications in brain structure and functions associated with cognitive deficit [17,18,21,32]. Moreover, gait data obtained from wearable accelerometers were able to discriminate different subtypes of dementia [21], thus suggesting that such devices might represent a useful tool for supporting the clinical diagnosis.

Several studies also attempted to correlate trunk accelerations features, acquired during walking tests, with clinical characteristics of older adults with cognitive impairments. Their main findings can be summarized as follows: in comparison with unaffected individuals, older adults with cognitive impairments exhibit significant reduced value of the root mean square (RMS) and structure variability of the medio-lateral trunk acceleration [33], significant association of trunk stability measures with the white matter lesions [34] and with cognitive performance [35]. In particular, the study of Ijmker and Lamoth [35] showed that the presence of a cognitive impairment is accompanied by a decrease in smoothness of gait along the walking direction (anteroposterior, AP), as indicated by the significantly reduced value of the corresponding HR. Moreover, HR AP was found to be significantly correlated with cognitive status, as expressed by the Mini Mental State Examination (MMSE) score. However, although innovative and interesting, such findings require further verification and extension, firstly owing to the limited size of the tested sample, as well as its unbalanced composition in terms of the men to women ratio (75 to 85% of the tested individuals were men). It is also noticeable that the role of HR in the ML direction has not been clarified, being found to increase in cognitively impaired individuals, contrary to expectations. Finally, HR in the V direction was not even considered.

Based on the aforementioned considerations, the main purpose of the present study was to analyze the spatio-temporal and smoothness of gait parameters for a cohort of older adults, with and without cognitive impairments. The main hypothesis to verify was if individuals with impaired cognitive performance are characterized by altered gait patterns and reduced smoothness of gait. As a secondary goal, the existence of possible relationships between the degree of cognitive impairment and the gait parameters investigated will also be explored.

## 2. Materials and Methods

### 2.1. Participants

In the period January 2020–February 2020, 90 elderly adults aged over 65, consecutively examined at the Center for Cognitive Disorders and Dementia (in collaboration with the Geriatric Unit, "SS. Trinità" General Hospital, Cagliari, Italy), were recruited for the study. All participants were free from other neurologic conditions (e.g., Parkinson's disease, multiple sclerosis and stroke), excluding cognitive decline. They were also free from orthopedic conditions able to interfere in mobility, and could walk independently without the need of any support, such as canes, walking frames, crutches etc. After a detailed explanation of the purposes and methodology of the study, they (or their family members/caregivers when necessary) signed an informed consent form. The study was conducted in accordance with the ethical standards of the institutional research committee, and with the 1964 Helsinki declaration and its later amendments.

## 2.2. Neuropsychologic Assessment

After an overall clinical and geriatric assessment, participants underwent a screening of their cognitive status carried out by means of: (1) the Italian version of the Mini Mental State Examination (MMSE, [36,37]) and (2) Addenbrooke's Cognitive Examination Revised (ACE-R, [38,39]). ACE-R is articulated across five cognitive domains, namely attention and orientation, memory, verbal fluency (related to cognitive abilities of executive function), visuospatial, and language. The overall ACE-R score ranges from 0 to 100, lower scores being indicative of greater cognitive impairment. We decided to employ both tests, because although MMSE is probably the most widespread rapid cognitive screening instrument and, as such, has a large amount of reference data available, it also suffers from several drawbacks which are partly overcome by ACE-R.

Participants were stratified into 3 groups, according to their MMSE score, based on the cut-offs proposed by Isella et al. [40], as follows:

- Healthy controls (HC): MMSE score ≥ 24 (n = 34)
- Early cognitive decline (ECD): 18 ≤ MMSE score < 24 (n = 37);
- Advanced cognitive decline (ACD): MMSE score < 18 (n = 19);

Their anthropometric and clinical features are reported in Table 1.

**Table 1.** Anthropometric and clinical features of participants. Values are expressed as mean ± SD.

|  | Healthy Controls (HC) | Early Cognitive Decline (ECD) | Advanced Cognitive Decline (ACD) |
|---|---|---|---|
| **Participants # (F, M)** | 34 (22 F, 12 M) | 37 (22 F, 15 M) | 19 (12 F, 7 M) |
| **Female/Male Ratio** | F 65%, M 35% | F 60%, M 40% | F 63%, M 37% |
| **Age (years)** | 79.1 ± 3.9 | 78.8 ± 5.8 | 78.9 ± 4.6 |
| **Body Mass (kg)** | 64.1 ± 13.5 | 62.5 ± 12.9 | 62.6 ± 17.1 |
| **Height (cm)** | 159.9 ± 8.6 | 159.3 ± 8.8 | 158.1 ± 9.7 |
| **Mini Mental State Examination (MMSE)** | 27.6 ± 1.7 | 22.0 ± 1.5 | 11.8 ± 5.1 |
| **Addenbrooke's Cognitive Examination Revised (ACE-R)** | 77.8 ± 11.1 | 55.5 ± 9.7 | 25.0 ± 15.4 |

## 2.3. Instrumental Gait Analysis

Gait patterns were investigated based on trunk accelerations collected using a miniaturized wearable inertial sensor (G-Sensor®, BTS Bioengineering, Italy), previously employed in studies involving the elderly [41,42]. The sensor was attached to participants' lower back, at approximately the S1 vertebrae level, using a dedicated semi-elastic belt. After a brief familiarization phase, participants were requested to walk along a 30-m hallway following a straight trajectory at a self-selected speed, and in the most natural manner. During the trial, the sensor acquired, at 100 Hz frequency, the accelerations along three orthogonal axes, namely: antero-posterior (AP) corresponding to the walking direction, medio-lateral (ML), and supero-inferior (V). In order to reduce the error possibly introduced by the initial misalignment of the sensor (particularly with regards to the V direction), the participants were asked to stand still for 10 s before starting the walking trial, and the local reference system of the device was rotated in such a way as to align its vertical axis with the gravity vector [43]. Acquired data were sent in real-time via Bluetooth to a Personal Computer, where they were subsequently processed with a custom Matlab® routine to calculate:

- spatio-temporal parameters of gait (namely gait speed, stride length, cadence, duration of stance, swing and double support phase expressed as a percentage of the gait cycle). The identification of the gait cycle and the subsequent extraction of such parameters was carried out by means of a peak-detection algorithm, according to the procedure described by Zijlstra [44];

- HRs for AP, ML and V directions.

The calculation of the HRs was carried out according to the procedure proposed by Menz, Lord and Fitzpatrick [31]. In short, the raw accelerometric signal is processed in the frequency domain using a finite Fourier series, and the HRs for the AP and V directions (see Equation (1)) are calculated as the ratio between the sum of the amplitudes (A) of the first ten even harmonics (which are representative of the in-phase components of the signal) and the sum of the amplitudes of the first ten odd harmonics (associated with the out-of-phase components), the latter being minimized as gait symmetry improves. Instead, the HR in the ML direction (see Equation (2)) is obtained by dividing the sum of the amplitudes of the odd harmonics by the sum of the amplitudes of the even harmonics, since the acceleration pattern exhibits one peak per stride, thus resulting in the dominance of the first harmonic and subsequent odd harmonics.

$$HR_{AP-V} = \frac{\sum A_{even\ harmonics}}{\sum A_{odd\ harmonics}} \tag{1}$$

$$HR_{ML} = \frac{\sum A_{odd\ harmonics}}{\sum A_{even\ harmonics}} \tag{2}$$

The interpretation of the HR values is quite straightforward, as lower values indicate a less smooth/symmetrical gait. Reference values for healthy older adults lie in the range 3–4 (for AP and V directions) and 2.1–2.6 for the ML direction [26,45–48].

*2.4. Statistical Analysis*

The existence of possible differences introduced in spatio-temporal parameters and HRs by participants' cognitive status was assessed using a one-way multivariate analysis of variance (MANOVA) and a one-way multivariate analysis of covariance (MANCOVA), respectively. In the latter case, gait speed was included in the analysis as a covariate, given its influence on HR values [46]. The independent variable was the participant's status (e.g., HC, ECD or ACD) and the dependent variables were the 6 spatio-temporal parameters and the 3 HRs. In both cases, the level of significance was set at $p = 0.05$, and the effect sizes were assessed using the eta-squared ($\eta^2$) coefficient. Univariate ANOVA was carried out as a post-hoc test, by reducing the level of significance to $p = 0.008$ (0.05/6) for spatio-temporal parameters and $p = 0.016$ (0.05/3) for HRs, after a Bonferroni correction for multiple comparisons. The relationship between spatio-temporal gait parameters and cognitive status (as indicated by both MMSE and ACE-R scores) was explored using Spearman's rank correlation coefficient rho, by setting the level of significance at $p < 0.05$. Rho values of 0.1, 0.3, and 0.5 were assumed to be representative of small, moderate, and large correlations respectively, according to Cohen's guidelines [49]. In the case of HR, we used partial correlation coefficients, checking for gait speed. All analyses were carried out using the IBM SPSS Statistics v.23 software (IBM, Armonk, NY, USA).

## 3. Results

*3.1. Spatio-Temporal Parameters of Gait and Harmonic Ratio*

The results of the experimental test are summarized in Table 2 (comparison of the spatio-temporal and HR values across the three groups) and in Table 3 (correlation analysis between gait parameters and MMSE/ACE-R scores).

**Table 2.** Spatio-temporal and smoothness-of-gait parameters calculated for the three groups of elderly. Values are expressed as mean ± SD.

| Gait Parameter | Healthy Controls (HC) | Early Cognitive Decline (ECD) | Advanced Cognitive Decline (ACD) |
|---|---|---|---|
| Gait speed (m s$^{-1}$) | 0.92 ± 0.23 | 0.68 ± 0.30 [a] | 0.63 ± 0.25 [a] |
| Stride length (m) | 1.03 ± 0.23 | 0.81 ± 0.32 [a] | 0.73 ± 0.23 [a] |
| Cadence (steps min$^{-1}$) | 107.3 ± 8.5 | 100.6 ± 12.6 | 101.7 ± 11.8 |
| Stance phase (% GC) | 61.3 ± 2.0 | 61.9 ± 2.3 | 62.0 ± 1.6 |
| Swing phase (% GC) | 38.8 ± 1.9 | 38.1 ± 2.3 | 37.4 ± 2.8 |
| Double support phase (% GC) | 22.3 ± 2.0 | 23.8 ± 2.3 | 24.0 ± 1.7 |
| Harmonic ratio (HR) anteroposterior (AP) direction * | 3.37 ± 0.69 | 2.49 ± 0.88 [a] | 2.31 ± 0.76 [a] |
| HR mediolateral (ML) direction * | 2.40 ± 0.72 | 2.03 ± 0.53 | 2.05 ± 0.55 |
| HR vertical (V) direction * | 3.70 ± 0.93 | 2.60 ± 0.87 [a] | 2.60 ± 0.88 [a] |

[a] significant difference vs. HC after Bonferroni correction; * controlled for gait speed; GC: Gait Cycle.

**Table 3.** Spearman's coefficients for correlations between spatial-temporal and smoothness of gait parameters and scores obtained from the neuropsychological assessment.

| Gait Variables | | MMSE | ACE-R |
|---|---|---|---|
| Spatial-temporal parameters | Gait speed | 0.449 [††] | 0.430 [††] |
| | Stride length | 0.446 [††] | 0.422 [††] |
| | Cadence | 0.199 | 0.191 |
| | Stance phase | −0.156 | −0.143 |
| | Swing phase | 0.192 | 0.182 |
| | Double support phase | −0.153 | −0.149 |
| Harmonic Ratio | HR AP direction * | 0.323 [††] | 0.303 [††] |
| | HR ML direction * | 0.213 [†] | 0.251 [†] |
| | HR V direction * | 0.259 [†] | 0.207 [†] |

[†] $p < 0.05$; [††] $p < 0.01$; * controlled for gait speed; ACE-R: Addenbrooke's Cognitive Examination (Revised); MMSE: Mini Mental State Examination; AP: antero-posterior; ML: medio-lateral; V: vertical.

MANOVA detected a significant main effect of group on the spatio-temporal parameters of gait [$F(12, 164) = 2.17$, $p = 0.016$, Wilks $\lambda = 0.74$, $\eta^2 = 0.14$], but the post-hoc analysis revealed that only gait speed and stride length actually differed across the tested groups. In particular, individuals with both ECD and ACD exhibited a significant reduced gait speed (0.68 and 0.63 m/s respectively vs. 0.92 m/s of HC, $p = 0.001$ in both cases) and stride length (0.81 and 0.73 m vs. 1.03 m of HC, $p < 0.01$ in both cases) with respect to unaffected participants.

Trends of the HR, calculated using the two methods previously described, are reported in Figure 1.

After controlling for gait speed, MANCOVA detected a significant main effect of individuals' status on HR values [$F(6168) = 3.42$, $p = 0.003$, Wilks $\lambda = 0.79$, $\eta^2 = 0.11$], and the post-hoc analysis revealed that HR in the AP and V directions differed significantly across the tested groups. For both directions in particular, individuals of the ECD and ACD groups exhibited HR values that were significantly lower that healthy controls, while no differences were found between the two groups of cognitively impaired elderly.

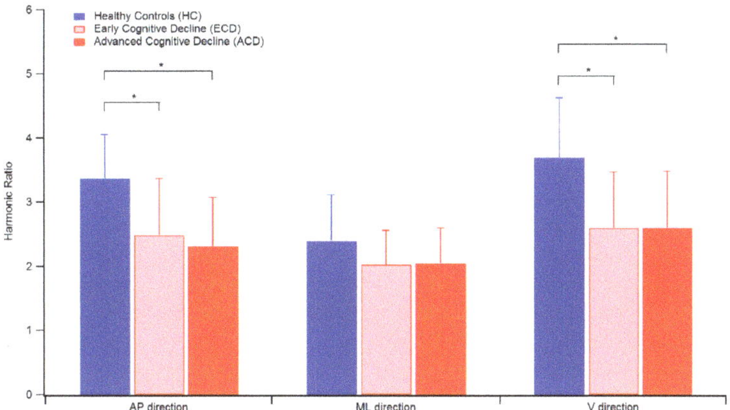


**Figure 1.** Trend of HR values for the three groups of tested elderly. The symbol * denotes a statistically significant difference after Bonferroni correction ($p < 0.016$)

### 3.2. Correlation between Gait Parameters and Cognitive Impairment

Gait speed and stride length were positively correlated with both measures of cognitive status with coefficients similar in magnitude, while no correlations were found with the remaining gait parameters. When we checked speed on the relationship between MMSE scores and HRs, we found a significant positive partial correlation, with rho ranging from 0.21 (ML direction) to 0.32 (AP direction). Similar results were obtained in the case of ACE-R, where the coefficients varied between 0.21 (HR V direction) and 0.30 (HR in AP direction).

## 4. Discussion

### 4.1. General Considerations

The aim of this study was to quantitatively investigate the alterations of gait patterns consequent to the presence of a cognitive impairment of different severity, using a wearable inertial sensor in a clinical setting, and to explore the existence of possible relationships between gait parameters and the degree of impairment. To this end, we employed the typical spatio-temporal parameters of gait, and trunk acceleration-based measures such as HR, which provide a different point of the view of gait alterations associated with overall body stability. In particular, we attempted to extend the previous limited findings by calculating HR for all three directions (AP, ML and V), enlarging the tested sample and analyzing the correlations of HR with two different measures of cognitive performance, namely MMSE and ACE-R.

At first, consistent with most existing studies, our data confirm that the existence of cognitive impairment, even mild, is associated with significant reductions in gait speed and stride length, while cadence and phase subdivision of the gait cycle appear to be less altered. The speed reductions of individuals with cognitive impairments with respect to unaffected controls is clinically meaningful and, particularly in the case of ECD, in very good agreement with the values recently reported by Peel et al. [50], in a meta-analysis, summarizing the results of 36 studies, involving more than 29,000 participants. Participants with more advanced impairment showed slower speed (−8%) and shorter stride length (−11%) with respect to individuals with ECD, but such a change was not found to be significant and, as such, should rather be considered as a trend. Taken together, the reduction in gait speed and stride length indicate that cognitive decline influences gait strategy, through the adoption of a cautious approach that probably reflects the diminished efficiency of sensory and motor systems and attempts to achieve a more stable locomotion to reduce the risk of falls [51].

### 4.2. Smoothness of Gait

Firstly, it is to be noted that the HR values calculated in the present study for our reference group of healthy older adults are consistent, even from a quantitative point of view, with those reported in previous studies involving individuals of the same age range [45–48]. This demonstrates that, despite the variability in terms of equipment and measurement protocols, the approach based on HR analysis is reliable and robust. As regards the values observed in individuals with cognitive impairment, even after checking for gait speed (which is known to have a direct influence on HR) our data show a substantial significant decrease in smoothness for all three directions considered, although only in the case of the AP and V directions were the variations statistically significant.

In the last two decades, several studies have employed HR to investigate gait performance in older adults for different purposes, such as characterizing the changes associated with either aging [27,45] or the presence of neurological diseases [26], assessing the risk of falls [41,44] and verifying the differences between overground and treadmill walking [43]. In short, their main findings indicate that older adults feature lower HR values with respect to young individuals. Moreover, further reductions have been observed in those suffering either from recurrent falls or in the presence of neurologic conditions known to affect balance and stability, such as Parkinson's disease [26], stroke [28] and multiple sclerosis [29]. However, only in the study by Ijmker and Lamoth [35] was the analysis of smoothness of gait applied to a small cohort of individuals with cognitive impairments, and it thus represents the only term of comparison for the findings derived by the present study. Consistently with our results, they observed a significant lower value for HR in the AP direction in cognitively impaired individuals, with respect to the unaffected elderly. In contrast, individuals with dementia exhibited higher values of HR ML with respect to both unaffected elderly and young subjects. This is contrary to the findings of our study, as participants in the ECD and ACD groups had significantly lower HR ML values than unaffected controls. The findings by Imjker and Lamoth [35] are actually quite surprising, as higher HR values indicate better smoothness of gait and stability, while most literature reports that dementia is accompanied by poor stability, especially in the ML direction [33,52,53]. Nevertheless, since this result was not discussed in detail by the authors, we can only speculate that factors such as a different composition of the sample (i.e., presence/different proportion of individuals with Alzheimer's disease or vascular cognitive impairment and a different woman/man ratio), as well as environmental and socio-economic backgrounds of the countries in which the studies were performed, might partly explain such a discrepancy. Generally speaking, the reduction of smoothness of gait can be attributed to alterations in limb dynamics and overall function, which can be present even in the early stages of cognitive impairment [26,54], as well as in trunk stability, especially in the presence of brain structural changes such as severe white matter lesions [34]. Moreover, individuals at increased risk of falls, such as those with cognitive impairment [55,56], have difficulties in controlling the rhythmic displacements of the trunk during gait [45], which is thus another factor able to worsen the overall smoothness of gait.

Interestingly, the most relevant changes in HR are evident already from the early stages of cognitive decline, which is the case of ECD, while further worsening appears not to be accompanied by a corresponding deterioration in gait smoothness. This suggests that the impact of cognitive decline on gait performance is already relevant during its early or even prodromal stages, a fact that is consistent with previous observations that pointed out how the deterioration of walking abilities precedes cognitive decline and the presence of dementia [57].

### 4.3. Correlation between Cognitive Status and Gait Parameters

The results of the correlation analysis between cognitive status and spatio-temporal parameters of gait confirm its relevant role in mobility performance [58,59]. In particular, the significant moderate correlations found between cognitive scores and gait speed (0.43 for ACE-R and 0.45 for MMSE) and stride length (0.42 for ACE-R and 0.45 for MMSE) are consistent with the findings of previous studies which reported coefficient values from 0.36 to 0.60 for gait speed (vs. ACE-R [60]; vs. MMSE [17]) and 0.59 for stride length (vs. MMSE, [17]).

There is instead a scarcity of data regarding the relationship between HR and cognitive measures, even though a number of studies have investigated the alterations of trunk accelerations in cognitively impaired people using a variety of metrics, including some quite similar in principle to HR [33–57], concluding that gait outcomes related to speed, regularity, predictability, and stability of trunk accelerations may suitably integrate other physical, cognitive, and behavioral measures, to better identify the extent of a cognitive impairment in the elderly. To the best of our knowledge, only Ijmker and Lamoth [35] attempted to investigate the existence of a possible relationship between HRs and MMSE score. They found a moderate positive correlation between HR AP and MMSE, similar to the observations of the present study, although slightly larger in magnitude (rho = 0.48 vs. 0.32). In contrast, Ijmker and Lamoth [35] found no significant correlation for the ML direction and did not consider the V direction. Possible reasons for the discrepancies with our findings are: (1) the fact that they did not consider the effect of gait speed, which may have some effect on HR values, as demonstrated by Lowry et al. [46]; (2) the different number of participants, which was less than a half with respect to our sample; (3) the unbalanced composition of the groups, which were predominantly composed of men.

Overall, our data suggest that gait parameters (both spatio-temporal and smoothness) are similarly influenced by the cognitive status, regardless of the way in which it is assessed, since the coefficients of correlation did not differ greatly. This would imply that while ACE-R, given its superior sensitivity, may be beneficial in better identifying the presence of dementia with respect to the MMSE, the latter appears to have sufficient capabilities for detecting the cognitive impairments associated with alterations in mobility.

What are the clinical implications of the findings obtained in the present study? Previous research demonstrated that HR is a metric more sensitive to subtle alterations in locomotor mechanisms, with respect to spatio-temporal parameters like speed or stride length [25]. Some examples of this phenomenon were observed in individuals in the early stages of Parkinson's disease [26] and multiple sclerosis [29]. In aging, recent research demonstrated that reductions in gait speed predicts incident dementia and cognitive decline [60], thus it is likely that the regular monitoring of trunk accelerations would probably allow the detection of changes in HR that are likely to occur earlier, with respect to those of walking speed. If such hypothesis would be confirmed by further longitudinal studies, the information provided by HR would support clinicians in the diagnosis of suspected cognitive impairment, allowing the planning of timely interventions.

### 4.4. Possible Issues Associated with the Use of IMU to Assess Gait Parameters and HR

As previously mentioned, IMU is a very appealing tool to perform the quick and inexpensive assessment of gait in a clinical setting, especially to test people with cognitive impairment, because, unlike more sophisticated equipment like optoelectronic motion capture system (which represents the gold-standard for the quantitative analysis of human movement), the test does not require a specific preparation of the individual for marker positioning and can be performed having him/her fully dressed. However, it must be noted that the validity and reliability of gait data obtained by IMU are influenced by several factors which should be considered. At first, the estimation of gait parameters could be affected by changes in sensor orientation, which may change during walking. Therefore, vertical acceleration may exhibit components in the remaining two axes which alter their actual value.

Specific issues are also associated with the calculation of the HR, which in some cases has been criticized for poor reliability, which is not associated with the methodology by itself, but rather with a poor standardization of the measurement protocols [43]. In particular, the approach proposed by Menz et al. [31] used in the present study (which is probably the most widespread) considers the first 20 harmonics of the accelerometric signal in the frequency domain. However, as pointed out by Bellanca et al. [25], such value is justified and adequate for "regular" cadences (i.e., approximately in the range 80–135 steps/min), because very slow walking may cut a significant part of the power spectrum, thus altering the HR value. Although, in our sample, all participants satisfied this criterion, in studies involving older adults with more severe cognitive decline, who also usually exhibit significantly reduced gait speed, such an aspect should be carefully considered.

### 4.5. Limitations of the Study

Some limitations of the study are to be acknowledged, beside the technical issues previously mentioned. Although it significantly extends the amount of available data, in terms of participants tested, the number of HRs considered, and the neuropsychologic tools used to explore the relationship between gait and cognitive status, some important factors have not been included in our analysis. Firstly, we did not consider education, wealth and occupational status, which are all known to have some influence on mobility performance [61–63]. Thus, the generalization of the results presented here considering different socio-economic contexts should be performed cautiously. Secondly, since a non-negligible percentage of the participants were overweight or obese (31% and 13% respectively), such conditions may have introduced alterations in gait parameters, especially for their HR values [64].

## 5. Conclusions

In the present study, we have attempted to clarify the relationship between smoothness of gait and cognitive performance in a cohort of the Italian elderly, using trunk acceleration-based data acquired in a clinical setting by means of a wearable inertial sensor. The results confirm the existence of gait pattern alterations in terms of slower speed and shorter stride length, as well as a decrease of HR in all the directions investigated, which were already evident in individuals with ECD. Instead, no further worsening of smoothness of gait was detected in the presence of a more severe cognitive impairment. All the aforementioned alterations were found to be moderately correlated with the extent of the cognitive impairment in a similar way, regardless of the use of different neuropsychologic screening tools such as MMSE and ACE-R.

Based on these findings, it is possible to state that the smoothness of gait parameters may represent a metric potentially useful in detecting subtle changes in gait possibly present in prodromal stages of dementia, but not evident from the analysis of spatio-temporal parameters alone. Such data might support the clinician in performing a more accurate diagnosis of cognitive impairment as well, in verifying the effectiveness of all those interventions targeted to overcome any possible mobility limitations in cognitively impaired individuals.

**Author Contributions:** M.P.: Conceptualization, Methodology, Formal Analysis, Writing—Original Draft; I.M. (Ilaria Mulas): Data Curation, Formal Analysis; V.P., G.A. (Gesuina Asoni), D.V., I.M. (Irene Mameli): Clinical Assessment of the Participants; B.L.: Formal Analysis, Software; G.A. (Gilles Allali): Writing—Review and Editing. All authors have read and agreed to the published version of the manuscript.

**Funding:** This research received no external funding.

**Acknowledgments:** The experimental tests of this study were completed a few weeks before the COVID-19 outbreak in Italy, which has caused the death of more than 25,000 adults aged over 70. This paper is dedicated to their memory.

**Conflicts of Interest:** The authors declare no conflict of interest.

## References

1. Hardy, S.E.; Perera, S.; Roumani, Y.F.; Chandler, J.M.; Studenski, S.A. Improvement in Usual Gait Speed Predicts Better Survival in Older Adults. *J. Am. Geriatr. Soc.* **2007**, *55*, 1727–1734. [CrossRef]
2. Studenski, S.; Perera, S.; Patel, K.; Rosano, C.; Faulkner, K.; Inzitari, M.; Brach, J.S.; Chandler, J.; Cawthon, P.; Connor, E.B.; et al. Gait Speed and Survival in Older Adults. *JAMA* **2011**, *305*, 50–58. [CrossRef]
3. Sargent-Cox, K.; Anstey, K.J.; Luszcz, M.A. The relationship between change in self-perceptions of aging and physical functioning in older adults. *Psychol. Aging* **2012**, *27*, 750–760. [CrossRef] [PubMed]
4. Rosso, A.; Taylor, J.A.; Tabb, L.P.; Michael, Y.L. Mobility, disability, and social engagement in older adults. *J. Aging Heal.* **2013**, *25*, 617–637. [CrossRef] [PubMed]

5.  Davis, J.C.; Bryan, S.; Best, J.R.; Li, L.C.; Hsu, C.L.; Gomez, C.; Vertes, K.A.; Liu-Ambrose, T. Mobility predicts change in older adults' health-related quality of life: Evidence from a Vancouver falls prevention prospective cohort study. *Heal. Qual. Life Outcomes* **2015**, *13*, 101. [CrossRef] [PubMed]

6.  Rantakokko, M.; Mänty, M.; Rantanen, T. Mobility Decline in Old Age. *Exerc. Sport Sci. Rev.* **2013**, *41*, 19–25. [CrossRef] [PubMed]

7.  Iosa, M.; Fusco, A.; Morone, G.; Paolucci, S. Development and Decline of Upright Gait Stability. *Front. Aging Neurosci.* **2014**, *6*, 14. [CrossRef] [PubMed]

8.  Ebeling, P.R.; Cicuttini, F.; Scott, D.; Jones, G. Promoting mobility and healthy aging in men: A narrative review. *Osteoporos. Int.* **2019**, *30*, 1911–1922. [CrossRef]

9.  Prince, F.; Corriveau, H.; Hébert, R.; Winter, D.A. Gait in the elderly. *Gait Posture* **1997**, *5*, 128–135. [CrossRef]

10. Giladi, N.; Herman, T.; Reider-Groswasser, I.I.; Gurevich, T.; Hausdorff, J.M. Clinical characteristics of elderly patients with a cautious gait of unknown origin. *J. Neurol.* **2005**, *252*, 300–306. [CrossRef]

11. Cohen, J.A.; Verghese, J.; Zwerling, J.L. Cognition and gait in older people. *Maturitas* **2016**, *93*, 73–77. [CrossRef] [PubMed]

12. Beauchet, O.; Annweiler, C.; Montero-Odasso, M.; Fantino, B.; Herrmann, F.; Allali, G. Gait control: A specific subdomain of executive function? *J. Neuroeng. Rehabil.* **2012**, *9*, 12. [CrossRef] [PubMed]

13. Montero-Odasso, M.; Verghese, J.; Beauchet, O.; Hausdorff, J.M. Gait and cognition: A complementary approach to understanding brain function and the risk of falling. *J. Am. Geriatr. Soc.* **2012**, *60*, 2127–2136. [CrossRef] [PubMed]

14. Amboni, M.; Barone, P.; Hausdorff, J.M. Cognitive contributions to gait and falls: Evidence and implications. *Mov. Disord.* **2013**, *28*, 1520–1533. [CrossRef]

15. Iosa, M.; Picerno, P.; Paolucci, S.; Morone, G. Wearable Inertial Sensors for Human Movement Analysis. *Expert Rev. Med. Devices* **2016**, *13*, 641–659. [CrossRef]

16. Culhane, K.M.; O'Connor, M.; Lyons, D.; Lyons, G.M. Accelerometers in rehabilitation medicine for older adults. *Age Ageing* **2005**, *34*, 556–560. [CrossRef]

17. Maquet, D.; Lekeu, F.; Warzee, E.; Gillain, S.; Wojtasik, V.; Salmon, E.; Petermans, J.; Croisier, J.L. Gait analysis in elderly adult patients with mild cognitive impairment and patients with mild Alzheimer's disease: Simple versus dual task: A preliminary report. *Clin. Physiol. Funct. Imaging* **2010**, *30*, 51–56. [CrossRef]

18. Choi, J.-S.; Oh, H.-S.; Kang, D.-W.; Mun, K.-R.; Choi, M.-H.; Lee, S.-J.; Yang, J.-W.; Chung, S.-C.; Mun, S.-W.; Tack, G.-R. Comparison of gait and cognitive function among the elderly with Alzheimer's disease, Mild Cognitive Impairment and Healthy. *Int. J. Precis. Eng. Manuf.* **2011**, *12*, 169–173. [CrossRef]

19. Mirelman, A.; Weiss, A.; Buchman, A.S.; Bennett, D.A.; Giladi, N.; Hausdorff, J.M. Association between performance on Timed Up and Go subtasks and mild cognitive impairment: Further insights into the links between cognitive and motor function. *J. Am. Geriatr. Soc.* **2014**, *62*, 673–678. [CrossRef]

20. Grimm, B.; Bolink, S. Evaluating physical function and activity in the elderly patient using wearable motion sensors. *EFORT Open Rev.* **2016**, *1*, 112–120. [CrossRef]

21. Mc Ardle, R.; Del Din, S.; Galna, B.; Thomas, A.; Rochester, L. Differentiating dementia disease subtypes with gait analysis: Feasibility of wearable sensors? *Gait Posture* **2020**, *76*, 372–376. [CrossRef] [PubMed]

22. Brognara, L.; Palumbo, P.; Grimm, B.; Palmerini, L. Assessing Gait in Parkinson's Disease Using Wearable Motion Sensors: A Systematic Review. *Diseases* **2019**, *7*, 18. [CrossRef] [PubMed]

23. Jarchi, D.; Pope, J.; Lee, T.K.M.; Tamjidi, L.; Mirzaei, A.; Sanei, S. A Review on Accelerometry-Based Gait Analysis and Emerging Clinical Applications. *IEEE Rev. Biomed. Eng.* **2018**, *11*, 177–194. [CrossRef] [PubMed]

24. Smidt, G.; Deusinger, R.; Arora, J.; Albright, J. An automated accelerometry system for gait analysis. *J. Biomech.* **1977**, *10*, 367–375. [CrossRef]

25. Bellanca, J.; Lowry, K.A.; VanSwearingen, J.; Brach, J.S.; Redfern, M. Harmonic ratios: A quantification of step to step symmetry. *J. Biomech.* **2013**, *46*, 828–831. [CrossRef] [PubMed]

26. Lowry, K.A.; Carrel, A.J.; Kerr, J.P.; Smiley-Oyen, A.L. Walking stability using harmonic ratios in Parkinson's disease. *Mov. Disord.* **2009**, *24*, 261–267. [CrossRef]

27. Brach, J.S.; McGurl, D.; Wert, D.; VanSwearingen, J.M.; Perera, S.; Cham, R.; Studenski, S. Validation of a Measure of Smoothness of Walking. *J. Gerontol. Ser. A Boil. Sci. Med. Sci.* **2010**, *66*, 136–141. [CrossRef]

28. Iosa, M.; Bini, F.; Marinozzi, F.; Fusco, A.; Morone, G.; Koch, G.; Cinnera, A.M.; Bonnì, S.; Paolucci, S. Stability and Harmony of Gait in Patients with Subacute Stroke. *J. Med. Boil. Eng.* **2016**, *36*, 635–643. [CrossRef]

29. Pau, M.; Mandaresu, S.; Pilloni, G.; Porta, M.; Coghe, G.; Marrosu, M.G.; Cocco, E. Smoothness of gait detects early alterations of walking in persons with multiple sclerosis without disability. *Gait Posture* **2017**, *58*, 307–309. [CrossRef]

30. Yack, H.J.; Berger, R.C. Dynamic Stability in the Elderly: Identifying a Possible Measure. *J. Gerontol.* **1993**, *48*. [CrossRef]

31. Menz, H.; Lord, S.R.; Fitzpatrick, R.C. Acceleration patterns of the head and pelvis when walking on level and irregular surfaces. *Gait Posture* **2003**, *18*, 35–46. [CrossRef]

32. Byun, S.; Han, J.W.; Kim, T.H.; Kim, K.; Kim, T.H.; Park, J.Y.; Suh, S.W.; Seo, J.Y.; So, Y.; Lee, K.H.; et al. Gait Variability Can Predict the Risk of Cognitive Decline in Cognitively Normal Older People. *Dement. Geriatr. Cogn. Disord.* **2018**, *45*, 251–261. [CrossRef] [PubMed]

33. Lamoth, C.J.C.; Van Deudekom, F.J.; Van Campen, J.P.; Appels, B.A.; De Vries, O.J.; Pijnappels, M. Gait stability and variability measures show effects of impaired cognition and dual tasking in frail people. *J. Neuroeng. Rehabil.* **2011**, *8*, 2. [CrossRef] [PubMed]

34. Doi, T.; Shimada, H.; Makizako, H.; Tsutsumimoto, K.; Hotta, R.; Nakakubo, S.; Suzuki, T. Effects of white matter lesions on trunk stability during dual-task walking among older adults with mild cognitive impairment. *AGE* **2015**, *37*, 120. [CrossRef]

35. Ijmker, T.; Lamoth, C.J.C. Gait and cognition: The relationship between gait stability and variability with executive function in persons with and without dementia. *Gait Posture* **2012**, *35*, 126–130. [CrossRef]

36. Folstein, M.F.; Folstein, S.E.; McHugh, P.R. Mini-mental state. *J. Psychiatry Res.* **1975**, *12*, 189–198. [CrossRef]

37. Magni, E.; Binetti, G.; Bianchetti, A.; Rozzini, R.; Trabucchi, M. Mini-Mental State Examination: A normative study in Italian elderly population. *Eur. J. Neurol.* **1996**, *3*, 198–202. [CrossRef]

38. Mioshi, E.; Dawson, K.; Mitchell, J.; Arnold, R.; Hodges, J.R. The Addenbrooke's Cognitive Examination Revised (ACE-R): A brief cognitive test battery for dementia screening. *Int. J. Geriatr. Psychiatry* **2006**, *21*, 1078–1085. [CrossRef]

39. Pigliautile, M.; Ricci, M.; Mioshi, E.; Ercolani, S.; Mangialasche, F.; Monastero, R.; Croce, M.; Federici, S.; Mecocci, P. Validation Study of the Italian Addenbrooke's Cognitive Examination Revised in a Young-Old and Old-Old Population. *Dement. Geriatr. Cogn. Disord.* **2011**, *32*, 301–307. [CrossRef]

40. Isella, V.; Villa, M.; Frattola, L.; Appollonio, I. Screening cognitive decline in dementia: Preliminary data on the Italian version of the IQCODE. *Neurol. Sci.* **2002**, *23*, s79–s80. [CrossRef]

41. Viscione, I.; D'Elia, F.; Vastola, R.; Sibilio, M. The Correlation between Technologies and Rating Scales in Gait Analysis. *J. Sports Sci.* **2016**, *4*, 119–123. [CrossRef]

42. Pau, M.; Porta, M.; Pilloni, G.; Corona, F.; Fastame, M.C.; Hitchcott, P.K.; Penna, M.L.F. Texting While Walking Induces Gait Pattern Alterations in Healthy Older Adults. *Proc. Hum. Factors Ergon. Soc. Annu. Meet.* **2018**, *62*, 1908–1912. [CrossRef]

43. Pasciuto, I.; Bergamini, E.; Iosa, M.; Vannozzi, G.; Cappozzo, A. Overcoming the limitations of the Harmonic Ratio for the reliable assessment of gait symmetry. *J. Biomech.* **2017**, *53*, 84–89. [CrossRef] [PubMed]

44. Zijlstra, W. Assessment of spatio-temporal parameters during unconstrained walking. *Graefe's Arch. Clin. Exp. Ophthalmol.* **2004**, *92*, 39–44. [CrossRef]

45. Menz, H.; Lord, S.R.; Fitzpatrick, R.C. Acceleration patterns of the head and pelvis when walking are associated with risk of falling in community-dwelling older people. *J. Gerontol. Ser. A Boil. Sci. Med. Sci.* **2003**, *58*, M446–M452. [CrossRef]

46. Lowry, K.A.; Lokenvitz, N.; Smiley-Oyen, A. Age- and speed-related differences in harmonic ratios during walking. *Gait Posture* **2012**, *35*, 272–276. [CrossRef]

47. Lazzarini, B.S.R.; Kataras, T.J. Treadmill walking is not equivalent to overground walking for the study of walking smoothness and rhythmicity in older adults. *Gait Posture* **2016**, *46*, 42–46. [CrossRef]

48. Asai, T.; Misu, S.; Sawa, R.; Doi, T.; Yamada, M. The association between fear of falling and smoothness of lower trunk oscillation in gait varies according to gait speed in community-dwelling older adults. *J. Neuroeng. Rehabil.* **2017**, *14*, 5. [CrossRef]

49. Cohen, J. Statistical Power Analysis. *Curr. Dir. Psychol. Sci.* **1992**, *1*, 98–101. [CrossRef]

50. Peel, N.M.; Alapatt, L.J.; Jones, L.V.; Hubbard, R.E. The Association between Gait Speed and Cognitive Status in Community-Dwelling Older People: A Systematic Review and Meta-analysis. *J. Gerontol. Ser. A Boil. Sci. Med. Sci.* **2018**, *74*, 943–948. [CrossRef]

51. Salzman, B.E. Gait and balance disorders in older adults. *Am. Fam. Physician* **2010**, *82*, 61–68. [PubMed]

52. Allan, L.M.; Ballard, C.; Burn, D.; Kenny, R.A. Prevalence and Severity of Gait Disorders in Alzheimer's and Non-Alzheimer's Dementias. *J. Am. Geriatr. Soc.* **2005**, *53*, 1681–1687. [CrossRef] [PubMed]

53. Beauchet, O.; Allali, G.; Berrut, G.; Hommet, C.; Dubost, V.; Assal, F. Gait analysis in demented subjects: Interests and perspectives. *Neuropsychiatr. Dis. Treat.* **2008**, *4*, 155–160. [CrossRef] [PubMed]

54. Eggermont, L.H.; Gavett, B.E.; Volkers, K.M.; Blankevoort, C.G.; Scherder, E.J.; Jefferson, A.L.; Steinberg, E.; Nair, A.; Green, R.C.; Stern, R. Lower-Extremity Function in Cognitively Healthy Aging, Mild Cognitive Impairment, and Alzheimer's Disease. *Arch. Phys. Med. Rehabil.* **2010**, *91*, 584–588. [CrossRef]

55. Shaw, F.E.; Bond, J.; Richardson, D.A.; Dawson, P.; Steen, I.N.; McKeith, I.G.; Kenny, R.A. Multifactorial intervention after a fall in older people with cognitive impairment and dementia presenting to the accident and emergency department: Randomised controlled trial. *BMJ* **2003**, *326*, 73. [CrossRef]

56. Zhang, W.; Low, L.-F.; Schwenk, M.; Mills, N.; Gwynn, J.D.; Clemson, L. Review of Gait, Cognition, and Fall Risks with Implications for Fall Prevention in Older Adults with Dementia. *Dement. Geriatr. Cogn. Disord.* **2019**, *48*, 17–29. [CrossRef]

57. Kikkert, L.H.J.; Vuillerme, N.; Van Campen, J.P.; Hortobágyi, T.; Lamoth, C.J.C. Walking ability to predict future cognitive decline in old adults: A scoping review. *Ageing Res. Rev.* **2016**, *27*, 1–14. [CrossRef]

58. Morris, R.; Lord, S.; Bunce, J.; Burn, D.; Rochester, L. Gait and cognition: Mapping the global and discrete relationships in ageing and neurodegenerative disease. *Neurosci. Biobehav. Rev.* **2016**, *64*, 326–345. [CrossRef]

59. Demnitz, N.; Esser, P.; Dawes, H.; Valkanova, V.; Johansen-Berg, H.; Ebmeier, K.P.; Sexton, C. A systematic review and meta-analysis of cross-sectional studies examining the relationship between mobility and cognition in healthy older adults. *Gait Posture* **2016**, *50*, 164–174. [CrossRef]

60. Grande, G.; Triolo, F.; Nuara, A.; Welmer, A.-K.; Fratiglioni, L.; Vetrano, D.L. Measuring gait speed to better identify prodromal dementia. *Exp. Gerontol.* **2019**, *124*, 110625. [CrossRef] [PubMed]

61. Zaninotto, P.; Sacker, A.; Head, J. Relationship between Wealth and Age Trajectories of Walking Speed among Older Adults: Evidence from the English Longitudinal Study of Ageing. *J. Gerontol. Ser. A Boil. Sci. Med. Sci.* **2013**, *68*, 1525–1531. [CrossRef] [PubMed]

62. Busch, T.; Duarte, Y.A.D.O.; Nunes, D.P.; Lebrão, M.L.; Naslavsky, M.S.; Rodrigues, A.D.S.; Amaro, E. Factors associated with lower gait speed among the elderly living in a developing country: A cross-sectional population-based study. *BMC Geriatr.* **2015**, *15*, 35. [CrossRef] [PubMed]

63. Weber, D. Differences in physical aging measured by walking speed: Evidence from the English Longitudinal Study of Ageing. *BMC Geriatr.* **2016**, *16*, 2326. [CrossRef] [PubMed]

64. Cimolin, V.; Cau, N.; Sartorio, A.; Capodaglio, P.; Galli, M.; Tringali, G.; Leban, B.; Porta, M.; Pau, M. Symmetry of Gait in Underweight, Normal and Overweight Children and Adolescents. *Sensors* **2019**, *19*, 2054. [CrossRef] [PubMed]

*Systematic Review*

# A Systematic Review of Diagnostic Accuracy and Clinical Applications of Wearable Movement Sensors for Knee Joint Rehabilitation

Robert Prill [1,*], Marina Walter [2], Aleksandra Królikowska [3] and Roland Becker [1]

[1] Center of Orthopaedics and Traumatology, Brandenburg Medical School, University Hospital Brandenburg/Havel, 14770 Brandenburg an der Havel, Germany; roland.becker@mhb-fontane.de
[2] Hasso-Plattner-Institut, University of Potsdam, 14469 Potsdam, Germany; mail@marinawalter.de
[3] Ergonomics and Biomedical Monitoring Laboratory, Department of Physiotherapy, Faculty of Health Sciences, Wroclaw Medical University, 50-367 Wrocław, Poland; aleksandra.krolikowska@umw.edu.pl
* Correspondence: robert.prill@mhb-fontane.de

**Abstract:** In clinical practice, only a few reliable measurement instruments are available for monitoring knee joint rehabilitation. Advances to replace motion capturing with sensor data measurement have been made in the last years. Thus, a systematic review of the literature was performed, focusing on the implementation, diagnostic accuracy, and facilitators and barriers of integrating wearable sensor technology in clinical practices based on a Preferred Reporting Items for Systematic Reviews and Meta-Analyses (PRISMA) statement. For critical appraisal, the COSMIN Risk of Bias tool for reliability and measurement of error was used. PUBMED, Prospero, Cochrane database, and EMBASE were searched for eligible studies. Six studies reporting reliability aspects in using wearable sensor technology at any point after knee surgery in humans were included. All studies reported excellent results with high reliability coefficients, high limits of agreement, or a few detectable errors. They used different or partly inappropriate methods for estimating reliability or missed reporting essential information. Therefore, a moderate risk of bias must be considered. Further quality criterion studies in clinical settings are needed to synthesize the evidence for providing transparent recommendations for the clinical use of wearable movement sensors in knee joint rehabilitation.

**Keywords:** wearable movement sensor; IMU; motion capture; reliability; clinical; orthopedic

**Citation:** Prill, R.; Walter, M.; Królikowska, A.; Becker, R. A Systematic Review of Diagnostic Accuracy and Clinical Applications of Wearable Movement Sensors for Knee Joint Rehabilitation. *Sensors* **2021**, *21*, 8221. https://doi.org/10.3390/s21248221

Academic Editors: Paolo Capodaglio and Veronica Cimolin

Received: 28 October 2021
Accepted: 7 December 2021
Published: 9 December 2021

**Publisher's Note:** MDPI stays neutral with regard to jurisdictional claims in published maps and institutional affiliations.

## 1. Introduction

Knee joint problems are widespread and may occur throughout a patient's lifespan. Given the high incidence across the age continuum and the frequent need for surgical repair and long-term rehabilitation, knee injuries present one of the highest clinical and public health injury-related burdens [1,2]. Ligament damage to the knee, including the most frequently injured anterior cruciate ligament (ACL), is more common than any other type of knee injury pathology [3,4]. Additionally, knee osteoarthritis (KOA), with its global prevalence, amounts to almost 23% in individuals aged 40 and over [3], and accounts for nearly four-fifths of OA burden worldwide [5]. The incidence of KOA is 203 per 100,000 person-years in individuals aged 20 and over, and it increases with age to peak at 70–79 years old [6]. Although end-stage KOA can be effectively treated with total knee arthroplasty (TKA), the procedure is related to substantial health costs [7,8].

Patients with knee disorders of different natures require a dedicated follow-up involving physicians, nurses, physical therapists, and other medical staff. Therefore, the healthcare sector is facing challenges regarding the rapidly growing elderly population, rising cost pressure, and limited temporal resources of medical staff. New postoperative protocols are well established and have significantly reduced the time of hospitalization.

Cost explosion has induced an increasingly shorter inpatient care of surgical patients, which often induces restrictions in rehabilitation and follow-up quality.

Sensing technology is widely used in orthopedics nowadays. Most commonly it is established in intraoperative care and basic science on human movement [9–12]. Since wearable technology nowadays possesses the capacity for monitoring and diagnostic functionality, this technology might help solve some of the challenges the healthcare sector faces. Current research has indicated that wearable sensing technology can benefit patients' care. This device helps physiotherapists and orthopedic surgeons detect movement pattern problems, such as asymmetrical limb loading after anterior cruciate ligament reconstruction (ACLR) [13,14] and quantification of varus thrust in patients with KOA [15]. For patients with total knee replacement, some parameters were used to describe the progress of certain selected parameters relevant for rehabilitation, even if not evaluated for this setting. General gait analysis [16], stance and swing phase development [17], range of motion [18], and knee instability before [19] and after [20] were evaluated before and after total knee arthroplasty (TKA).

So far, no wearable sensing technology system has been successfully incorporated into everyday clinical practice or in a hospital or rehabilitative setting. The feasibility of clinical implementation and the possibility of reimbursement by health insurance companies largely depend on usability, cost-effectiveness, availability, and, most important, diagnostic accuracy. To account for this current gap in knowledge, reviews that focus on these aspects would be helpful.

To date, reviews that have tackled the topic of sensor technology in the medical field have investigated the issue from a broader view. A review from 2012 by Patel et al. focused on wearable sensors and systems with applications in rehabilitation [21]. This review provided an overview of different sensing technologies, such as built-in smartphone sensors, ambient home sensory sensors, fabric electrodes, and various types of wearable devices, to measure blood glucose levels, respiratory rate, ECG, etc. Additionally, potential use cases of telemonitoring in the aging population were discussed. Sensing technology and biomedical markers are commonly used nowadays in various fields of medicine, such as stroke rehabilitation [22] or ankle joint power [23], and rehabilitation issues, such as hand-finger orientation, have already been considered [24].

In 2018, Porciuncula et al. provided what they called a "focused discussion" about current sensor technologies and their clinical applications [25]. They did not provide a comprehensive systematic review but provided an overview of clinical applications used in patients with neurological and musculoskeletal diagnoses, which could potentially benefit from wearable sensors during their rehabilitation. They included different sorts of sensors, such as phone-based sensors or those included in shoes or wristbands for activity recognition, identification of pathologic motor features, falls management, and other clinical applications. The most recent scoping review from 2019 provided by Small assessed the current methodology and clinical application of accelerometers and inertial measurement units (IMUs) to evaluate a patient's activity and functional recovery after knee arthroplasty [26].

The reviews mentioned above provide a broad scope of the topic. However, apart from the review by Small et al. (2019), the issues of patients with knee pathologies have only been covered to a limited extent. Therefore, the current review focuses on diagnostic accuracy and the different approaches of wearable sensing technology used for monitoring knee and lower limb motion in clinical practice.

**Highlights:**

- Promising IMU quality criterion data exist for describing knee joint status
- No wearable sensing technology assessing knee joint rehabilitation issues has been incorporated successfully into clinical practice
- No consensus about added value from IMUs and quality criterion parameter statistics to be reported

- IMUs are currently used to raise the efficiency of established tests but have high potentials for new parameters with higher validity for function

## 2. Materials and Methods

A systematic review was conducted using Preferred Items for Systematic Review and Meta-Analysis (PRISMA) and accordance with recently published author guidelines for Systematic Reviews and Meta-Analyses [27]. The protocol was preregistered at the open science framework: 10.17605/OSF.IO/DQEAX. To be included in this review, papers must report on the use of at least one IMU for assessing knee joint kinematics, knee stability, or gait analysis. Optimally, studies include validation against a gold standard. Included studies were conducted either in a hospital, ambulatory, or gait laboratory setting. The study population underwent either TKA or ACLR as some of the most commonly performed knee surgeries. Wearable sensing technology has become smaller, more efficient, less obtrusive, and increasingly affordable due to advanced technology. This also leads to an increased number of scientific studies in the field in the last months and years. Nevertheless, to capture all potentially relevant research for this very specific systematic research, PUBMED, Prospero, Cochrane database, and EMBASE were screened for papers from 1980 to 13 March 2021. For identification of relevant studies in the English language, a literature search with the keywords "knee" AND "sensors" OR "IMU" OR "inertial measurement unit" in those electronic databases was conducted.

Due to various methodologies among different journals, a comprehensible guideline for inclusion or exclusion criteria was required, as provided in Table 1. Review articles were excluded but examined for potentially relevant research articles. Exclusion criteria included the use of intraoperative sensor technology to enhance surgical outcomes, app-based intervention, and telerehabilitation studies that did not use wearable sensor technology.

**Table 1.** Inclusion and exclusion criteria.

| Inclusion Criteria | Exclusion Criteria |
| --- | --- |
| Studies including patients with knee osteoarthritis, total knee arthroplasty, or anterior cruciate ligament reconstruction | Studies including intraoperative sensors for enhancing surgical outcomes, such as using pressure sensors for total knee replacement |
| Studies including patients investigated with at least one IMU | Studies that perform postoperative digital interventions or telerehabilitation without using wearable sensing technology |
| Studies including body-mounted sensors | Cadaveric studies |
| Some form of quality measurement of the data needs to be provided | Studies including patients with neurological or rheumatic diseases that impaired balance or ability to walk |
| | Study protocols |

Two independent reviewers screened the manuscript titles and abstracts. Exclusion and inclusion criteria, as presented in Table 1, were discussed among reviewers before the title and abstract screening. After searching and title screening the online database resources, duplicates were removed. For the manuscripts that both reviewers included, a full-text search was performed to decide upon inclusion for the review. Exclusion and inclusion criteria were discussed among reviewers before the title and abstract screening. The full-text screening was performed accordingly.

Following the relevant items of the STARD for reliability checklist, data from the included papers were summarized in a data extraction spreadsheet independently by both reviewers. Disagreements were solved via discussion. Data extraction was grouped by patients' demographics, type of sensing technology, outcome variables, and diagnostic accuracy criteria. An overview of the different testing protocols was included. A COSMIN Risk of Bias tool was used to examine the quality in a systematic and transparent manner [28]. No ethical approval was required since only existing peer-reviewed literature sources were accepted for evaluation. No data registration plan was needed.

## 3. Results

The initial database research with the previously defined search string yielded 2368 results. After the title and abstract search, 84 manuscripts remained and underwent full-text assessment, of which 78 were excluded according to the criteria specified in Table 1. Therefore, six manuscripts remained for inclusion in the qualitative synthesis. Figure 1 shows a PRISMA flow diagram detailing the results of the literature search and review.

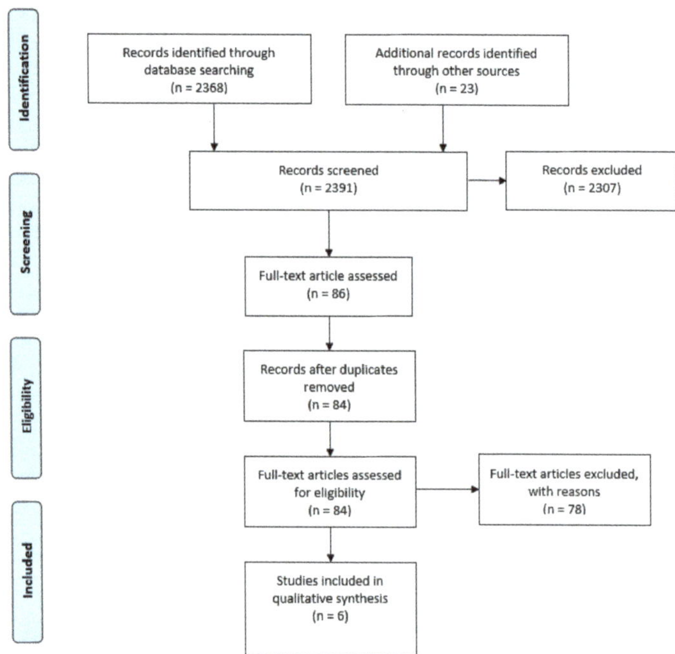


**Figure 1.** PRISMA flow diagram detailing the results of the literature search and review.

Promising study protocols that assess the practical clinical usability of sensing technology have been registered in the last two years. Still, since no results have been published yet, they were excluded from this review. Comparability of studies was limited since various methodological approaches existed. Due to a lack of standardization and an abundance of proprietary solutions, the studies differed regarding sensing technology, dedicated analysis software, sensor placement, testing protocols, and measured outcome variables. Three studies investigated patients who received TKA surgery, and the other three focused on patients after ACLR. The most widely used reference system was the optoelectronic motion capturing system, often not reported in detail, and sometimes complemented with additional force plates.

Different outcome variables were used for patient evaluation after TKA. Temporospatial parameters of gait were measured (cycle time, stance time, and swing time) by De Vroey et al. [29] and knee flexion angles by Roberts et al. [20]. For leg swings, joint instability acceleration-based parameters were measured by Huang et al. [30]. Outcome measures for the ACLR population included gait analysis in one study [14] and knee loading asymmetries with a single limb loading (SLL) task in two other studies [13,31]. Table 2 presents the baseline characteristics of the included studies.

**Table 2.** Baseline characteristics.

| | TKA ($n$) | TKA % Female | ACLR ($n$) | ACLR % Female | Healthy Controls | Healthy Contr. % FEM. | 2–3 Months *p.* Surgery | 4–6 Months *p.* Surgery | >6 Months *p.* Surgery | Gait Analysis | SLL | Other | Mocap | Force Plates | Other |
|---|---|---|---|---|---|---|---|---|---|---|---|---|---|---|---|
| De Vroey | 16 | 50 | | | | | | | √ | √ | | | √ | | |
| Huang | 8 | 75 | | | 16 | 50 | | | √ | | | √ | | | √ |
| Pratt a | | | 21 | 57 | | | | √ | | | √ | | √ | √ | |
| Pratt b | | | 21 | 57 | | | | √ | | | √ | | | | |
| Roberts | 27 | 59 | | | 18 | 61 | | | √ | | | √ | | | |
| Sigward | | | 19 | 74 | | | √ | | | √ | | | √ | √ | |

ACLR, anterior cruciate ligament reconstruction; $n$, number of individuals in a given sample; TKA, total knee arthroplasty.

De Vroey et al. [29] used wearable sensing technology to analyze the temporal parameters of gait in a TKA population. The objective was to investigate the agreement between an IMU and a camera-based motion capturing system. Sixteen patients included one year after TKA were asked to perform three gait trials with a self-selected speed along a six-meter walkway. The sensors were placed at the anteromedial facet of the tibia at the left and right lower leg, approximately 5 cm below the knee joint line. Inertial measurement sensor data and optoelectronic camera motion data were collected simultaneously during the gait trials. Custom-made software was used to identify gait events from the gyroscope data. From these data, cycle time, stance time, and swing time were derived. The kinematic data from the camera system were analyzed based on a coordinate-based algorithm. Both sets of temporal variables were compared by calculating intra-class correlation coefficients (ICCs), mean errors, and root mean squared errors. De Vroey et al. found very good to excellent ICC values (0.826–0.972) between the sensor-based and optoelectric motion-based method. The root mean squared errors between both methods ranged from 0.036 to 0.055. Overall, all observed variables showed high levels of agreement. The findings of De Vroey et al. indicated that IMUs can be used in clinical settings to assess temporal gait parameters in the knee arthroplasty population. However, no studies have been published so far proving the usage of the sensors in daily clinical practice.

In a study on monitoring knee flexion angles for rehabilitation purposes in a total knee replacement population, Huang et al. used wearable sensing technology. They compared the measured range of motion between inertial measurement sensors and the Cybex® isokinetic dynamometer (Cybex NORM; Lumex, Inc., Ronkonkoma, NY, USA). The sensor comprised an ATMEGA328 microcontroller, a MPU6050 triaxial accelerometer and gyroscope module, an Arduino Bluetooth module, a lithium battery (9 V, 650 mAh), and a smartphone. The smartphone was used to receive signals transmitted by the Bluetooth module from the accelerometer and gyroscope. The two sensor devices were worn on the thigh and ankle. Thirty-five subjects were enrolled in the experiments, comprising 16 healthy controls and eight patients post total knee replacement. The testing protocol of Huang et al. comprised three indices used as metrics to measure knee rehabilitation progress: number of swings, maximum knee flexion angle, and duration of practice each time. Each subject wore one sensor device on the right shank, and angular speeds of 25, 60, and 180°/s were used, while the swing phase was driven by the Cybex®. The system's accuracy was calculated based on the difference between the detected angle of the sensors and the ROM of Cybex. Huang et al. found that the correlation coefficients between the two measurements at the three angular speeds mentioned above were 0.975, 0.969, and

0.967, respectively. The results indicated high consistency between the sensor-based system and the Cybex reference standard. Correlation coefficients for the TKA subjects, under the same measurement conditions, were calculated to be 0.993, 0.982, and 0.986, again based on three different angular speeds of 25, 60, and 180°/s. Again, this implies a high correlation between the sensor-based system and Cybex. They also found that the average absolute swing errors for the TKA patients were between 1.65° and 3.27°, resulting in accuracies between 96.16% and 98.09%, depending on angular speeds, while accuracies decreased with higher angular speeds of Cybex. Huang et al. concluded that inertial measurement sensors are comparable with professional equipment and, therefore, can be deployed in a clinical setting [30].

Roberts et al. attached a single IMU at the level of the tibial tubercle in patients after TKA and healthy controls. They measured the linear acceleration of the knee joint during several activities of daily living. A direct tibia-mounted accelerometer was compared with a rubber skin-mounted accelerometer in a cadaveric study to ensure skin-mounted devices accuracy. Bland-Altman analysis of acceleration profiles indicated limits of agreement of −0.600 to 1.252 between the two methods. The healthy controls and the TKA cohort were analyzed for statistically significant differences regarding their general activity level, pain for each activity, and instability for each activity. They developed a testing protocol that included five activities of daily living, which were then evaluated with the IMUs and compared against self-reported instability levels. Controls and patients with TKA were found to be comparable regarding general activity scores. Twenty-four out of 38 patients with TKA reported instability during the exercises, with instability depending significantly on the activity performed ($p = 0.015$). Stepping up and down was the most prone to experiencing instability. Furthermore, this was the only activity in which any patient reported severe instability. None of the parameters concerning pain or instability were clinically relevant. Parameters in the y-plane seem most promising, showing extremes in movement [20].

Pratt et al. used wearable sensing technology following ACLR to detect knee power deficits. Their objective was to determine the diagnostic accuracy of inertial sensor thigh angular velocities to detect asymmetrical knee loading. Pratt et al. used two inertial sensors equipped with triaxial accelerometers, gyroscopes, and magnetometers (manufactured by Opal brand, APDM Inc., Portland, OR, USA). The sensors were placed bilaterally on the mid-lateral thighs. Twenty-one individuals following ACLR performed three trials of SLL tasks on each leg while being recorded with a wearable sensor system. Concurrently, the subjects were monitored using an optoelectronic motion capturing system with additional force plates. Pratt et al. calculated between limb ratios for knee power in ACL-reconstructed and contralateral legs based on motion-capturing data. Furthermore, thigh angular velocity was extracted from the inertial sensors, and their ratio was used to diagnose asymmetrical knee loading with receiver operating characteristic curve (ROC) analysis. Asymmetrical knee loading was defined as knee power deficits exceeding 15%. Thigh angular velocity symmetry ratio was discriminated between asymmetrical and symmetrical knee power with high specificity (100%) and sensitivity (81.2%). The study's findings underlined the feasibility of thigh angular velocities extracted from inertial sensors for clinical detection of knee power asymmetries in individuals following ACLR, allowing for clinical quantification of dynamic knee loading deficits [13]. Furthermore, the authors aimed to prove that knee loading deficits can be identified more easily and with less clinical expenditure using inertial sensor technology. They tried to deduce information about knee moment/knee power (KMom/KPow) during dynamic tasks based on angular velocity measurements with inertial sensors in a cohort of post-ACLR patients. ICCs exceeded 0.947 ($p < 0.001$) for all variables [31].

Sigward et al. explored knee loading asymmetries in individuals after unilateral ACLR using sensor technology too. The authors analyzed the relationship between shank angular velocity and knee extensor moment during a gait trial using an IMU, while validating against a motion-capturing system with force plates. Sigward et al. used two calibrated and synchronized inertial sensors equipped with tri-axial accelerometers, gyroscopes, and magnetometers manufactured by Mobility Lab software, APDM Inc., Portland, Oregon, USA. The inertial sensors were placed bilaterally on the lateral shanks. If the IMU position coincided with that of the MOCAP tracking marker cluster, the IMUs were fixed firmly on top with adhesive tape. Nineteen individuals were instructed to walk 10 m at a self-selected speed. Three trials for each limb were collected. The symmetry between the limbs was calculated using the ratio of peak knee extensor moments of the surgical knee relative to the non-surgical knee. Three trials were averaged for analysis. Sigward et al. found no differences between the limbs regarding stance ($p = 0.132$) and swing ($p = 0.840$) times. However, the peak knee extensor moment and peak shank angular velocity in the ACL-reconstructed knee markedly exceeded those of the contralateral knee ($p < 0.001$). The authors found a strong positive correlation between knee extensor moment and shank angular velocity. Shank angular velocities measured by wearable IMUs can therefore be used to calculate knee extensor moments, while the in-between limb ratios were identified as indicators of knee extensor moment deficits. These findings make wearable IMUs feasible for detecting gait impairment after ACLR. It was concluded that spatiotemporal gait parameters, such as stance and swing time, in an ACLR population normalize sooner than knee loading deficits. What is more, the study indicates that observation of gait deviations by clinicians may not be sufficient to detect rehabilitation progress in subjects following ACLR. Wearable IMUs can account for this gap in rehabilitation progress detection [14]. The extracted data are presented in Table 3.

**Table 3.** Data extraction, sensor information, and results.

|  | **Sensor Information and Application** | **Knee-Joint Measurement Method** | **Results** |
|---|---|---|---|
| De Vroey (2018) | Gyroscope data: Three gait trials 6 m walk; TKA patients | Shank worn IMUs | ICC = 0.826–0.972 RMSE = 0.036–0.055 |
| Huang (2020) | Three axial accelerometer and gyroscope data: Number of swings, ROM knee flex, duration, TKA patients, and controls | MPU6050, ATMEGA328  Cybex | Measurement error = 1.65°–3.27° |
| Pratt (2018a) | Shank gyroscope, maker-based motion and force plate data: Sagittal plane peak knee power absorption, ACLR patients | Opal APDM, Qualisis AB, AMTI | 81%, Specificity 100% for asymmetrical knee loading |
| Pratt (2018b) | Shank gyroscope, knee moments, knee power (angular velocity): single limb loading tasks, ACLR patients | OPAL APDM, Qualisis | ICCs (>0.947); r = 0.81 for thigh and r = 0.54 for knee velocity |
| Roberts (2013) | Tibial tuberositas IMU; joint acceleration, Jerk: Joint stability, 5 activities on one leg and the other, TKA patients and controls | Motion Nod, gyroscope | Differences ($p > 0.05$) in 22 IMU parameters between patients and controls |
| Sigwards (2016) | Shank angular velocity and knee extensors movement during gait | Opal APDM gyroscope, Qualisis, AMTI | Peak velocity and knee extensor movement correlate with r = 0.75 |

ICC = intraclass correlation, RMSE = root mean square errors, ROM = range of motion.

The risk of bias assessment using the COSMIN Risk of Bias tool is presented in Table 4, showing, on average, a moderate risk of bias for included studies.

**Table 4.** Risk of bias assessment (consensus results).

| | Design Requirements | Stability of the Patients | Time Interval | Similarity of Measurement | Administration without Knowledge of Scores | Score Assignment or Determination of Values | Other important Flaws | Statistical Methods | For continuous Data ICC | For Ordinal: Kappa | For Nominal: Kappa for each Category | Final Rating |
|---|---|---|---|---|---|---|---|---|---|---|---|---|
| De Vroey | | √ | NA | √ | √ | √ | - | | √ | - | - | A |
| Huang | | √ | NA | √ | √ | √ | - | | √ | - | - | A |
| Pratt a | | √ | NA | (√) | √ | √ | - | | NA | | | DF |
| Pratt b | | √ | NA | √ | √ | √ | - | | √ | | | A |
| Roberts | | √ | NA | (√) | √ | √ | - | | NA | | | DF |
| Sigward | | √ | NA | √ | √ | √ | - | | NA | | | DF |

NA = not available or wrong, (√) = correct, but unclear, A = adequate, DF = doubtful.

## 4. Discussion

All the analyzed studies used commercially available sensor technology, apart from Huang et al., who developed a proprietary solution based on Arduino technology [30]. For data analysis, commercial software was complemented with proprietary solutions, often based on MATLAB, for data analysis purposes. Findings from the studies indicate that IMU usage in rehabilitating the knee surgery population provides reliable data compared to the motion capturing gold standard. Due to various study designs and the resulting methodological differences, a synthesis of evidence is not possible. In Table 5, additionally a summary of sensor issues is provided.

**Table 5.** Sensor summary.

| | Number of Wearable Sensors | Accelerometer | Gyroscope | Magnetometer | Additional Force Platform | Not Reported | 100–200 Hz | 50–100 Hz | Commercial Software | Proprietary Solution | Not Described | Leg | Hip | Not Described | Commercial Sensor | Proprietary Sensor |
|---|---|---|---|---|---|---|---|---|---|---|---|---|---|---|---|---|
| De Vroey | 2 | | | | | | | √ | | √ | | | √ | | √ | |
| Huang | 2 | √ | √ | | | √ | | | | | √ | √ | | | | √ |
| Pratt a | 2 | √ | √ | √ | √ | | | √ | √ | √ | | | √ | | √ | |
| Pratt b | 4 | √ | √ | √ | | √ | | | √ | √ | | | | √ | √ | |
| Roberts | 1 | √ | √ | √ | | | | √ | | √ | | √ | | | √ | |
| Sigward | 2 | √ | √ | √ | √ | | √ | | √ | √ | | √ | | | √ | |

Usability: Experience from De Vroey et al. showed that using IMUs drastically reduces the time needed for data collection and processing. Placement of motion capture (MOCAP) markers took them, on average, 20 min, while sensor placement took 3 min only. Data processing from MOCAP markers took, on average, 40 min per subject and trial. Concurrently, computing gait events from IMU data (with the proposed algorithm) required another 10 min per subject and trial [29]. Huang et al. found that IMUs were well usable since they can be worn without spatiotemporal constraints; they can reduce the frequency of patients needing to return to the hospital for inpatient services and thus save medical expenses. Furthermore, they provide accuracy in monitoring the rehabilitation progress. The sensor devices presented in the study can be easily worn on the thigh and ankle with Velcro and an elastic band, and the number of swings and ROM from each rehabilitation course can be recorded and tracked by users or potentially shared with other medical staff [30]. Roberts et al. underlined the advantages of IMU portability and ease of adaptation to space limitations inherent in clinical follow-up visits after TKA surgery. Furthermore, IMUs are less expensive than other diagnostic tools, such as gait analysis systems and fluoroscopy [20]. According to Pratt et al., IMU utilization should be limited. Although they are less expensive than gold-standard motion capturing systems, they still require a computer and expertise to operate and analyze the data accordingly, leaving the need to develop clinician-friendly technology, especially for placement and calibration [13].

Resolving shortcomings in current rehabilitation practice with IMUs: IMUs offer the potential to extend the existing range of rehabilitation measurements. Knee joint instability after TKA is one of the leading causes of further surgical intervention. Quantification of knee joint instability still lacks objectively quantifiable parameters and is evaluated instead through patient history and physical examination [32]. Roberts et al. identified activities of daily life that can help quantify self-perceived instability in the TKR population using a single tibial-worn IMU and supported Khan et al. [33]. Furthermore, according to a systematic review by Barber-Westin et al., in patients after ACLR, the timing of return to unrestricted sports activities still lacks objective assessment [34]. General recommendations are based on the quantification of muscle strength, stability, neuromuscular control, and general function. Furthermore, there is evidence that, in individuals following ACLR, unilateral deficits may be masked during double-limb performance activities and therefore be overseen in conventional clinical assessments. Isolation of the involved limb with unilateral tasks, such as hopping, should be used to identify deficits in performance [35,36]. Unilateral limb monitoring is hard to accomplish in a clinical context without using MOCAP technology. A possible solution is presented with the IMU-assisted detection of knee loading impairment proposed by Sigward et al. The surgical limb can be separately monitored and, therefore, may offer a new criterion for returning to sporting activities in the ACLR population.

Influencing factors and confounders: Joint angle calculation based on inertial measurement data for human motion analysis remains challenging. In IMU-based human motion analysis, the common problem is that the IMU's local coordinate axes are not aligned with any physiologically meaningful axis. Within the scope of this review, the decision about the optimal sensor set and sensor position remains unclear [37]. Data from the review showed that sensor placement between the studies varied significantly. Some gave detailed instructions for placement regarding specific anatomical landmarks, while others just vaguely mentioned the body part to which the sensor was attached to. Sometimes additional pictures clarify the sensor placement, but reproducibility is not necessarily provided. Previous studies have shown that the placement of sensors is critical for detecting temporal gait events [37,38]. DeVroey et al. mounted the sensor on the anteromedial surface of the tibia to reduce the chance of soft-tissue artifacts. Compared to other studies, where sensors were mounted to the foot or waist, three authors reported it beneficial for gait event detection when the sensor is placed on the shank since gyroscopic data from shank-worn IMUs show a very distinctive pattern. However, single IMUs attached to the pelvis were shown to miss gait events [29]. Pratt et al. placed IMU sensors on top of MOCAP marker clusters, which provided some standardization concerning their placement. They stated that if their findings are translated to the clinical setting, there will be a need to develop a placement protocol to reproduce sensor placement without these marker cluster plates [13]. Suggestions regarding sensor placement have been made before. Rueterbories et al. published a meta-analysis of sensors and sensor combinations capable of analyzing gait in ambulatory settings and showed a comprehensive overview of sensor devices at different body parts [37]. Furthermore, Storm et al. proposed methods that avoid assuming specific orientations in which the sensors are mounted regarding body segments [38]. To achieve comparability of results, in future research, the standardization of sensor placement should be considered carefully.

Testing protocols contained in this review included different activities, such as gait, ROM measurements, SLL tasks, alongside the performance of various activities of daily life, and therefore impaired comparability as well. Roberts et al. found their testing protocol best suited to detecting significant differences between patients and controls in the sagittal plane since most movement parameters of their tested activities projected to this plane [20]. Sigward et al., who detected impaired knee loading in the ACLR population, used between-limb ratios for their assessment—a widely used method for comparing gait mechanics after surgery. They stated that this is feasible assuming that the non-surgical limb demonstrates normal gait mechanics, which may not be accurate. Nevertheless, this provides the best available frame of reference. Furthermore, Sigward et al. noted that gait mechanics are related to walking velocity and likely influenced by other factors, such as shoe wear or walking surface, which might provide further potential for standardization of testing protocols [14]. Testing protocols, population characteristics, and intervention times differed, leading to a lack of comparability of results, although all the studies induced higher accuracy of sensing devices than the standard measurement methods.

Accuracy issues were verified by Huang et al., especially the accuracy of the sensor devices regarding Cybex when detecting lower limb flexion, and they identified potential reasons for inaccuracies during measurements. Sensor data reception issues arose due to sensors not being worn tightly enough to the leg and, therefore, slide during swings. Another reason is the possible inadequacies of the sampling rate. Their sensor device transmitted their measurements with a frequency of 100 Hz to the connected smartphone, which might induce missed capturing of swing angle and overhead of the smartphone memory due to the trade-off between sampling rate and overhead of the smartphone memory. As a second reason, they discussed vibration from the participant's leg in cases where they tried to resist their leg being passively swung by the Cybex device. This can induce errors in sensor devices [30]. De Vroey et al., who assessed the temporal gait parameters in the TKA population, traced back measurement deviations to the algorithm used to analyze IMU data. The algorithm showed some variability in detecting gait events compared to actual kinetic detection, likely a consequence of flexion and extension of the metatarsal-phalangeal joints [29]. Nevertheless, these errors in timing estimations were small enough not to be of clinical relevance. Adding to the choice of sensing equipment and placement, different algorithms exist to extract gait events from kinematic IMU data. While most algorithms show good accuracy in normal gait, care has to be taken in the gait-impaired population, where the selection of the appropriate algorithm makes a difference [37]. Bruening et al. compared different algorithms for detecting gait events from kinematic data. They suggested that algorithm choice depended on whether the foot's motion in terminal swing was more horizontal or vertical for foot strike events. They concluded that algorithms match actual gait events best when selected according to visually distinct gait patterns [39]. Their findings can be applied to routine clinical practice since they identified the most appropriate algorithm for each specific gait pattern. Nevertheless, within this review, only a few authors mention their choice of applied algorithm. Huge varieties of IMU gait analysis algorithms and the lack of consensus for their validation make it difficult for researchers to assess the algorithms' reliability for specific use cases [40].

The following parameters would help to raise acceptance and facilitate IMU implementation in clinical practice: comparable algorithms, bigger sample sizes for powering the conclusions, strong methods for bias reduction, such as standardized marker application, test–retest designs, the inclusion of more testers and different settings in different stages, and, consequently, reporting Intra Class Correlations and Limits of Agreement.

One of the key problems not solved so far is a valid description and detection of the most relevant parameters for measurement to collect from IMUs for describing rehabilitation after knee surgery. Most benefits from the IMU data will be provided for describing the domain "function". The gold standard for this domain is the use patient-reported outcome measurements (PROMs) and performance based measures so far, but with few correlation in the early rehabilitation [41]. According to Bolink (2015), PROMs and performance-based outcome measures are, for example, only moderately correlated one year after TKA, probably due to capturing a different dimension of function [42]. As shown in this review, there is widespread usage of sensors for detecting changes in knee joint rehabilitation. Still, many of them were only partly evaluated for quality criteria, probably caused by a lack of consensus on relevant parameters for describing the function of the knee joint and related rehabilitation progress. It seems obvious that, in clinical practice, sensors have mostly been used to express existing tests and parameters in an easier or faster way. Bigger advantage from implementing sensors will probably be given when developing new parameters. It might be of value to expand the scope of potentially relevant parameters first, highlighting the value of wearable sensing technology unlike standard performance-based measures in the past. New parameters such as "whole day knee joint angle movement", "all day stairs used", or "average limb loading while walking" are currently not reported and might provide higher correlation to patient-reported function and, therefore, broader acceptance among stakeholders, thereby inducing more explicit quality criteria studies in the field. This might lead to consensus discussions and the establishment of core domain sets for this field in addition to existing outcome sets for total knee arthroplasty [43].

## 5. Limitations

Although the studies included in this review showed a wide variety in their approaches, test protocols, and study population, some valuable information can be derived from them. Multiple studies mentioned limited applicability due to the relatively small sample size [13,31]. Furthermore, Pratt et al. emphasized restricted applicability of results since their established testing paradigm can only be applied to individuals four to six months post-surgery who are progressing back to running. Other phases of rehabilitation remain unexplored. Translation of Pratt's findings cannot be assumed to be widely translatable to different tasks than SLL, which leaves the need to assess other dynamic tasks, such as running. Roberts et al. drew limitations regarding the assessment of tibia and femur motion, respectively, and proposed using two IMUs for better characterization of the relative motion between the two bones to assess the movement of the knee implant parts. They stated that these dynamics might differ in patients with bilateral TKA, unlike those with unilateral one [20]. Meta-analysis was inappropriate because studies were not similar enough from a methodological and clinical viewpoint. No grading of evidence for a specific outcome was possible because of the different topics covered in the included studies. From our point of view, many studies close to the topic had to be excluded, caused by strict inclusion criteria. Reviews on similar topic especially on new potential parameters should be performed.

## 6. Conclusions

The present review shows that IMUs offer sufficient accuracy to replace, combine, and extend the existing range of rehabilitation devices. IMUs can subsume different measures for rehabilitation by assessing outcomes that would typically be measured individually, such as ROM, gait analysis, and detection of asymmetric knee loading, while adding new rehabilitation hallmarks, such as quantification of instability. IMUs can replace time-consuming equipment such as motion-capturing systems and force platforms in the knee surgery population. Developing clinician-friendly, standardized applications of IMUs for clinical practice is imperative. However, all the data provided were collected in a laboratory environment. Furthermore, studies regarding sensing technology utilization in clinical practice remain lacking. Since this technology provides evidence to benefit patients and healthcare providers, its translation into clinical rehabilitation practice is imperative. Some interesting work was done to clarify the diagnostic accuracy of wearable movement sensors for knee joint rehabilitation. Still, in the current stage, comparable quality criterion studies are lacking for an evidence summary of potential measurement bias and clear recommendations for using wearable movement technology in quantifying knee injuries in clinical settings. Developing a core measurement set for quality criterion studies on IMUs for medical use might help harmonize research in knee joint rehabilitation. Generally, within the scope of this review, although there are distinct limitations of sensor usage in rehabilitating knee surgery populations, the potential of these devices is obvious.

**Author Contributions:** Conceptualization: R.B. and R.P.; methodology: R.P.; Literature search: M.W.; Title/Abstract/Fulltext Screening: M.W. and R.P., Data extraction: M.W. and R.P., A.K.; formal analysis: M.W. and R.P.; Manuscript writing: M.W., R.P., A.K. and R.B., formatting: R.P., final revision: A.K., R.B. All authors have read and agreed to the published version of the manuscript.

**Funding:** We acknowledge funding from the MHB Open Access Publication Fund supported by the German Research Association (DFG).

**Institutional Review Board Statement:** Not applicable.

**Informed Consent Statement:** Not applicable. No patients have been involved in this review.

**Data Availability Statement:** All extracted data are included in the manuscript.

**Conflicts of Interest:** The authors declare no conflict of interest.

## References

1. Gage, B.E.; McIlvain, N.M.; Collins, C.L.; Fields, S.K.; Comstock, R.D. Epidemiology of 6.6 million knee injuries presenting to United States emergency departments from 1999 through 2008. *Acad. Emerg. Med.* **2012**, *19*, 378–385. [CrossRef]
2. Majewski, M.; Susanne, H.; Klaus, S. Epidemiology of athletic knee injuries: A 10-year study. *Knee* **2006**, *13*, 184–188. [CrossRef] [PubMed]
3. Bollen, S. Epidemiology of knee injuries: Diagnosis and triage. *Br. J. Sports Med.* **2000**, *34*, 227–228. [CrossRef]
4. Bram, J.T.; Magee, L.C.; Mehta, N.N.; Patel, N.M.; Ganley, T.J. Anterior Cruciate Ligament Injury Incidence in Adolescent Athletes: A Systematic Review and Meta-analysis. *Am. J. Sports Med.* **2021**, *49*, 1962–1972. [CrossRef]
5. Global Burden of Disease Study Collaborators. Global, regional, and national incidence, prevalence, and years lived with disability for 301 acute and chronic diseases and injuries in 188 countries, 1990–2013: A systematic analysis for the Global Burden of Disease Study 2013. *Lancet* **2015**, *386*, 743–800. [CrossRef]
6. Cui, A.; Li, H.; Wang, D.; Zhong, J.; Chen, Y.; Lu, H. Global, regional prevalence, incidence and risk factors of knee osteoarthritis in population-based studies. *EClinicalMedicine* **2020**, *29–30*, 100587. [CrossRef]
7. Palazzo, C.; Nguyen, C.; Lefevre-Colau, M.M.; Rannou, F.; Poiraudeau, S. Risk factors and burden of osteoarthritis. *Ann. Phys. Rehabil. Med.* **2016**, *59*, 134–138. [CrossRef]
8. Bannuru, R.R.; Osani, M.C.; Vaysbrot, E.E.; Arden, N.K.; Bennell, K.; Bierma-Zeinstra, S.M.A.; Kraus, V.B.; Lohmander, L.S.; Abbott, J.H.; Bhandari, M.; et al. OARSI guidelines for the non-surgical management of knee, hip, and polyarticular osteoarthritis. *Osteoarthr. Cartil.* **2019**, *27*, 1578–1589. [CrossRef] [PubMed]
9. Begon, M.; Andersen, M.S.; Dumas, R. Multibody kinematic optimization for the estimation of upper and lower limb human joint kinematics: A systematic review. *J. Biomech. Eng.* **2018**, *140*, 030801. [CrossRef] [PubMed]
10. Shull, P.B.; Jirattigalachote, W.; Hunt, M.A.; Cutkosky, M.R.; Delp, S.L. Quantified self and human movement: A review on the clinical impact of wearable sensing and feedback for gait analysis and intervention. *Gait Posture* **2014**, *40*, 11–19. [CrossRef] [PubMed]
11. Lam, M.H.; Fong, D.T.; Yung, P.S.; Chan, K.M. Biomechanical techniques to evaluate tibial rotation. A systematic review. *Knee Surg. Sports Traumatol. Arthrosc. Off. J. ESSKA* **2012**, *20*, 1720–1729. [CrossRef]
12. Batailler, C.; Lording, T.; Naaim, A.; Servien, E.; Cheze, L.; Lustig, S. No difference of gait parameters in patients with image-free robotic-assisted medial unicompartmental knee arthroplasty compared to a conventional technique: Early results of a randomized controlled trial. *Knee Surg. Sports Traumatol. Arthrosc.* **2021**, *11*, 1–11. [CrossRef] [PubMed]

13. Pratt, K.A.; Sigward, S.M. Detection of Knee Power Deficits Following Anterior Cruciate Ligament Reconstruction Using Wearable Sensors. *J. Orthop. Sports Phys. Ther.* **2018**, *48*, 895–902. [CrossRef] [PubMed]

14. Sigward, S.M.; Chan, M.M.; Lin, P.E. Characterizing knee loading asymmetry in individuals following anterior cruciate ligament reconstruction using inertial sensors. *Gait Posture* **2016**, *49*, 114–119. [CrossRef]

15. Costello, K.E.; Eigenbrot, S.; Geronimo, A.; Guermazi, A.; Felson, D.T.; Richards, J.; Kumar, D. Quantifying varus thrust in knee osteoarthritis using wearable inertial sensors: A proof of concept. *Clin. Biomech.* **2020**, *80*, 105232. [CrossRef] [PubMed]

16. Bravi, M.; Gallotta, E.; Morrone, M.; Maselli, M.; Santacaterina, F.; Toglia, R.; Foti, C.; Sterzi, S.; Bressi, F.; Miccinilli, S. Concurrent validity and inter trial reliability of a single inertial measurement unit for spatial-temporal gait parameter analysis in patients with recent total hip or total knee arthroplasty. *Gait Posture* **2020**, *76*, 175–181. [CrossRef]

17. Chapman, R.M.; Moschetti, W.E.; Van Citters, D.W. Stance and swing phase knee flexion recover at different rates following total knee arthroplasty: An inertial measurement unit study. *J. Biomech.* **2019**, *84*, 129–137. [CrossRef]

18. Chiang, C.-Y.; Chen, K.-H.; Liu, K.-C.; Hsu, S.J.-P.; Chan, C.-T. Data Collection and Analysis Using Wearable Sensors for Monitoring Knee Range of Motion after Total Knee Arthroplasty. *Sensors* **2017**, *17*, 418. [CrossRef] [PubMed]

19. Na, A.; Buchanan, T.S. Validating Wearable Sensors Using Self-Reported Instability among Patients with Knee Osteoarthritis. *PM R* **2021**, *13*, 119–127. [CrossRef]

20. Roberts, D.; Khan, H.; Kim, J.H.; Slover, J.; Walker, P.S. Acceleration-based joint stability parameters for total knee arthroplasty that correspond with patient-reported instability. *Proc. Inst. Mech. Eng. Part H* **2013**, *227*, 1104–1113. [CrossRef]

21. Patel, S.; Park, H.; Bonato, P.; Chan, L.; Rodgers, M. A review of wearable sensors and systems with application in rehabilitation. *J. Neuroeng. Rehabil.* **2012**, *9*, 21. [CrossRef] [PubMed]

22. Hussain, I.; Park, S.-J. Prediction of Myoelectric Biomarkers in Post-Stroke Gait. *Sensors* **2021**, *21*, 5334. [CrossRef]

23. Jiang, X.; Gholami, M.; Khoshnam, M.; Eng, J.J.; Menon, C. Estimation of Ankle Joint Power during Walking Using Two Inertial Sensors. *Sensors* **2019**, *19*, 2796. [CrossRef]

24. Yang, Z.; Van Beijnum, B.-J.F.; Li, B.; Yan, S.; Veltink, P.H. Estimation of Relative Hand-Finger Orientation Using a Small IMU Configuration. *Sensors* **2020**, *20*, 4008. [CrossRef] [PubMed]

25. Porciuncula, F.; Roto, A.V.; Kumar, D.; Davis, I.; Roy, S.; Walsh, C.J.; Awad, L.N. Wearable Movement Sensors for Rehabilitation: A Focused Review of Technological and Clinical Advances. *PM R* **2018**, *10*, S220–S232. [CrossRef]

26. Small, S.R.; Bullock, G.S.; Khalid, S.; Barker, K.; Trivella, M.; Price, A.J. Current clinical utilisation of wearable motion sensors for the assessment of outcome following knee arthroplasty: A scoping review. *BMJ Open* **2019**, *9*, e033832. [CrossRef]

27. Prill, R.; Karlsson, J.; Ayeni, O.R.; Becker, R. Author guidelines for conducting systematic reviews and meta-analyses. *Knee Surg. Sports Traumatol. Arthrosc.* **2021**, *29*, 2739–2744. [CrossRef]

28. Mokkink, L.B.; Boers, M.; Van Der Vleuten, C.P.M.; Bouter, L.M.; Alonso, J.; Patrick, D.L.; de Vet, H.C.W.; Terwee, C.B. COSMIN Risk of Bias tool to assess the quality of studies on reliability or measurement error of outcome measurement instruments: A Delphi study. *BMC Med. Res. Methodol.* **2020**, *20*, 293. [CrossRef] [PubMed]

29. De Vroey, H.; Staes, F.; Weygers, I.; Vereecke, E.; Vanrenterghem, J.; Deklerck, J.; Van Damme, G.; Hallez, H.; Claeys, K. The implementation of inertial sensors for the assessment of temporal parameters of gait in the knee arthroplasty population. *Clin. Biomech.* **2018**, *54*, 22–27. [CrossRef]

30. Huang, Y.P.; Liu, Y.Y.; Hsu, W.H.; Lai, L.J.; Lee, M.S. Progress on Range of Motion After Total Knee Replacement by Sensor-Based System. *Sensors* **2020**, *20*, 1703. [CrossRef]

31. Pratt, K.A.; Sigward, S.M. Inertial Sensor Angular Velocities Reflect Dynamic Knee Loading during Single Limb Loading in Individuals Following Anterior Cruciate Ligament Reconstruction. *Sensors* **2018**, *18*, 3460. [CrossRef] [PubMed]

32. Cottino, U.; Sculco, P.K.; Sierra, R.J.; Abdel, M.P. Instability After Total Knee Arthroplasty. *Orthop. Clin. N. Am.* **2016**, *47*, 311–316. [CrossRef]

33. Khan, H.; Walker, P.S.; Zuckerman, J.D.; Slover, J.; Jaffe, F.; Karia, R.J.; Kim, J.H. The potential of accelerometers in the evaluation of stability of total knee arthroplasty. *J. Arthroplast.* **2013**, *28*, 459–462. [CrossRef] [PubMed]

34. Barber-Westin, S.D.; Noyes, F.R. Factors used to determine return to unrestricted sports activities after anterior cruciate ligament reconstruction. *Arthroscopy* **2011**, *27*, 1697–1705. [CrossRef]

35. Myer, G.D.; Schmitt, L.C.; Brent, J.L.; Ford, K.R.; Barber Foss, K.D.; Scherer, B.J.; Heidt, R.S., Jr.; Divine, J.G.; Hewett, T.E. Utilization of modified NFL combine testing to identify functional deficits in athletes following ACL reconstruction. *J. Orthop. Sports Phys. Ther.* **2011**, *41*, 377–387. [CrossRef]

36. Sigward, S.M.; Lin, P.; Pratt, K. Knee loading asymmetries during gait and running in early rehabilitation following anterior cruciate ligament reconstruction: A longitudinal study. *Clin. Biomech.* **2016**, *32*, 249–254. [CrossRef]

37. Rueterbories, J.; Spaich, E.G.; Larsen, B.; Andersen, O.K. Methods for gait event detection and analysis in ambulatory systems. *Med. Eng. Phys.* **2010**, *32*, 545–552. [CrossRef]

38. Storm, F.A.; Buckley, C.J.; Mazza, C. Gait event detection in laboratory and real life settings: Accuracy of ankle and waist sensor based methods. *Gait Posture* **2016**, *50*, 42–46. [CrossRef]

39. Bruening, D.A.; Ridge, S.T. Automated event detection algorithms in pathological gait. *Gait Posture* **2014**, *39*, 472–477. [CrossRef] [PubMed]

40. Zhou, L.; Tunca, C.; Fischer, E.; Brahms, C.M.; Ersoy, C.; Granacher, U.; Arnrich, B. Validation of an IMU Gait Analysis Algorithm for Gait Monitoring in Daily Life Situations. In Proceedings of the 2020 42nd Annual International Conference of the IEEE Engineering in Medicine & Biology Society (EMBC), Montreal, QC, Canada, 20–24 July 2020; 2020, pp. 4229–4232. [CrossRef]
41. Prill, R.; Becker, R.; Schulz, R.; Michel, S.; Hommel, H. No correlation between symmetry-based performance measures and patient-related outcome prior to and after total knee arthroplasty. *Knee Surg. Sports Traumatol. Arthrosc.* **2021**, 1–7. [CrossRef] [PubMed]
42. Bolink, S.A.; Grimm, B.; Heyligers, I.C. Patient-reported outcome measures versus inertial performance-based outcome measures: A prospective study in patients undergoing primary total knee arthroplasty. *Knee* **2015**, *22*, 618–623. [CrossRef] [PubMed]
43. Prill, R.; Singh, J.A.; Seeber, G.H.; Nielsen, S.M.; Goodman, S.; Michel, S.; Kopkow, C.; Schulz, R.; Choong, P.; Hommel, H. Patient, physiotherapist and surgeon endorsement of the core domain set for total hip and total knee replacement in Germany: A study protocol for an OMERACT initiative. *BMJ Open* **2020**, *10*, e035207. [CrossRef] [PubMed]

MDPI

St. Alban-Anlage 66

4052 Basel

Switzerland

Tel. +41 61 683 77 34

Fax +41 61 302 89 18

www.mdpi.com

*Sensors* Editorial Office

E-mail: sensors@mdpi.com

www.mdpi.com/journal/sensors